Springer Geophysics

The Springer Geophysics series seeks to publish a broad portfolio of scientific books, aiming at researchers, students, and everyone interested in geophysics. The series includes peer-reviewed monographs, edited volumes, textbooks, and conference proceedings. It covers the entire research area including, but not limited to, applied geophysics, computational geophysics, electrical and electromagnetic geophysics, geodesy, geodynamics, geomagnetism, gravity, lithosphere research, paleomagnetism, planetology, tectonophysics, thermal geophysics, and seismology.

More information about this series at http://www.springer.com/series/10173

Yuanxue Liu · Yingren Zheng

Plastic Mechanics of Geomaterial

Science Press
Beijing

Springer

Yuanxue Liu
Institute of Geotechnical Engineering
Logistical Engineering University
Chongqing, China

Yingren Zheng
Institute of Geotechnical Engineering
Logistical Engineering University
Chongqing, China

ISSN 2364-9119 ISSN 2364-9127 (electronic)
Springer Geophysics
ISBN 978-981-13-3752-9 ISBN 978-981-13-3753-6 (eBook)
https://doi.org/10.1007/978-981-13-3753-6

Jointly published with Science Press, Beijing, China

The print edition is not for sale in China Mainland. Customers from China Mainland please order the print book from: Science Press.

Library of Congress Control Number: 2018964049

This Springer imprint is published by the registered company Springer Nature Singapore Pte Ltd.
The registered company address is: 152 Beach Road, #21-01/04 Gateway East, Singapore 189721, Singapore

Preface

Plastic theory is widely used to describe the mechanical behaviors of geomaterial in civil engineering. This book is the summary of the author's experiences in teaching and research of plastic mechanics of geomaterial, closely focusing on the difficulty and confusion of the graduate students in the learning process. In this book, the general theory of plasticity is concisely stated and special attention is paid to the mechanical response and constitutive description for geomaterial. Besides, simple and easy-to-understand language is used to introduce the basic concepts, theories, methods and models, and the overall logic and systematic elaboration is guaranteed. We hope this book can enable the colleagues and graduate students of geotechnical engineering to gain in-depth understanding of basic mechanical properties and constitutive models of geomaterial, and to select and establish the most reasonable constitutive model after evaluating the actual engineering situation.

This book content is organized as follows: the basic concept of plastic theory of geomaterial; The continuum mechanics basis of plastic theory: stress analysis, strain analysis, and basic equations; The basic mechanical properties and their descriptions of geomaterial; Nonlinear theory, as well as the typical nonlinear constitutive model for geomaterial; The foundation of classic plastic theory and the classical plastic theory; Research progress of the basic problem in constitutive theory of geomaterial, and development of plastic theory of geomaterial in yield surface theory, hardening theory, flow rule and loading–unloading criterion; Representative static constitutive model for geomaterial; The computing theory for rotation effect of principal stress axes; Basic dynamic mechanical properties and typical dynamic constitutive models for geomaterial; Limit analysis in geotechnical engineering and its latest progress-FEM limit analysis. Each chapter ends with some questions and references which will help the readers to understand further and read extensively.

In the process of writing this book, we obtain the help of Ming Hu, Zhongyou Li, Jiawu Zhou, Peiyong Wang, Changbing Shan, Yizhong Tan, Xiaoliang Chen, Chenyu Qiu, and Shangyi Zhao. The nice figures come from the hard work of Penghui Chu, Hongwei Li, Chaoguang Qin, Can Li, and Meichen Shen. The proofreading and revision of English are finished by teacher Ling Du.

Chongqing, China

刘元雪
Yuanxue Liu
郑颖人
Yingren Zheng

Contents

1 Introduction . . . 1
1.1 Basic Concept . . . 1
1.1.1 Plastic Deformation . . . 1
1.1.2 Plastic Mechanics . . . 2
1.1.3 Plastic Mechanics of Geomaterial . . . 3
1.2 The Basic Hypothesis of Plastic Mechanics of Geomaterial . . . 4
1.3 The Constitutive Model . . . 4
1.3.1 What Is Model . . . 4
1.3.2 Model in Classical Soil Mechanics . . . 5
1.4 Development History for Plastic Mechanics of Geomaterial . . . 6
References . . . 8

2 Stress and Strain and Its Basic Equations . . . 11
2.1 Continuum Model . . . 11
2.2 Stress Tensor . . . 12
2.3 Decomposition of Stress Tensor and Its Invariants . . . 15
2.3.1 Decomposition of Stress Tensor . . . 15
2.3.2 Other Representation of Stress Invariant . . . 16
2.4 Deformation and Strain . . . 19
2.5 The Invariant of Strain Tensor . . . 22
2.6 Decomposition of Strain Tensor and Its Invariants . . . 23
2.6.1 Decomposition of Strain Tensor . . . 23
2.6.2 Other Representation of Strain Invariant . . . 24
2.7 Stress Path and Strain Path . . . 27
2.7.1 Expression of Stress Path . . . 27
2.7.2 The Realization of Stress Path . . . 29
2.7.3 Total Stress Path and Effective Stress Path . . . 29
2.7.4 Strain Path . . . 30

2.8 Basic Equations of Plastic Mechanics of Geomaterial 30
2.8.1 Basic Equations 31
2.8.2 Boundary Condition and Initial Value 33
References 34

3 The Basic Mechanical Characteristics of the Geomaterial 35
3.1 Pressure-Hardening 35
3.2 Yield Caused by Hydrostatic Pressure 36
3.3 Dilatancy 37
3.4 Plastic Deformation Dependent on Stress Path 39
3.5 Other Important Characteristics 41
3.6 Mechanical Characteristic at Small Strain of Geomaterial 43
3.7 Mechanical Difference for the Natural and Remolded Soil. 45
References 47

4 The Elastic Model of Geomaterial 49
4.1 Nonlinear Elastic Theory 51
4.1.1 Variable Elasticity Theory 51
4.1.2 Hyperelastic Theory 51
4.1.3 Hypoelastic Theory 52
4.2 The Anisotropic Elastic Theory 52
4.2.1 Isotropic Elastic Constitutive Model 54
4.2.2 The Elastic Constitutive Model with Cross-Anisotropy 57
4.3 The Isotropic Nonlinear Elastic Model of Geomaterial 57
4.3.1 The Basic Principle of Duncan-Chang Model. 58
4.3.2 Two Elastic Function of Duncan-Chang Model 59
4.3.3 Review of Duncan-Chang Model 63
4.4 The Elastic Model with Transverse Isotropy 64
4.4.1 Xiao-nan Gong Model 64
4.4.2 Graham Model 65
References 71

5 Classical Plastic Theory 73
5.1 Potential Function and Thermodynamics 73
5.1.1 First Law of Thermodynamics. 73
5.1.2 Second Law of Thermodynamics 74
5.1.3 Thermodynamics Potential and Dissipative Inequality. 75
5.2 Plastic Postulate 78
5.2.1 Drucker's Stability Postulate 78
5.2.2 Inference of Drucker's Postulate 79

5.3 The Constitutive Model Based on the Classic Plastic Theory 82
5.3.1 The Framework of the Classic Plastic Theory 82
5.3.2 Commonly Used Models 83
References 86

6 The Development of the Plastic Theory of Geomaterial 87
6.1 Study of Several Basic Problems in Plastic Theory of Geomaterial 87
6.1.1 Proving that Drucker Postulate Is Unsuitable for Geomaterial 88
6.1.2 Proving that the Classic Plastic Theory Is Unsuitable for Geomaterial 91
6.1.3 Study of Several Key Problems in the Plastic Theory of Geomaterial 93
6.2 Development of the Yield Surface for Geomaterial 97
6.2.1 Significance of Yield Surface 97
6.2.2 Yield of Geomaterial 97
6.2.3 The Shear Yield Surface 98
6.2.4 Volumetric Yield Criterion 106
6.2.5 Yield Surface of Overconsolidated Soil 114
6.2.6 Part Yield 116
6.3 Hardening Laws 117
6.3.1 Hardening Theory 117
6.3.2 Hardening Model 118
6.3.3 Isotropic Hardening 119
6.3.4 Kinematic Hardening 121
6.3.5 Mixed Hardening 124
6.3.6 The General Form of Hardening Model 125
6.4 Plastic Flow Rule 128
6.4.1 Associated Flow Rule 129
6.4.2 Nonassociated Flow Rule 129
6.4.3 Mixed Flow Rule 129
6.5 Loading and Unloading Rule 131
6.5.1 Loading–Unloading Rule Based on Yield Surface 131
6.5.2 Loading–Unloading Rule with Stress Type 132
6.5.3 Loading–Unloading Rule with Strain Style 132
References 134

7 The Static Elastoplastic Model for Geomaterial 137
7.1 Cam-Clay and Modified Cam-Clay Model 138
7.1.1 The Concept of Critical States 138
7.1.2 Cam-Clay Model 139

7.1.3 Modified Cam-Clay Model 144
7.1.4 Comment on Cam-Clay Model 145
7.2 Lade Model 146
7.2.1 Components of Constitutive Model 146
7.2.2 Elastic Behavior 146
7.2.3 Failure Criterion 147
7.2.4 Plastic Potential and Flow Rule 148
7.2.5 Yield Criterion and Work-Hardening/Softening Relation 150
7.2.6 Determination of Material Parameters 153
7.2.7 Model Comments 160
7.3 A Unified Hardening Constitutive Model for Soils 161
7.3.1 Introduction 161
7.3.2 The Unified Hardening Parameter Which Has Nothing to Do with the Stress Path 162
7.3.3 Unified Hardening Model for Natural Consolidation Soil 168
7.3.4 The Unified Hardening Model of the Normal Consolidated Soil 170
7.3.5 The Stress–Strain Relationship 175
7.3.6 Model Comments 177
References 179

8 Generalized Plastic Mechanics Considering the Rotation of Principal Axis of Stress 181
8.1 Decomposition of the Stress Increment 182
8.1.1 Decomposition of Two-Dimensional Stress Increment 182
8.1.2 Decomposition of Three-Dimensional Stress Increment 184
8.2 Generalized Plastic Potential Theory Considering Rotation of Principal Stress Axis 186
8.3 Complete Stress Increment Expression of Elastoplastic Stress–Strain Relationship for Geomaterial 187
8.4 Plastic Deformation Caused by Coaxial Stress Increment 191
8.5 Plastic Deformation Caused by the Rotational Increment of Stress 193
8.5.1 Plastic Deformation Caused by Stress Increment $d\sigma_{r1}$ 194
8.5.2 Plastic Deformation Caused by Rotational Stress Increment $d\sigma_{r2}$ and $d\sigma_{r3}$ 198
8.6 Elastoplastic Stress–Strain Relations Considering Rotation of Principal Stress Axis 200

8.6.1 Elastic Compliance Matrix 200
8.6.2 Coaxial Plastic Compliance Matrix 200
8.6.3 Rotating Plastic Compliance Matrix 201
8.7 Example 203
References 206

9 The Dynamic Constitutive Model of Geomaterial 207
9.1 Basic Characteristics of Dynamic Stress–Strain Relationship of Geomaterial 207
9.2 Empirical Model for Dynamic Stress–Strain Relationship 210
9.3 Equivalent Dynamic Linear Viscoelastic Model of Soil 213
9.3.1 Viscoelastic Model 213
9.3.2 Parameter for Viscoelastic Model 217
9.4 Viscoelastoplastic Dynamic Constitutive Model of Geomaterial 219
9.4.1 Framework for Viscoelastic Plasticity Model 219
9.4.2 The Computation of Viscoelastic Part 219
9.4.3 The Computation of Elastoplastic Part 220
9.4.4 The Computation of Total Part 222
9.5 Viscoplastic Cap Models 222
9.5.1 The Perzyna-Type Viscoplastic Cap Model 223
9.5.2 Solution Algorithms 225
9.5.3 The Duvant–Lions Type Viscoplastic Cap Model 227
9.5.4 Illustration Example 228
References 230

10 Limit Analysis for Geotechnical Engineering 231
10.1 Overview 231
10.2 The Basic Equations for Limit Analysis 231
10.3 The Characteristic Line Method in Limit Analysis 232
10.3.1 Slip Line of Stress 232
10.3.2 Prandtl Solution of Stress Characteristic Line for Ultimate Load of a Half-Infinite Plane Body 236
10.3.3 Velocity Slip Line Field for Plane Strain Problem 239
10.4 Principle of Limit Analysis and Approximate Method 239
10.4.1 Limit Analysis Theorem 240
10.4.2 Example of Limit Analysis Principle 240
10.5 Numerical Limit Analysis 243
10.5.1 The Basic Principle of FEM Limit Analysis 244
10.5.2 Definition of Safety Factor 244
10.5.3 Criterion for Limit State 245
10.5.4 Example for the Calculation of Slope Safety Factor 247
References 250

Abstract

This book is the coagulation of the author's painstaking efforts in the teaching and research work of the plastic mechanics of geomaterial for a long time. It is an attempt to set forth clearly the basic concepts of geomaterial plasticity, the basic static and dynamic mechanical properties, nonlinear and classical plastic theory, geomaterial yield surface theory, hardening model, flow rule and loading–unloading criterion, the calculation theory of principal stress axe rotation, and geomaterial limit analysis and its latest progress-FEM limit analysis. In addition, the typical static and dynamic constitutive model of geomaterial is introduced in detail too, hopefully this book will help the readers acquire in-depth understanding of the basic mechanical properties and constitutive model for geomaterial, and establish and select the most reasonable constitutive model according to the specific engineering problems and geomaterial characteristics. This book is suitable for the designers and researchers in the related fields of geotechnical engineering, and it can also be used as a textbook for the relevant graduate courses.

Chapter 1
Introduction

The solution of complicated geotechnical engineering problems (stability, deformation) is inseparable from the computer, and the core of numerical analysis is the constitutive model of geomaterial. If the model is wrong, the calculation result is rubbish. Many geotechnical workers do not know that their calculation conclusion is dependent on the constitutive model that they selected. The basis of the constitutive model of geomaterial is plastic mechanics of geomaterial.

1.1 Basic Concept

1.1.1 Plastic Deformation

The mechanical response of materials can be divided into three phases: elastic deformation stage, deformation can be recovered after unloading of stress; plastic deformation stage, part or all deformation cannot be recovered after unloading of stress; and damage stage, the stress reaches the ultimate strength of materials, causing failure of material structure or macroscopic failure.

Elastic deformation is generally regarded as a linear model, and plastic deformation is generally regarded as a nonlinear model. Let us have a look at Fig. 1.1 and identify the elasticity, plasticity, and nonlinear.

The loading and unloading lines overlap in Fig. 1.1a. It means that there is no residual deformation although the lines are curves, namely, nonlinear. Figure 1.1b is not the same, the loading path is different from the unloading path, and the residual deformation appears after unloading although these paths are straight lines.

In general, elastic deformation of materials is described as a linear model, while the nonlinear characteristic is obvious for the plastic deformation. However, nonlinear is not the essential difference between elastic materials and plastic ones, for nonlinear characteristic is also displayed in the elastic deformation of some materials such as

Y. Liu and Y. Zheng, *Plastic Mechanics of Geomaterial*, Springer Geophysics,
https://doi.org/10.1007/978-981-13-3753-6_1

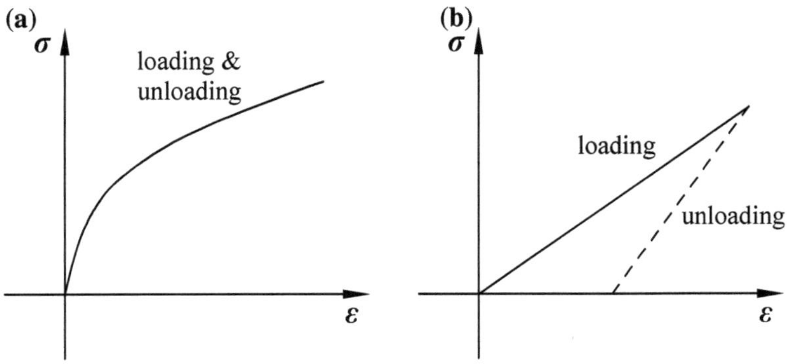

Fig. 1.1 The schematic diagram of loading–unloading (solid line indicates loading; dashed line shows unloading; and loading and unloading lines overlap in the left figure)

rubber: the stress–strain curve is nonlinear, but deformation can be fully recovered once unloaded. The material mechanical behavior is linear in Fig. 1.1b, but there is residual deformation after unloading (plastic deformation). Of course, this kind of material is imaginary, and I hope it will be found or created.

The essential difference between elastic material and plastic material is whether any difference exists in loading and unloading path, and whether the residual deformation (plastic deformation) exists after unloading. Thus, Fig. 1.1a is called elastic behavior, while Fig. 1.1b is elastoplastic behavior.

The plastic deformation is the irreversible distortion, namely, the deformation cannot restore completely after the stress eliminates.

Certain stress corresponds to certain strain in elastic deformation, and it is called the whole total relationship. The loading path and unloading path are different in the plastic deformation stage, and the non-corresponding relationship exists for stress and strain under plastic state (Fig. 1.2). The total relationship can only be applied to some simple loading cases, while incremental relationship should be adopted in other cases.

1.1.2 Plastic Mechanics

Plastic Mechanics is classified under mechanics. Mechanics is a science that studies the law of mechanical motion. The mechanical movement is the simplest and most basic movement forms in nature. The location of an object relative to another one, or the location of one part of an object relative to other parts, the change process of these locations with time is called the mechanical movement.

Plastic Mechanics is the subject that studies the deformation and stability of object in the plastic deformation phase. It is a branch of Continuum Mechanics. Continuum Mechanics studies the general laws of mechanics which the deformable

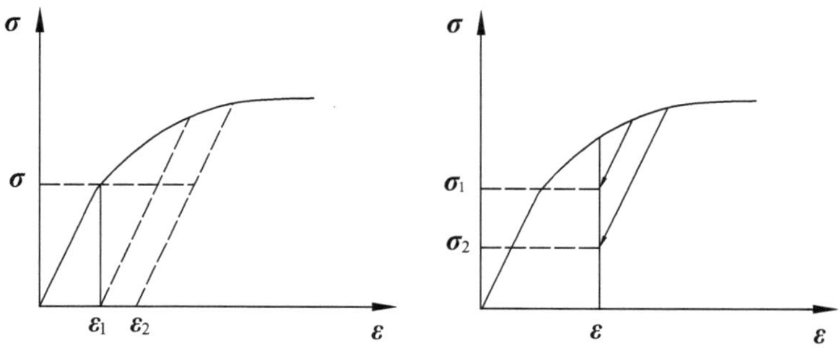

(a) The samestress for different strains (b) The samestrain corresponding to different stresses

Fig. 1.2 The non-corresponding relationship of stress and strain under plastic state

objects follow with mass of continuous distribution. For example, the conservation of mass, momentum and angular momentum theorem, energy conservation, etc.

The basic equations for Continuum Mechanics could be divided into three categories:

1. Geometric equations: to describe the geometric relationship between deformation and movement of the object (relation of displacement and strain);
2. Conservation equation: conservation of mass, angular momentum, momentum, and energy;
3. Physical equations (constitutive equation): to depict the relation between physical state and mechanical properties.

The difference among the branches of Continuum Mechanics (such as Elastic Mechanics, Plastic Mechanics, and Fluid Mechanics) is the different constitutive equations (the third category equation). The task of Plastic Mechanics is to build this kind of equation and solve engineering problems in the plastic deformation stage.

1.1.3 Plastic Mechanics of Geomaterial

Plastic Mechanics of Geomaterial is a subject that studies the deformation and the stability of geomaterial in plastic deformation stage. A classic content of the traditional plastic mechanics is to seek the analytical solution of the elastoplastic problems with boundary value or initial value. But the main task of the Plastic Mechanics of Geomaterial is to establish the constitutive equation of geomaterial and determine the condition of initial value and boundary value, and then the engineering problems are usually solved by numerical simulation.

1.2 The Basic Hypothesis of Plastic Mechanics of Geomaterial

Rock and soil is the research object of Plastic Mechanics of Geomaterial. During the whole geological history of geomaterial formation and existence, it has been subjected to various complicated geological processes, so it has a complex structure and environment of geostress field. The engineering properties of different types of rock are often quite different in different regions due to the different geological processes. When rocks are exposed to the surface, they are formed by weathering. They are either retained in situ or deposited in other places through erosion and transport by wind, water, and glaciers. For different weathering environment and the dynamic conditions of transportation and deposition of various geological periods in different regions, soil display not only complex engineering properties but also strong regional and individualities. Practical stress–strain relationship is very complicated for geomaterial, such as nonlinearity, elasticity, plasticity and viscosity, dilatancy, and anisotropy. At the same time, a lot of factors have an effect on it, for instance, the stress path, strength, composition, structure, temperature, etc. Because geomaterial is too complex and has many influencing factors, the following five basic hypotheses are introduced to grasp the main contradiction in Plastic Mechanics of Geotechnical:

1. Ignoring the influence of temperature.
2. Ignoring the influence of the time.
3. Continuity hypothesis.
4. Small deformation hypothesis.
5. Principle of effective stress.

1.3 The Constitutive Model

1.3.1 What Is Model

Scientific understanding proceeds through constructing and analyzing models of the segments or aspects of reality under study.

The purpose of these models is not to give a mirror image of reality, not to include all its elements in their exact sizes and proportions, but rather to single out and make available for intensive investigation those elements which are decisive.

We abstract from nonessentials, blow out the unimportant to get an unobstructed view of the important, and magnify in order to improve the range and accuracy of our observation.

A model is and must be unrealistic in the sense in which the word is most commonly used. Nevertheless, paradoxically, if it is a good model, it provides the key to understanding reality.

Engineering is concerned with understanding, analyzing, and predicting the way in which real devices, structures, and pieces of equipment will behave in use. It is rarely possible to perform an analysis in which full knowledge of the object being analyzed permits a complete and accurate description of the object to be incorporated in the analysis.

This is particularly true for geotechnical engineering. The soil conditions under a foundation or embankment can be discovered only at discrete locations by retrieving samples of soil from boreholes or performing in situ tests; soil conditions between such discrete locations can be deduced only by informed interpolation.

This is a major difference between geotechnical engineering and structural or mechanical engineering, in which it is feasible to specify and control the properties of the steel, concrete, or other material from which a structural member or mechanical component is to be manufactured.

Not only is it rarely possible to perform such an analysis, but it is also rarely desirable. Understanding the behavior of real objects is improved if intelligent simplifications of reality are made, and analyses are performed using simplified models.

The objective of using conceptual models is to focus attention on the important features of a problem and to leave aside features which are irrelevant. The choice of model depends on the application. For example, a spacecraft can be considered as a point mass, etc.

1.3.2 Model in Classical Soil Mechanics

Classic soil mechanics makes much implicit use of idealized stress–strain relationships. A typical stress–strain relationship denoted by a curve is illustrated in Fig. 1.3. Two groups of calculations are regularly performed in geotechnical engineering: stability calculation and settlement calculation.

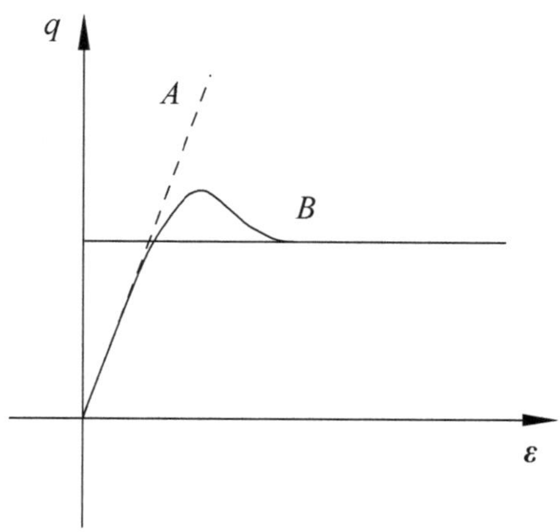

Fig. 1.3 Observed and idealized behavior of soils for classic soils mechanics

Settlement calculation is concerned with the stiffness of soil masses under applied loads. An obvious idealization of the stress–strain curve is to assume that over the range of stresses applied under working loads, the stress–strain behavior is linear and elastic, represented by dashed line *A* in Fig. 1.3.

Stability calculation is concerned with the complete failure of soil masses, with large deformation occurring on complete failure of soil masses, accompanied by the collapse of geotechnical structures. If the deformation is large, the precise shape of the early stage of the stress–strain curve is of little importance, and the stress–strain behavior can be idealized as rigid-perfectly plastic, represented by line *B* in Fig. 1.3.

Traditional model is two extreme cases in soil mechanics.

Establishing model is the simplification of highlighting the key factors. The key factors are decided by the purpose of application. The constitutive model is also set up on the same principle.

1.4 Development History for Plastic Mechanics of Geomaterial

Any object must experience three stages generally from deformation to failure: elasticity, plasticity, and failure. The elastic theory should be adopted to calculate the stress and deformation in elastic stage. In this stage, the force corresponds to the deformation monotonously, and that the deformation will completely restore after the force removes. The Plastic Mechanics of Geomaterial is used to analyze the stress and deformation of the material in plastic stage. In this stage, the stress–strain relation is influenced by load condition, stress level, stress history, and stress path.

The plastic mechanics comes of Coulomb failure criterion, and now it is developed to Mohr–Coulomb criterion. In 1857, the slip plane concept was put forward by Rankine [1] after the semi-infinite equilibrium was investigated. The slip line method was established by Kotter. Then the method of limiting equilibrium was proposed by Fellenius. In 1943, the Fellenius theory was developed by Terzaghi [2], which was adopted to solve various stability problems in soil mechanics. The method of limiting equilibrium was developed by Drucker and Prager [3–7], and later a lot of achievements were obtained by Chen [8]. All the above methods could only solve the limit bearing capacity for geotechnical engineering, and the stress–strain relation was uncared for in the analysis.

In the late 1950s, the Plastic Mechanics of Geomaterial became an independent subject developed from classical plastic mechanics, modern soil mechanics, rock mechanics, and numerical simulation, such as the finite element method. In 1957, Drucker [9] pointed out that the mean stress or strain could cause the volumetric yield of geomaterial, thus a hat-style yield surface should be added in the Mohr–Coulomb cone-shaped yield space. In 1958, the concept of critical state for clay was put forward by Roscoe [10] at Cambridge University. Then the elastoplastic constitutive model was proposed for the Cam-Clay in 1963 [11], within which the characteristics of elastoplastic deformation of geomaterials were described properly. It was a precedent of the practical computation model for geomaterial. From 1970 to now, the research of the constitutive model for geomaterial has been extremely active in the world [12–16].

The monographs of Plastic Mechanics of Geomaterial have been published one after another. In 1968, *Critical State Soil Mechanics* was published by Schofield and Wroth [17]. In 1982, *Constitutive Relations for Engineering Material* was written by Chen [18]. In 1982, the concept of generalized plastic mechanics was proposed by Zienkiewicz [19], in which it pointed out that the Plastic Mechanics of Geomaterial was the promotion of classical plastic mechanics. In 1984, *Constitutive Law for Engineering Material* was fulfilled by Desai [20]. In 2002, *Plasticity and Geomechanics* was published by Davis and Selvadurai [21].

In China, the Qinghua model, Zhujiang Shen model, the model with double yield surface or multiple yield surface appeared one after another in 1980s [22–27]. In 1980, Qian [28] published the first edition of *Geotechnical Principle and Calculation*. In 1983, Huang [29] fulfilled *Engineering Nature of Soil*. In 1987, Qu [30] finished *Plastic Mechanics of Soil*. In 1989, *The Basis for Plastic Mechanics of Geomaterial* was completed by Zheng [31]. In 1990, Gong [32] published *Plastic Mechanics of Soil*. Gong [33] accomplished *Constitutive Equations of Engineering Materials*. Qian and Yin [34] achieved *Geotechnical Principle and Calculation (second edition)*. In 2000, Shen [35] presented *Theoretical Soil Mechanics*. In 2002, *The Principle of the Plastic Mechanics of Geomaterial-the Generalized Plastic Mechanics* was published by Zheng [36]. In 2004, Zhang [37] put forward *Fundamentals of Plastic Mechanics of Geomaterial*, and Li [38] brought up *Advance Soil Mechanics*. In 2007, Jiang [39] fulfilled *Constitutive Models of Soil*, and Yang [40] finished *Generalized Potential Theory of Soil Constitutive Model and Its Application*. The uniform hardening model was proposed for soil by Yao [41]. In 2010, *Plastic Mechanics* of *Geomaterial* was published by Zheng [42]. However, the current plastic theory of geomaterial is far from perfect, some basic concepts are not clear, and some theories and the models lack the confirmation of scientific experiment. Now the plastic theory of geomaterial is in the developing stage, and the developing direction of Plastic Mechanics of Geomaterial would include the following points:

1. The current geomaterial model could not reflect the deformation mechanism of geomaterial, and some of them lack the strict theoretic basis. Thus, the urgent matter is to clarify the basic concept of Plastic Mechanics of Geomaterial and establish the system of generalized plastic mechanics which is adaptable to the deformation mechanism of geomaterial.
2. The accuracy of numerical simulation is decided not only by the strict scientific theory but also the mechanical parameters conforming to the actual situation. Thus, in the developing stage of the geomechanics, we must persist in combining the theory, the experiment and the engineering practice, and improving the measure instrument and method.

3. The deep level plastic theory of geomaterial should be further developed and the practical model should be established for the complex load, aeolotropy, and unsaturated soil.
4. We should explore the new theory and the new model, introduce the new theory such as the damage mechanics, continuum mechanics and the intelligent algorithm for Plastic Mechanics of Geomaterial, and start a new generation constitutive model of geomaterial by combining macroscopical scale with microcosmic scope.
5. Stability, strain soften, damage, strain localization, and shear zone of the geomaterial should be researched; then, the real destruction process could be described properly. Although these researches have just started, they are very important in judging the stabilization and the damage in the geotechnical engineering, and these theories would certainly become the important constituents in the Plastic Mechanics of Geomaterials.

Questions

1. What is the essence of plastic deformation?
2. What are the basic principles of establishing the constitutive model?

References

1. Rankine WJM (1857) On the stability of loose earth. Philos Trans R Soc Lond 147:9–27
2. Terzaghi K (1943) Theoretical soil mechanics. Wiley, New York
3. Drucker DC, Prager W (1952) Soil mechanics and plastic analysis on limit design. J Appl Math 10(2):157–165
4. Drucker DC, Prager W (1952) Soil mechanics and plastic analysis or limit design. Q Appl Math 10(2):157–164
5. Drucker DC (1953) Coulomb friction, plasticity, and limit loads. J Appl Math 21(1):71–74
6. Prager W (1954) Limit analysis and design. Appl Mech Rev 7:421–423
7. Brady WG, Drucker DC (1953) An experimental investigation and limit analysis of net area in tension. Trans ASCE 120:1133–1154
8. Chen WF (1975) Limit analysis and soil plasticity. Elsevier Scientific Publish Company
9. Drucker DC, Gibson RE, Henkel DD (1957) Soil mechanics and work hardening theories of plasticity. Trans ASCE 122:338–346
10. Roscoe KH, Schofield AN, Wroth CP (1958) On the yielding of soils. Geotechnique 8(1):22–53
11. Roscoe KH, Schofield AN, Thurairajah A (1963) Yielding of clays in states wetter than critical. Geotechnique 13(3):211–240
12. Duncan JM (1970) Nonlinear analysis of stress and strain in soils. J Soil Mech Found Div (ASCE) 96(5):1629–1653
13. Lade PV (1977) Elasto-plastic stress-strain theory for cohesionless soil with curved yield surfaces. Int J Solids Struct 13(11):1019–1035
14. Kim MK, Lade PV (1988) Single hardening constitutive model for frictional materials I. Plastic potential function. Comput Geotech 5(4):307–324
15. Lade PV, Kim MK (1988) Single hardening constitutive model for frictional materials II. Yield critirion and plastic work contours. Comput Geotech 6(1):13–29
16. Lade PV, Kim MK (1988) Single hardening constitutive model for frictional materials III. Comparisons with experimental data. Comput Geotech 6(1):31–47

17. Schofield A, Wroth P (1968) Critical state soil mechanics. McGraw-Hill, London
18. Chen WF, Saleeb AF, Dvorak GJ (1983) Constitutive equations for engineering materials, volume I: elasticity and modeling. J Appl Mech 50(3):269–271
19. Zienkiewicz OC (1982) Soils and other saturated media under transient, dynamic conditions: general formulation and the validity of various simplifying assumptions. In: Soil mechanics-transient and cyclic loads, pp 1–16
20. Desai CS, Siriwardane HJ (1984) Constitutive laws for engineering materials: with emphasis on geologic materials. Prentice Hall Incorporated, Englewood Cliffs
21. Davis RO, Selvadurai APS (2002) Plasticity and geomechanics. Cambridge University Press, UK
22. Shen ZJ (1980) The rational form of stress-strain relationship of soils based on elastoplastic theory. Chin J Geotech Eng 2(2):11–19
23. Shen ZJ (1984) A stress-strain model for soils with three yield surfaces. Acta Mech Solida Sin 2:163–174
24. Shen ZJ (1985) Elastoplastic analysis of consolidation deformation of soft soil foundation. Chin Sci (Ser A) 15(11):1049–1060
25. Li GX (1985) Study and verification of three dimensional constitutive relationship of soils Doctoral Dissertation, Tsinghua University, Beijing
26. Pu JL, Li GX (1986) The constitutive relationship of soil and its verification and application. J Geotech Eng 8(1):47–82
27. Yin ZZ (1988) A double yield surface stress-strain model of soil. Chin J Geotech Eng 8(4):64–71
28. Qian JH (1981) Geotechnical principle and calculation. China Hydraulic Press, Beijing
29. Huang WX (1983) Engineering nature of soil. China WaterPower Press, Beijing
30. Qu ZJ (1987) Plastic mechanics of soil. Chengdu University of Science and Technology Press, Chengdu
31. Zheng YR, Gong XN (1989) Fundamentals of plastic mechanics of geomaterial. China Architecture & Building Press, Beijing
32. Gong XN (1990) Plastic mechanics of soil. Zhejiang University Press, Hangzhou
33. Gong XN (1995) Constitutive equations of engineering materials. China Architecture & Building Press, Beijing
34. Qian JH, Yin ZZ (1994) Geotechnical principle and calculation, 2nd edn. China WaterPower Press, Beijing
35. Shen ZJ (2000) Theoretical soil mechanics. China WaterPower Press, Beijing
36. Zheng YR, Shen ZJ, Gong XN (2002) Principle of plastic mechanics of geomaterial—generalized plastic mechanics. China Architecture & Building press, Beijing
37. Zhang XY, Yan SW (2004) Fundamentals of plastic mechanics of geomaterial. Tianjin University Press, Tianjin
38. Li GX (2004) Advance soil mechanics. Tsinghua University Press, Beijing
39. Jiang PN (2007) Constitutive models of soil. Science Press, Beijing
40. Yang GH, Li GX, Jie YX (2007) Generalized potential theory of soil constitutive model and its application. China WaterPower Press, Beijing
41. Yao YP, Hou W (2008) Uniform hardening model of overconsolidated soil. J Geotech Eng 30(3):316–322
42. Zheng YR, Kong L (2010) Plastic mechanics of geomatrial. China Architecture & Building Press, Beijing

Chapter 2
Stress and Strain and Its Basic Equations

This chapter is the basis of mechanics analysis and constitutive modeling for geomaterial. It mainly introduces the description of stress state, strain state and its basic equations, and the basic concepts for the constitutive description (physical equation).

2.1 Continuum Model

Even the most casual inspection of any real soil shows clearly the random, particulate, and disordered character we associate with natural materials of geologic origin. The soil will be a mixture of particles of varying mineral (and possibly organic) content, with the pore space between particles being occupied by either water, or air, or both. Modern theories have it that model particulate behavior does directly exist, but in nearly all engineering applications, we idealize soil as a continuum: a body that may be subdivided indefinitely without altering its properties.

Relying on the continuum assumption, we can attribute familiar properties to all points in a soil body. For example, we can associate with any point x in the body a mass density ρ. In continuum mechanics, we define ρ as the limiting ratio of an elemental mass Δm and volume ΔV.

$$\rho = \lim_{\Delta V \to 0} \frac{\Delta m}{\Delta V} \tag{2.1.1}$$

Of course, we realize that we should shrink the elemental volume ΔV to zero in a real soil, and we would find a highly variable result depending on whether the point coincides with the position occupied by a particle, or by water, or by air. Thus, we interpret the density in Eq. (2.1.1) as a representative average value, as if the volume remains finite and of sufficient size to capture the salient qualities of the soil as a whole in the region of our point.

Y. Liu and Y. Zheng, *Plastic Mechanics of Geomaterial*, Springer Geophysics,
https://doi.org/10.1007/978-981-13-3753-6_2

Similar notions apply to other quantities of engineering interest. For example, there will be forces acting in the interior of the soil mass. In reality, they will be unwieldy combinations of interparticle contact forces and hydrostatic forces. We will consider appropriate average forces and permit them to be supported by continuous surfaces. We can then consider the ratio of an elemental force in an elemental area and define stresses within the soil, which is one of elementary concepts that we wish to elaborate in this chapter.

Although the concept of a continuum is elementary, it represents a powerful artifice, which enables the mathematical treatment of physical and mechanical phenomena in materials with complex internal structure such as soils. It allows us to take advantage of many mathematical tools in formulating theories of material behavior for practical engineering applications.

2.2 Stress Tensor

First, we need to identify the orientation of the surface we are interested in. This is accomplished by the construction of a unit vector $\boldsymbol{n}$ normal to the surface. Stress tensor, which in the infinitesimal area with normal at any internal point p in a continuum, is the ratio of all external traction $\boldsymbol{F}$ (including the force and moment) and its area.

$$\boldsymbol{\sigma} = \frac{\mathrm{d}\boldsymbol{F}}{\mathrm{d}s} \tag{2.2.1}$$

where $\boldsymbol{\sigma}$, $\mathrm{d}\boldsymbol{F}$, and $\mathrm{d}s$ are stress, external traction, and area, respectively.

Stress is clearly associated with the area of the specific direction. All independent direction vectors and its corresponding stress are called stress state at the point (Figs. 2.1 and 2.2).

Stress state is very complex and its number is infinite, but there is only nine independent stress components in 3D stress space.

$$\begin{cases} \boldsymbol{F}_x = \sigma_{11}\boldsymbol{i} + \sigma_{12}\boldsymbol{j} + \sigma_{13}\boldsymbol{k} \\ \boldsymbol{F}_y = \sigma_{21}\boldsymbol{i} + \sigma_{22}\boldsymbol{j} + \sigma_{23}\boldsymbol{k} \\ \boldsymbol{F}_z = \sigma_{31}\boldsymbol{i} + \sigma_{32}\boldsymbol{j} + \sigma_{33}\boldsymbol{k} \end{cases} \tag{2.2.2}$$

where $\boldsymbol{F}_x$, $\boldsymbol{F}_y$, and $\boldsymbol{F}_z$ are stresses acting on the surface with the normal identical to the coordinate axis; $\boldsymbol{i}, \boldsymbol{j}$, and $\boldsymbol{k}$ are the unit vectors of the coordinate axis; σ_{ij} is the stress component.

Fig. 2.1 The stress space

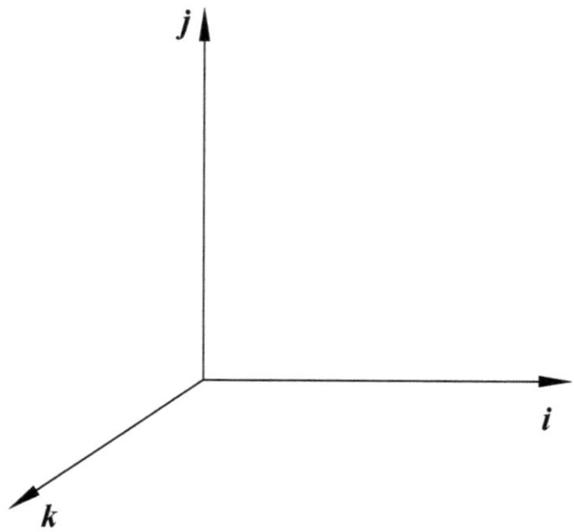

Fig. 2.2 Components of the stress matrix acting on a surface perpendicular to the $x(i)$ direction

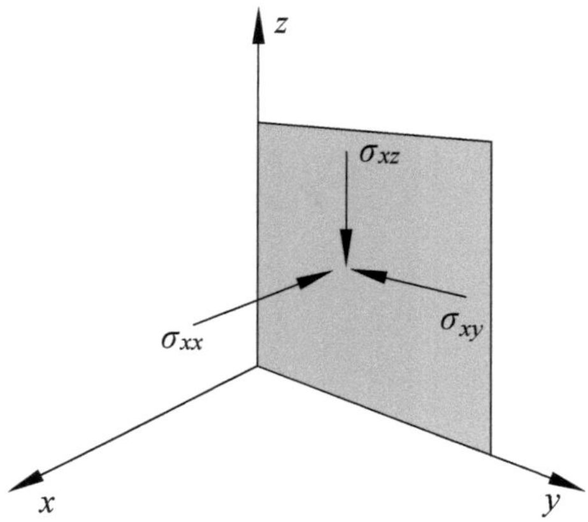

The stress vector acting on the infinitesimal area with the normal direction of $\boldsymbol{N}$ is

$$\begin{cases} f_x = n_1\sigma_{11} + n_2\sigma_{21} + n_3\sigma_{31} \\ f_y = n_1\sigma_{12} + n_2\sigma_{22} + n_3\sigma_{32} \\ f_z = n_1\sigma_{13} + n_2\sigma_{23} + n_3\sigma_{33} \end{cases} \tag{2.2.3}$$

where $\boldsymbol{N} = n_1\boldsymbol{i} + n_2\boldsymbol{j} + n_3\boldsymbol{k}$.

Equation (2.2.3) also could be expressed as

$$\boldsymbol{F}_{\mathrm{n}} = \begin{bmatrix} \sigma_{11} & \sigma_{12} & \sigma_{13} \\ \sigma_{21} & \sigma_{22} & \sigma_{23} \\ \sigma_{31} & \sigma_{32} & \sigma_{33} \end{bmatrix} \begin{bmatrix} n_1 \\ n_2 \\ n_3 \end{bmatrix} = \boldsymbol{\sigma} \cdot \boldsymbol{N} \tag{2.2.4}$$

That is to say that stress state can be expressed in a second-order tensor.

$$\boldsymbol{\sigma} = \begin{bmatrix} \sigma_{11} & \sigma_{12} & \sigma_{13} \\ \sigma_{21} & \sigma_{22} & \sigma_{23} \\ \sigma_{31} & \sigma_{32} & \sigma_{33} \end{bmatrix} \tag{2.2.5}$$

In general, stress tensor is symmetrical. So, there are six independent stress components.

$$\boldsymbol{\sigma} = \begin{bmatrix} \sigma_{11} & \sigma_{12} & \sigma_{13} \\ \sigma_{21} & \sigma_{22} & \sigma_{23} \\ \sigma_{31} & \sigma_{32} & \sigma_{33} \end{bmatrix} \qquad \begin{cases} \sigma_{12} = \sigma_{21} \\ \sigma_{13} = \sigma_{31} \\ \sigma_{23} = \sigma_{32} \end{cases}$$

There is such a special stress state that only normal stress acts on the surface with a particular normal direction, and no shear stress acts. Let the normal direction be $\boldsymbol{N}$.

Then, the stress vector on the surface is

$$\boldsymbol{F}_{\text{n}} = \boldsymbol{\sigma} \cdot \boldsymbol{N} = \begin{bmatrix} \sigma_{11} & \sigma_{12} & \sigma_{13} \\ \sigma_{21} & \sigma_{22} & \sigma_{23} \\ \sigma_{31} & \sigma_{32} & \sigma_{33} \end{bmatrix} \begin{bmatrix} n_1 \\ n_2 \\ n_3 \end{bmatrix} = \lambda \boldsymbol{N} \tag{2.2.6}$$

$$\boldsymbol{\sigma} \cdot \boldsymbol{N} - \lambda \boldsymbol{N} = 0$$
$$(\boldsymbol{\sigma} - \lambda \boldsymbol{I})\boldsymbol{N} = 0$$

$$\begin{bmatrix} \sigma_{11} - \lambda & \sigma_{12} & \sigma_{13} \\ \sigma_{21} & \sigma_{22} - \lambda & \sigma_{23} \\ \sigma_{31} & \sigma_{32} & \sigma_{33} - \lambda \end{bmatrix} = 0 \tag{2.2.7}$$

By solving this equation, we can get three solutions, σ_1, σ_2, and σ_3, which are called as the principal values of stress. Substituting the principal value into the Eq. (2.2.7), respectively, three principal directions can be obtained, known as the principal vector of stress, $\boldsymbol{N}_1$, $\boldsymbol{N}_2$, and $\boldsymbol{N}_3$.

In most cases, you do not need to consider the effect of the rotation of principal stress axes, thereafter the independent components of stress will be reduced to 3, such as gravity stress field.

The magnitude of the three principal values (σ_1, σ_2, and σ_3) has nothing to do with the orientation of rectangular coordinate system, also known as stress invariants, and other forms of stress invariants are as follows:

The first invariant of stress is

$$I_1 = \sigma_1 + \sigma_2 + \sigma_3 \tag{2.2.8a}$$

The second invariant of stress is

$$I_2 = -(\sigma_1\sigma_3 + \sigma_2\sigma_3 + \sigma_1\sigma_2) \tag{2.2.8b}$$

The third invariant of stress is

$$I_3 = \sigma_1\sigma_2\sigma_3 \tag{2.2.8c}$$

2.3 Decomposition of Stress Tensor and Its Invariants

2.3.1 Decomposition of Stress Tensor

In order to study conveniently, the stress tensor is decomposed into the spherical tensor and the deviatoric tensor.

$$\begin{bmatrix} \sigma_{11} & \sigma_{12} & \sigma_{13} \\ \sigma_{21} & \sigma_{22} & \sigma_{23} \\ \sigma_{31} & \sigma_{32} & \sigma_{33} \end{bmatrix} = \begin{bmatrix} p & 0 & 0 \\ 0 & p & 0 \\ 0 & 0 & p \end{bmatrix} + \begin{bmatrix} \sigma_{11} - p & \sigma_{12} & \sigma_{13} \\ \sigma_{21} & \sigma_{22} - p & \sigma_{23} \\ \sigma_{31} & \sigma_{32} & \sigma_{33} - p \end{bmatrix} \tag{2.3.1}$$

where $p = \frac{\sigma_{11}+\sigma_{22}+\sigma_{33}}{3}$ is the mean stress.

The former is called the spherical tensor of stress (namely, hydrostatic pressure, isotropic pressure) which is used to calculate the influence of isotropic component of stress. The latter is called the deviatoric tensor of stress, mainly reflecting the influence of shear stress.

The three principal values of deviatoric stress tensor are

$$S_1 = \sigma_1 - p, S_2 = \sigma_2 - p, S_3 = \sigma_3 - p \tag{2.3.2}$$

Its three invariants are

$$J_1 = S_1 + S_2 + S_3 = \sigma_1 + \sigma_2 + \sigma_3 - 3p = 0 \tag{2.3.3a}$$

$$\begin{aligned} J_2 &= -\left(S_1S_3 + S_2S_3 + S_1S_2\right) \\ &= \frac{(\sigma_1 - \sigma_2)^2 + (\sigma_3 - \sigma_2)^2 + (\sigma_1 - \sigma_3)^2}{6} \end{aligned} \tag{2.3.3b}$$

$$J_3 = S_1S_2S_3 = (\sigma_1 - p)(\sigma_2 - p)(\sigma_3 - p) \tag{2.3.3c}$$

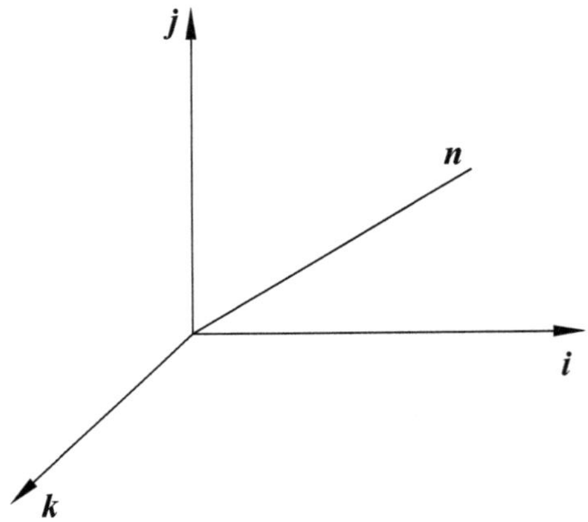

Fig. 2.3 The isoclinic lines in principal stress space

2.3.2 *Other Representation of Stress Invariant*

For convenience of use, or considering practices of different genres, different representations appear for stress invariants. The focus is the description of the sphere stress tensor and the generalized shear component. The main expression methods are octahedral plane and π plane expression.

1. **The octahedral stress**

The octahedral stress is the stress on the surface whose normal direction is identical to the isoclinic lines in principal stress space. The surface is called isoclinic surface.

The unit normal vector is as follows (Fig. 2.3).

$$\boldsymbol{n} = \begin{bmatrix} \frac{1}{\sqrt{3}} & \frac{1}{\sqrt{3}} & \frac{1}{\sqrt{3}} \end{bmatrix} \tag{2.3.4}$$

The stress on the surface would be

$$\boldsymbol{p} = \boldsymbol{\sigma} \cdot \boldsymbol{n} = \begin{bmatrix} \sigma_1 & 0 & 0 \\ 0 & \sigma_2 & 0 \\ 0 & 0 & \sigma_3 \end{bmatrix} \begin{bmatrix} \frac{1}{\sqrt{3}} \\ \frac{1}{\sqrt{3}} \\ \frac{1}{\sqrt{3}} \end{bmatrix} = \begin{bmatrix} \frac{\sigma_1}{\sqrt{3}} \\ \frac{\sigma_2}{\sqrt{3}} \\ \frac{\sigma_3}{\sqrt{3}} \end{bmatrix} \tag{2.3.5}$$

The normal stress on the surface would be the projection of the stress in its normal direction. The normal stress on the surface would be

$$\sigma_8 = \frac{\sigma_1 + \sigma_2 + \sigma_3}{3} = p \tag{2.3.6a}$$

The normal stress on isoclinic surface is equal to the magnitude of the spherical tensor of stress.

The shear stress orthogonal to the normal stress on the surface and the total stress is the synthesis of them. Then, the shear stress can be calculated by the Pythagorean Theorem.

$$\begin{aligned}
\tau_8 &= \sqrt{|\boldsymbol{p}|^2 - \sigma_8^2} \\
&= \sqrt{\frac{\sigma_1^2 + \sigma_2^2 + \sigma_3^2}{3} - p^2} \\
&= \frac{1}{3}\sqrt{(\sigma_1 - \sigma_2)^2 + (\sigma_2 - \sigma_3)^2 + (\sigma_1 - \sigma_3)^2} \\
&= \sqrt{\frac{2}{3}J_2}
\end{aligned} \tag{2.3.6b}$$

The octahedral normal stress describes the magnitude of the hydraulic tensor, and the octahedral shear stress is a characterization of the deviatoric stress tensor.

2. The generalized shear stress

The generalized shear stress is an overall reflection of the deviatoric stress tensor and is defined as

$$\begin{aligned}
q &= \frac{1}{\sqrt{2}}\sqrt{(\sigma_1 - \sigma_2)^2 + (\sigma_2 - \sigma_3)^2 + (\sigma_1 - \sigma_3)^2} \\
&= \sqrt{3J_2}
\end{aligned} \tag{2.3.7}$$

Under the conventional triaxial stress state ($\sigma_1 \neq \sigma_2 = \sigma_3$)

$$\begin{aligned}
q &= \frac{1}{\sqrt{2}}\sqrt{(\sigma_1 - \sigma_2)^2 + (\sigma_2 - \sigma_3)^2 + (\sigma_1 - \sigma_3)^2} \\
&= \sigma_1 - \sigma_3
\end{aligned} \tag{2.3.8}$$

3. The stress component on plane π

Plane π is a plane whose normal is the isoclinic lines in the principal stress space. The projection of principal stress vector on plane π is the shear stress on the plane and the projection of principal stress vector on the isoclinic line is the normal stress on plane π [1] (Fig. 2.4).

The principal stress vector is expressed as follows in principal stress space.

$$\boldsymbol{p}^{\mathrm{T}} = \begin{bmatrix} \sigma_1 & \sigma_2 & \sigma_3 \end{bmatrix}$$

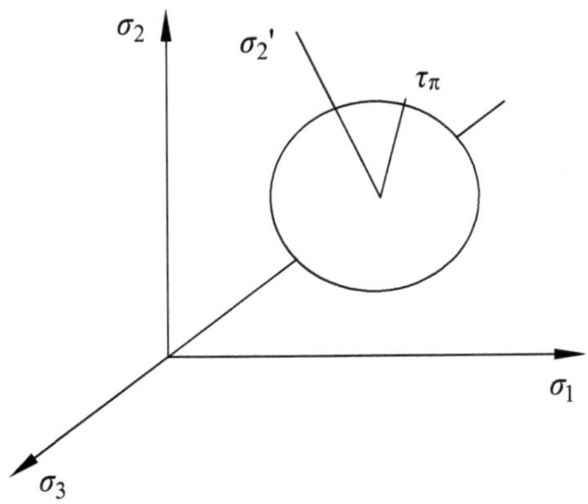

Fig. 2.4 The illustration of plane π

The normal stress on plane π is

$$\begin{aligned}\sigma_\pi &= \boldsymbol{p}\cdot\boldsymbol{n}^{\mathrm{T}} = \begin{bmatrix}\sigma_1\\ \sigma_2\\ \sigma_3\end{bmatrix}\begin{bmatrix}\frac{1}{\sqrt{3}} & \frac{1}{\sqrt{3}} & \frac{1}{\sqrt{3}}\end{bmatrix}\\ &= \frac{\sigma_1+\sigma_2+\sigma_3}{\sqrt{3}}\\ &= \sqrt{3}p \end{aligned}\tag{2.3.9}$$

Deviatoric stress on plane π is

$$\begin{aligned}\tau_\pi &= \sqrt{|\boldsymbol{p}|^2-\sigma_\pi^2}\\ &= \sqrt{\sigma_1^2+\sigma_2^2+\sigma_3^2-p^2}\\ &= \frac{1}{\sqrt{3}}\sqrt{(\sigma_1-\sigma_2)^2+(\sigma_2-\sigma_3)^2+(\sigma_1-\sigma_3)^2}\\ &= \sqrt{2J_2} = \sqrt{\frac{2}{3}}q\end{aligned}\tag{2.3.10}$$

The angle between the projection of principal stress vector ($\boldsymbol{\tau}_\pi$) and that of σ_2-axis ($\boldsymbol{\sigma}_2'$) on plane π is called Lode's Angle of stress (θ_σ).

$$\tan\theta_\sigma = \frac{2\sigma_2-\sigma_1-\sigma_3}{\sqrt{3}(\sigma_1-\sigma_3)}\tag{2.3.11a}$$

$$\theta_\sigma = \frac{1}{3}\sin^{-1}\left(\frac{-3\sqrt{3}}{2}\frac{J_3}{J_2^{\frac{3}{2}}}\right)\tag{2.3.11b}$$

Lode's Angle of stress (θ_σ) is a representation of the relative size of the intermediate principal stress.

4. **Conversion between two types of stress invariants**

Three principal values (σ_1, σ_2, and σ_3) and p, q, and θ_σ are commonly used stress invariants. What is the relationship among them?

(1) σ_1, σ_2, and σ_3 are expressed by p, q, and θ_σ

$$\begin{bmatrix} \sigma_1 \\ \sigma_2 \\ \sigma_3 \end{bmatrix} = \begin{bmatrix} p \\ p \\ p \end{bmatrix} + \frac{2}{3}q \begin{bmatrix} \sin(\theta_\sigma + \frac{2}{3}\pi) \\ \sin(\theta_\sigma) \\ \sin(\theta_\sigma - \frac{2}{3}\pi) \end{bmatrix} \tag{2.3.12}$$

(2) p, q, and θ_σ are expressed by $\sigma_1, \sigma_2, \sigma_3$

$$\begin{cases} p = \frac{\sigma_1+\sigma_2+\sigma_3}{3} \\ q = \frac{1}{\sqrt{2}}\sqrt{(\sigma_1-\sigma_2)^2+(\sigma_2-\sigma_3)^2+(\sigma_1-\sigma_3)^2} \\ \tan\theta_\sigma = \frac{2\sigma_2-\sigma_1-\sigma_3}{\sqrt{3}(\sigma_1-\sigma_3)} \end{cases} \tag{2.3.13}$$

2.4 Deformation and Strain

We begin by considering a continuum body with some generic shape similar to that as shown in Fig. 2.5. The body is placed in Fig. 2.5 in a reference system that we take to be a simple three-dimensional, rectangular Cartesian coordinate frame. A deformation of the body results in the movement from its the initial reference state to a new deformed state.

All deformations of a continuum are composed of two distinct parts. First, there are rigid motions. These are deformations for which the shape of the body is not changed in any way. Two categories of rigid motion are possible, rigid translation and rigid rotation. A rigid translation simply moves the body from one location in space to another without changing its attitude in relation to the coordinate directions. A rigid rotation changes the attitude of the body but not its position.

The second part of our deformation involves all the changes of shape of the body. It may be stretched, or twisted, or inflated or compressed. These sorts of deformations result in straining. Strains are usually the most interesting aspect of a deformation.

One way to characterize any deformation is to assign a displacement vector to every point in the body. The displacement vector joins the position of a point in the reference configuration to its position in the deformed configuration. We represent the vector by

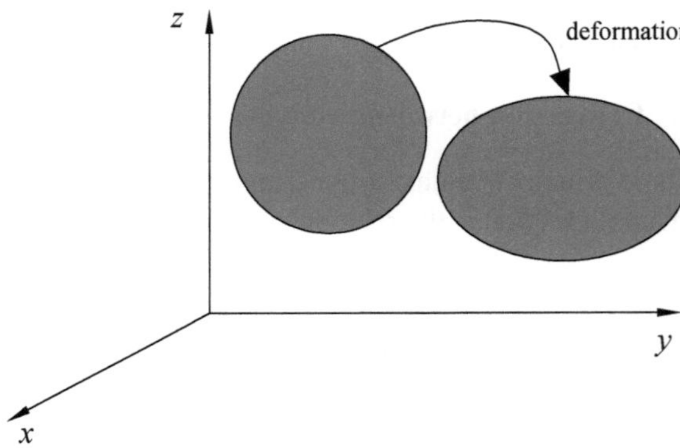

Fig. 2.5 Reference and deformed configurations of body

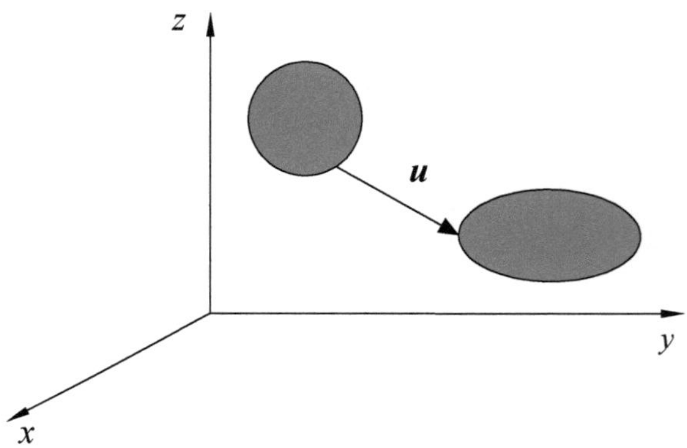

Fig. 2.6 The displacement vector

$$\boldsymbol{u} = \boldsymbol{u}(x, t)$$

where x is the position of any point within the body; t is time.

A typical displacement vector is shown in Fig. 2.6. Since there is a displacement vector associated with every point in the body, we say there is a displacement vector field covering the body. In our x, y, and z coordinate frame, $\boldsymbol{u}$ has components denoted by u_x, u_y, and u_z. Each component is, in general, a function of position and time, and according to our sign convention, components acting in negative coordinate directions will be considered to be positive [2].

If we know the displacement vector field, then we have complete knowledge of the deformation. Of course, part of the displacement field may be involved with rigid motions while the remainder results from straining. Our first task is to separate the two.

We begin by taking spatial derivatives of the components of the displacement vector. We arrange the derivatives into a 3×3 matrix called the displacement gradient matrix $\nabla \boldsymbol{u}$. If we are working in a three-dimensional rectangular Cartesian coordinate system, we can represent $\nabla \boldsymbol{u}$ in an array as follows.

$$\nabla \boldsymbol{u} = \begin{bmatrix} \frac{\partial u_x}{\partial x} & \frac{\partial u_x}{\partial y} & \frac{\partial u_x}{\partial z} \\ \frac{\partial u_y}{\partial x} & \frac{\partial u_y}{\partial y} & \frac{\partial u_y}{\partial z} \\ \frac{\partial u_z}{\partial x} & \frac{\partial u_z}{\partial y} & \frac{\partial u_z}{\partial z} \end{bmatrix} \tag{2.4.1}$$

where $\nabla = \frac{\partial}{\partial x}\boldsymbol{i} + \frac{\partial}{\partial y}\boldsymbol{j} + \frac{\partial}{\partial z}\boldsymbol{k}$, $\boldsymbol{i}, \boldsymbol{j}$, and $\boldsymbol{k}$ is the triad of unit base vectors.

Note the use of partial derivatives. Note also that the derivatives of $\boldsymbol{u}$ will not be affected by rigid translations. This might suggest we could use Eq. (2.4.1) as a measure of strain. But rigid rotations will give rise to nonzero derivatives of $\boldsymbol{u}$, so we need to introduce one more refinement. We use the symmetric part of $\nabla\boldsymbol{u}$. Let

$$\boldsymbol{\varepsilon} = \frac{1}{2}\left[\nabla\boldsymbol{u} + (\nabla\boldsymbol{u})^{\mathrm{T}}\right] \tag{2.4.2}$$

We call $\boldsymbol{\varepsilon}$ the strain matrix. Note that, the superscript T indicates the transpose of the displacement gradient matrix. Also note that $\boldsymbol{\varepsilon}$ is a symmetric matrix. As its name implies, $\boldsymbol{\varepsilon}$ represents the straining that occurs during our deformation. Just as the case with the displacement vector, $\boldsymbol{\varepsilon}$ is also a function of both position x and time t.

We write the components of $\boldsymbol{\varepsilon}$ as follows.

$$\boldsymbol{\varepsilon} = \begin{bmatrix} \varepsilon_{xx} & \varepsilon_{xy} & \varepsilon_{xz} \\ \varepsilon_{yx} & \varepsilon_{yy} & \varepsilon_{yz} \\ \varepsilon_{zx} & \varepsilon_{zy} & \varepsilon_{zz} \end{bmatrix} \tag{2.4.3}$$

The diagonal components of $\boldsymbol{\varepsilon}$ are referred to as extensional strains

$$\varepsilon_{xx} = \frac{\partial u_x}{\partial x}, \quad \varepsilon_{yy} = \frac{\partial u_y}{\partial y}, \quad \varepsilon_{zz} = \frac{\partial u_z}{\partial z} \tag{2.4.4}$$

Each of these represents the change in length per unit length of a material filament aligned in the appropriate coordinate direction.

The off-diagonal components of $\boldsymbol{\varepsilon}$ are called shear strains.

$$\varepsilon_{xy} = \varepsilon_{yx} = \frac{1}{2}\left(\frac{\partial u_x}{\partial y} + \frac{\partial u_y}{\partial x}\right), \quad \varepsilon_{yz} = \varepsilon_{zy} = \frac{1}{2}\left(\frac{\partial u_y}{\partial z} + \frac{\partial u_z}{\partial y}\right), \quad \varepsilon_{zx} = \varepsilon_{xz} = \frac{1}{2}\left(\frac{\partial u_z}{\partial x} + \frac{\partial u_x}{\partial z}\right) \tag{2.4.5}$$

These strains represent one-half the increase in the initially right angle between two material filaments aligned with the appropriate coordinate directions in the reference configuration (in solid mechanics, the shear strain represents the decrease in the right angle). We have the increase because of the assumption that compression is positive and our sign convention for displacements. For example, consider two filaments aligned with the x- and y-directions in the reference configuration as shown in Fig. 2.7. After the deformation, the attitude of the filaments may have changed

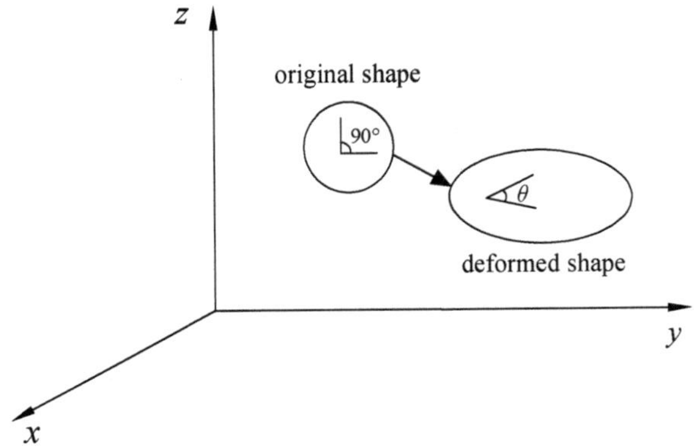

Fig. 2.7 Physical meaning of shearing strain

and the angle between them is θ now. Then $2\varepsilon_{xy} = 2\varepsilon_{yx} = \theta - \frac{\pi}{2}$. The presence of the factor of $\frac{1}{2}$ for shear strains is important to ensure that the strain matrix will give the correct measure of straining in different coordinate systems. Often, the change in an initially right angle (rather than one-half the change) is referred to as the engineering shear strain. It is usually denoted by the Greek letter gamma, γ. Obviously, if we know one of the shear strains defined in Eq. (2.4.5), we can determine the corresponding engineering shear strain.

2.5 The Invariant of Strain Tensor

Like stress, the magnitude of the three principal values of strain tensor (ε_1, ε_2, and ε_3) has nothing to do with the coordinate system, also known as strain invariants. The normal direction of the plane corresponding to three principal values is called the principal vector of strain.

In the principal strain space, the strain tensor is presented as

$$\boldsymbol{\varepsilon} = \begin{bmatrix} \varepsilon_1 & 0 & 0 \\ 0 & \varepsilon_2 & 0 \\ 0 & 0 & \varepsilon_3 \end{bmatrix} \tag{2.5.1}$$

The first, second, and third invariants of the strain are presented as follows.

$$I_1' = \varepsilon_1 + \varepsilon_2 + \varepsilon_3 \tag{2.5.2a}$$

$$I_2' = -(\varepsilon_1\varepsilon_3 + \varepsilon_2\varepsilon_3 + \varepsilon_1\varepsilon_2) \tag{2.5.2b}$$

$$I_3' = \varepsilon_1\varepsilon_2\varepsilon_3 \tag{2.5.2c}$$

The other invariant for the strain that is often defined as the volumetric strain, generalized shear strain, and strain Lode's angle, denoted by ε_v, ε_s, and θ_ε.

$$\varepsilon_v = \varepsilon_1 + \varepsilon_2 + \varepsilon_3 = \varepsilon_{xx} + \varepsilon_{yy} + \varepsilon_{zz} \tag{2.5.3}$$

$$\begin{aligned}\varepsilon_s &= \frac{\sqrt{2[(\varepsilon_1 - \varepsilon_2)^2 + (\varepsilon_1 - \varepsilon_3)^2 + (\varepsilon_3 - \varepsilon_2)^2]}}{3} \\ &= \frac{\sqrt{2[(\varepsilon_x - \varepsilon_y)^2 + (\varepsilon_y - \varepsilon_z)^2 + (\varepsilon_z - \varepsilon_x)^2 + 6(\varepsilon_{xy}^2 + \varepsilon_{xz}^2 + \varepsilon_{yz}^2)]}}{3}\end{aligned} \tag{2.5.4}$$

$$\tan\theta_\varepsilon = \frac{2\varepsilon_2 - \varepsilon_1 - \varepsilon_3}{\sqrt{3}(\varepsilon_1 - \varepsilon_3)} \tag{2.5.5}$$

2.6 Decomposition of Strain Tensor and Its Invariants

2.6.1 Decomposition of Strain Tensor

In the same way, the strain tensor is decomposed into the spherical tensor and the deviatoric tensor.

$$\begin{bmatrix} \varepsilon_{11} & \varepsilon_{12} & \varepsilon_{13} \\ \varepsilon_{21} & \varepsilon_{22} & \varepsilon_{23} \\ \varepsilon_{31} & \varepsilon_{32} & \varepsilon_{33} \end{bmatrix} = \begin{bmatrix} \varepsilon_m & 0 & 0 \\ 0 & \varepsilon_m & 0 \\ 0 & 0 & \varepsilon_m \end{bmatrix} + \begin{bmatrix} \varepsilon_{11} - \varepsilon_m & \varepsilon_{12} & \varepsilon_{13} \\ \varepsilon_{21} & \varepsilon_{22} - \varepsilon_m & \varepsilon_{23} \\ \varepsilon_{31} & \varepsilon_{32} & \varepsilon_{33} - \varepsilon_m \end{bmatrix} \tag{2.6.1}$$

where ε_m is the mean strain, and $\varepsilon_m = \frac{\varepsilon_{11}+\varepsilon_{22}+\varepsilon_{33}}{3}$.

The former is called the spherical tensor of strain which is used to calculate the influence of isotropic component of strain. The latter is called the deviatoric tensor of strain which mainly reflects the influence of shear strain.

The three principal values of deviatoric strain tensor are

$$S_1' = \varepsilon_1 - \varepsilon_m, S_2' = \varepsilon_2 - \varepsilon_m, S_3' = \varepsilon_3 - \varepsilon_m \tag{2.6.2}$$

Its first, second, and third invariants are

$$J_1' = S_1' + S_2' + S_3' = 0 \tag{2.6.3a}$$

$$\begin{aligned}J_2' &= -(S_1'S_3' + S_2'S_3' + S_1'S_2') \\ &= \frac{(\varepsilon_1 - \varepsilon_2)^2 + (\varepsilon_3 - \varepsilon_2)^2 + (\varepsilon_1 - \varepsilon_3)^2}{6}\end{aligned} \tag{2.6.3b}$$

$$J_3' = S_1'S_2'S_3' = (\varepsilon_1 - \varepsilon_m)(\varepsilon_2 - \varepsilon_m)(\varepsilon_3 - \varepsilon_m) \tag{2.6.3c}$$

2.6.2 Other Representation of Strain Invariant

For convenience of use, or considering practices of different schools, different representations appear for strain invariants. The focus is the description of the sphere strain tensor and the generalized shear component. The main expression methods are octahedral strain and π planar expression.

1. **The octahedral strain**

The octahedral stress is the stress on the surface whose normal direction is identical to the isoclinic lines in principal stress space. The surface is called isoclinic surface.

Its unit normal vector is (Fig. 2.8)

$$\boldsymbol{n} = \begin{bmatrix} \frac{1}{\sqrt{3}} & \frac{1}{\sqrt{3}} & \frac{1}{\sqrt{3}} \end{bmatrix} \tag{2.6.4}$$

The strain on the surface would be

$$\boldsymbol{\varepsilon}_8 = \boldsymbol{\varepsilon} \cdot \boldsymbol{n} = \begin{bmatrix} \varepsilon_1 & 0 & 0 \\ 0 & \varepsilon_2 & 0 \\ 0 & 0 & \varepsilon_3 \end{bmatrix} \begin{bmatrix} \frac{1}{\sqrt{3}} \\ \frac{1}{\sqrt{3}} \\ \frac{1}{\sqrt{3}} \end{bmatrix} = \begin{bmatrix} \frac{\varepsilon_1}{\sqrt{3}} \\ \frac{\varepsilon_2}{\sqrt{3}} \\ \frac{\varepsilon_3}{\sqrt{3}} \end{bmatrix} \tag{2.6.5}$$

The normal strain on the surface would be the projection of the strain on its normal direction. The normal strain on the surface would be

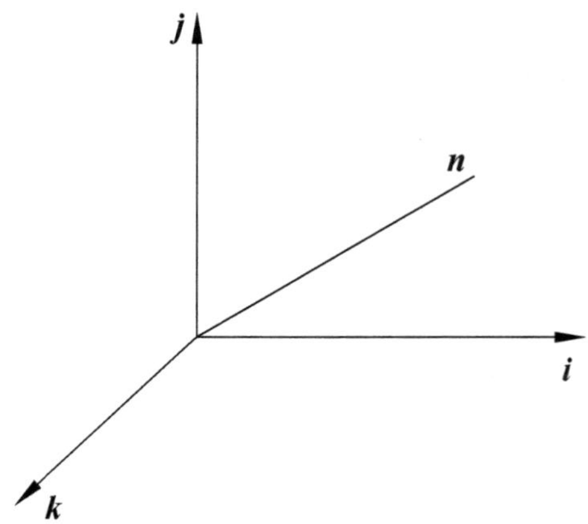

Fig. 2.8 The isoclinic lines in principal strain space

$$\boldsymbol{\varepsilon}_8 = \boldsymbol{\varepsilon}_{\mathrm{n}} = \boldsymbol{\varepsilon}_8 \cdot \boldsymbol{n}^{\mathrm{T}} = \begin{bmatrix} \frac{\varepsilon_1}{\sqrt{3}} \\ \frac{\varepsilon_2}{\sqrt{3}} \\ \frac{\varepsilon_3}{\sqrt{3}} \end{bmatrix} \begin{bmatrix} \frac{1}{\sqrt{3}} & \frac{1}{\sqrt{3}} & \frac{1}{\sqrt{3}} \end{bmatrix}$$
$$= \frac{\varepsilon_1 + \varepsilon_2 + \varepsilon_3}{3} = \varepsilon_{\mathrm{m}} \tag{2.6.6}$$

The normal strain on isoclinic surface is equal to the magnitude of the spherical tensor of strain.

Shear strain orthogonal to the normal strain on the surface, and total stress is the synthesis of them. Then, the shear strain could be calculated by the Pythagorean Theorem.

$$\gamma_8 = \sqrt{|\boldsymbol{\varepsilon}_8|^2 - \varepsilon_8^2}$$
$$= \frac{2}{3}\sqrt{(\varepsilon_1 - \varepsilon_2)^2 + (\varepsilon_2 - \varepsilon_3)^2 + (\varepsilon_1 - \varepsilon_3)^2}$$
$$= \frac{2\sqrt{2}}{\sqrt{3}}\sqrt{J_2'} \tag{2.6.7}$$

Octahedral normal stress describes the magnitude of the sphere strain tensor and octahedral shear strain is the reflection of the shear strain tensor.

2. **The generalized shear strain**

The generalized shear stress is an overall reflection of the deviatoric strain tensor and is defined as

$$\varepsilon_{\mathrm{s}} = \frac{\sqrt{2}}{3}\sqrt{(\varepsilon_1 - \varepsilon_2)^2 + (\varepsilon_2 - \varepsilon_3)^2 + (\varepsilon_1 - \varepsilon_3)^2}$$
$$= \frac{2}{\sqrt{3}}\sqrt{J_2'} \tag{2.6.8}$$

Under the conventional triaxial strain state ($\varepsilon_1 \neq \varepsilon_2 = \varepsilon_3$)

$$\varepsilon_{\mathrm{s}} = \frac{\sqrt{2}}{3}\sqrt{(\varepsilon_1 - \varepsilon_2)^2 + (\varepsilon_2 - \varepsilon_3)^2 + (\varepsilon_1 - \varepsilon_3)^2}$$
$$= \frac{2}{3}(\varepsilon_1 - \varepsilon_2) \tag{2.6.9}$$

3. **Strain component on plane π**

Plane π is a plane whose normal is the isoclinic lines in the principal strain space. The projection of principal strain vector on the plane is the shear strain on plane π, and the projection of principal strain vector on the isoclinic line is the normal strain on plane π (Fig. 2.9).

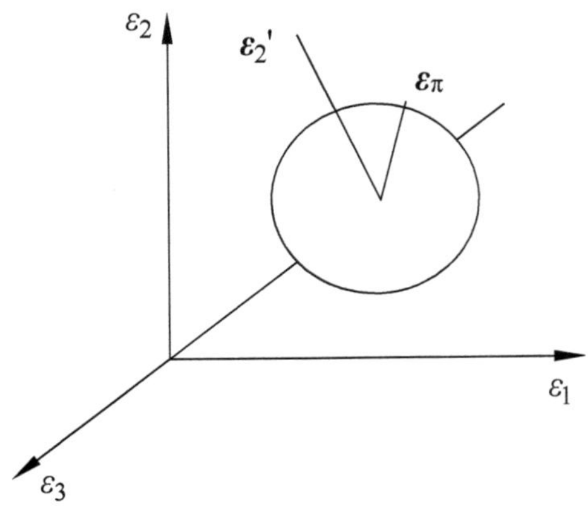

Fig. 2.9 The illustration of plane π of strain

The principal stress vector is expressed as follows in principal stress space.

$$\boldsymbol{\varepsilon}^{\mathrm{T}} = \begin{bmatrix} \varepsilon_1 & \varepsilon_2 & \varepsilon_3 \end{bmatrix}$$

Normal strain on the π plane is

$$\begin{aligned} \varepsilon_\pi &= \boldsymbol{\varepsilon} \cdot \boldsymbol{n}^{\mathrm{T}} = \begin{bmatrix} \varepsilon_1 \\ \varepsilon_2 \\ \varepsilon_3 \end{bmatrix} \begin{bmatrix} \frac{1}{\sqrt{3}} & \frac{1}{\sqrt{3}} & \frac{1}{\sqrt{3}} \end{bmatrix} \\ &= \frac{\varepsilon_1 + \varepsilon_2 + \varepsilon_3}{\sqrt{3}} \\ &= \sqrt{3}\varepsilon_{\mathrm{m}} \end{aligned} \tag{2.6.10}$$

Deviatoric strain on the π plane is

$$\begin{aligned} \gamma_\pi &= \sqrt{|\boldsymbol{\varepsilon}|^2 - \varepsilon_\pi^2} \\ &= \frac{1}{\sqrt{3}}\sqrt{(\varepsilon_1 - \varepsilon_2)^2 + (\varepsilon_2 - \varepsilon_3)^2 + (\varepsilon_1 - \varepsilon_3)^2} \\ &= \sqrt{6}\varepsilon_{\mathrm{s}} \end{aligned} \tag{2.6.11}$$

The angle is called Lode's Angle of strain (θ_ε) between the projection of principal strain vector ($\boldsymbol{\varepsilon}_\pi$) and that of ε_2-axis ($\boldsymbol{\varepsilon}_2'$) on π plane.

$$\tan\theta_\varepsilon = \frac{2\varepsilon_2 - \varepsilon_1 - \varepsilon_3}{\sqrt{3}(\varepsilon_1 - \varepsilon_3)} \tag{2.6.12}$$

Lode's angle of strain (θ_ε) is a representation of the magnitude of the intermediate principal strain.

4. **Conversion between two types of strain invariants**

Three principal values (ε_1, ε_2, and ε_3) and ε_m, ε_s, and θ_ε are commonly used strain invariants. What is the relationship between them?

(1) ε_1, ε_2, and ε_3 are expressed by ε_m, ε_s, and θ_ε

$$\begin{bmatrix} \varepsilon_1 \\ \varepsilon_2 \\ \varepsilon_3 \end{bmatrix} = \begin{bmatrix} \varepsilon_m \\ \varepsilon_m \\ \varepsilon_m \end{bmatrix} + \varepsilon_s \begin{bmatrix} \sin(\theta_\varepsilon + \frac{2}{3}\pi) \\ \sin(\theta_\varepsilon) \\ \sin(\theta_\varepsilon - \frac{2}{3}\pi) \end{bmatrix} \tag{2.6.13}$$

(2) ε_m, ε_s, and θ_ε are expressed by ε_1, ε_2, and ε_3

$$\begin{cases} \varepsilon_m = \frac{\varepsilon_1+\varepsilon_2+\varepsilon_3}{3} \\ \varepsilon_s = \frac{\sqrt{2}}{3}\sqrt{(\varepsilon_1-\varepsilon_2)^2+(\varepsilon_2-\varepsilon_3)^2+(\varepsilon_1-\varepsilon_3)^2} \\ \tan\theta_\varepsilon = \frac{2\varepsilon_2-\varepsilon_1-\varepsilon_3}{\sqrt{3}(\varepsilon_1-\varepsilon_3)} \end{cases} \tag{2.6.14}$$

2.7 Stress Path and Strain Path

The mechanical characteristic and constitutive relation of geomaterial depend on the changing process of stress state or strain state, so it is needed to describe the evolution process of stress and strain state in the process of loading. Stress path is described as change roadmap for the stress state, and the strain path is defined as the change routes for strain state. The stress path is widely applied on the geotechnical engineering at present.

2.7.1 Expression of Stress Path

In the most general case, there are six independent stress components, namely three principal values of stress and the rotation around three principal stress axes. It is difficult to express complex stress path in general stress space, and it is not necessary. The expression methods which are commonly used are:

1. **Expression in *p*-*q* plane**
   Common stress paths are often described in p (hydraulic stress)-q (generalized shear stress) plane. Two kinds of typical stress path are shown in Fig. 2.10.

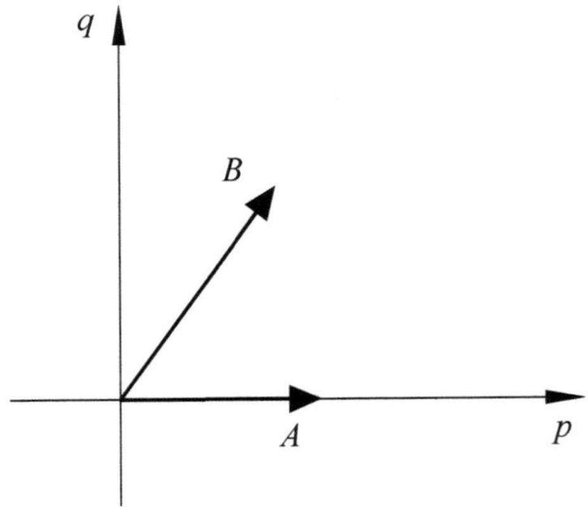

Fig. 2.10 The schematic diagram for stress path in *p*-*q* plane

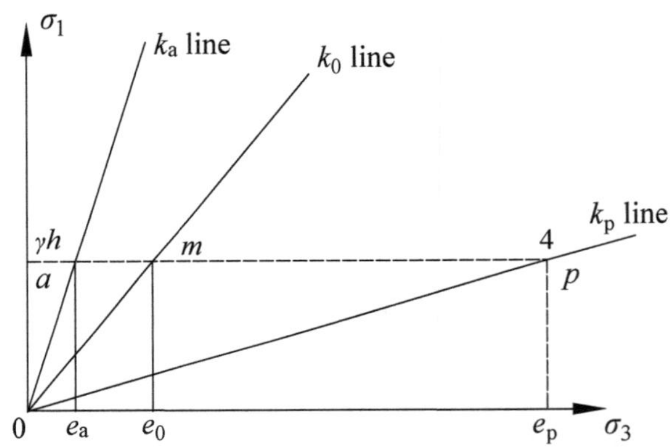

Fig. 2.11 Stress path of earth pressure of sand in principal stress plane σ_1–σ_3

(1) Hydraulic stress test (stress path *A*)
Under this stress path, $\sigma_1 = \sigma_2 = \sigma_3 \uparrow, p = \sigma_1 = \sigma_2 = \sigma_3 \uparrow, q = 0$, namely stress state changes along the *p*-axis.

(2) Pseudo triaxial test (stress path *B*)
Under this stress path, $c = \sigma_2 = \sigma_3 < \sigma_1 \uparrow$, $dq/dp = d\sigma_1/(d\sigma_1/3) = 3$, namely the stress path is a straight line with the slope of 3.

2. **Principal stress plane**
Many engineering problems can be simplified as an axisymmetric one, and generally can be simplified as stress state $c = \sigma_2 = \sigma_3 < \sigma_1$. It can be described in principal stress plane σ_1–σ_3. The stress paths are shown in Fig. 2.11 for earth pressure of sand, the active earth pressure, static earth pressure, and passive earth pressure, respectively.

2.7.2 *The Realization of Stress Path*

The influence of stress path has been a basic problem of geotechnical mechanics. It is the basis of revealing the effect of stress path to carry out the experimental research on the impact of stress path on the stress–strain relationship of geomaterial. In the general case, stress has six independent components. It is the most ideal that the experiment can achieve independent change of the six components, but in fact, the following methods can be achieved only at present.

1. **Uniaxial compression**

Uniaxial compression experiment is axial compression experiment under confining restriction. Only axial deformation exists under uniaxial compression. In general, it is viewed as the stress variation of a single direction. Actually, it is the change of the axial stress and horizontal stress at the same time. It is more suitable to regard it as the strain control for it is only the change of axial strain.

Under uniaxial compression, (horizontal stress) $\sigma_2 = \sigma_3 \uparrow < \sigma_1 \uparrow$ (axial stress), and (horizontal strain) $\varepsilon_2 = \varepsilon_3 = 0,\ \varepsilon_1 \uparrow$ (axial strain).

2. **Pseudo triaxial test**

Pseudo triaxial experiment can realize the change of the two principal stress components through the independent control of axial pressure and confining pressure.

Under different confining pressures, $c = \sigma_2 = \sigma_3 < \sigma_1 \uparrow$ (axial stress).

3. **True triaxial test**

It is realized for the change of the three principal stress components in true triaxial test through independent control of loads of three orthogonal directions.

Namely, three different principal stress values $\sigma_1 \neq \sigma_2 \neq \sigma_3$.

4. **The hollow torsional shear experiment**

Through independent control on the internal and external water pressure, axial pressure and the hoop torque on the hollow cylinder sample, the hollow torsional shear apparatus can realize the change of three principal stress values and the rotation around one principal stress axe.

Namely, three principal stress values $\sigma_1 \neq \sigma_2 \neq \sigma_3$, one rotational angle $\theta_1 \neq 0$.

2.7.3 *Total Stress Path and Effective Stress Path*

According to the Tersaghi principle of effective stress

$$\boldsymbol{\sigma}' = \boldsymbol{\sigma} - \boldsymbol{u} \tag{2.7.1}$$

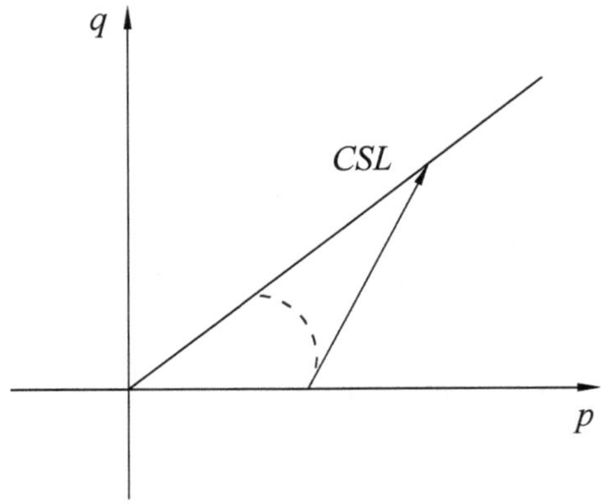

Fig. 2.12 The effective stress path and total stress path for undrained pseudo triaxial test

where $\boldsymbol{\sigma}'$, $\boldsymbol{\sigma}$, and $\boldsymbol{u}$ are the effective stress, total stress, and the pore water pressure, respectively.

It is consistent for total stress path and effective stress path under the condition of drainage, but it is different under undrained condition.

The stress path of the normal consolidated soil is shown in Fig. 2.12 for undrained pseudo triaxial shear test (the line for the total stress path, the dotted line for the effective stress path).

2.7.4 Strain Path

Strain path is the change process of strain state with time. It is always described in the principal strain space, ε_V–ε_S plane, or plane π of strain.

The advantage of the strain path description is of uniqueness and has nothing to do with drainage condition. The strain paths of sand earth pressure are shown in Fig. 2.13 in the plane of the principal strain.

In general geotechnical engineering, the deformation is deserved to calculate under a certain load. It is natural to get the stress changes at first, and then to determine the change process of the strain. So, the stress path description is more convenient.

2.8 Basic Equations of Plastic Mechanics of Geomaterial

Plastic Mechanics of Geomaterial is a branch of Continuum Mechanics, and the problem-solving framework is consistent with Continuum Mechanics. The computation is based on the three kinds of basic equation (conservation equation, geometric equation, and constitutive equation), and combining the boundary and initial conditions. These basic equations are described, respectively.

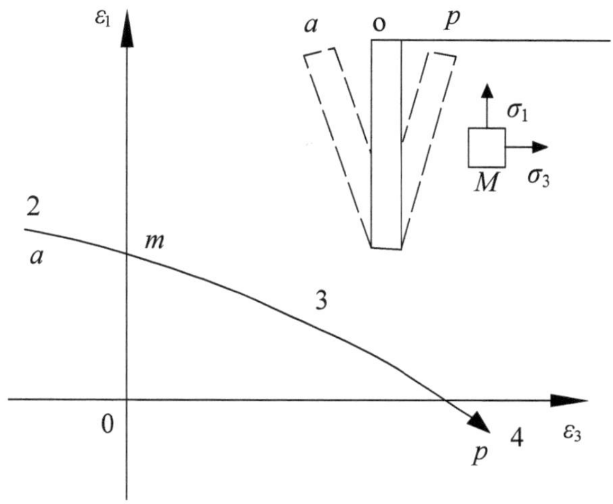

Fig. 2.13 The strain path of sand earth pressure in the plane of the principal strain

2.8.1 Basic Equations

There are three kinds of basic equations in Plastic Mechanics, conservation equation, geometric equation, and constitutive equation.

1. **Conservation equation**

All behavior of geotechnical material must satisfy all the physical conservation equation such as energy conservation, momentum conservation. The momentum conservation is used commonly in engineering, and is always presented as balance equation or equations of motion.

Equations of motion

$$\begin{cases} \dfrac{\partial \sigma_{xx}}{\partial x} + \dfrac{\partial \sigma_{xy}}{\partial y} + \dfrac{\partial \sigma_{xz}}{\partial z} + F_x = \rho \dfrac{\partial^2 u}{\partial t^2} \\ \dfrac{\partial \sigma_{yx}}{\partial x} + \dfrac{\partial \sigma_{yy}}{\partial y} + \dfrac{\partial \sigma_{yz}}{\partial z} + F_y = \rho \dfrac{\partial^2 v}{\partial t^2} \\ \dfrac{\partial \sigma_{zx}}{\partial x} + \dfrac{\partial \sigma_{zy}}{\partial y} + \dfrac{\partial \sigma_{zz}}{\partial z} + F_z = \rho \dfrac{\partial^2 w}{\partial t^2} \end{cases} \tag{2.8.1}$$

where F_x, F_y, and F_z are the body force; u, v, and w are displacement in the x-, y-, and z-direction; ρ is the density.

To the static problem, the equations of motion degenerate to the balance equation.

$$\begin{cases} \frac{\partial \sigma_{xx}}{\partial x} + \frac{\partial \sigma_{xy}}{\partial y} + \frac{\partial \sigma_{xz}}{\partial z} + F_x = 0 \\ \frac{\partial \sigma_{yx}}{\partial x} + \frac{\partial \sigma_{yy}}{\partial y} + \frac{\partial \sigma_{yz}}{\partial z} + F_y = 0 \\ \frac{\partial \sigma_{zx}}{\partial x} + \frac{\partial \sigma_{zy}}{\partial y} + \frac{\partial \sigma_{zz}}{\partial z} + F_z = o \end{cases} \tag{2.8.2}$$

2. **Geometric equation**

The geometric equations show the relationship between strain and displacement. In small strain stage, the geometric equations can be expressed as

$$\begin{cases} \varepsilon_x = \frac{\partial u}{\partial x} \\ \varepsilon_y = \frac{\partial v}{\partial y} \\ \varepsilon_z = \frac{\partial w}{\partial z} \\ \varepsilon_{xy} = \frac{\frac{\partial u}{\partial y} + \frac{\partial v}{\partial x}}{2} \\ \varepsilon_{xz} = \frac{\frac{\partial u}{\partial z} + \frac{\partial w}{\partial x}}{2} \\ \varepsilon_{yz} = \frac{\frac{\partial v}{\partial z} + \frac{\partial w}{\partial y}}{2} \end{cases} \tag{2.8.3}$$

An important concept with regard to deformation and strain is the idea of strain compatibility. In simplest terms, this is the physically reasonable requirement that when an intact body deforms, it does so without the development of gaps or overlaps. The French mathematician Saint-Venant solved the general problem of strain compatibility. He showed that the strain components must satisfy a set of six compatibility equations shown in Eq. (2.8.4). Equation (2.8.4) given below ensure that Eq. (2.8.3) can be integrated to yield single-valued and continuous displacements.

$$\begin{cases} \frac{\partial^2 \varepsilon_{xx}}{\partial y^2} + \frac{\partial^2 \varepsilon_{yy}}{\partial x^2} = 2\frac{\partial^2 \varepsilon_{xy}}{\partial x \partial y} \\ \frac{\partial^2 \varepsilon_{yy}}{\partial z^2} + \frac{\partial^2 \varepsilon_{zz}}{\partial y^2} = 2\frac{\partial^2 \varepsilon_{yz}}{\partial y \partial z} \\ \frac{\partial^2 \varepsilon_{zz}}{\partial x^2} + \frac{\partial^2 \varepsilon_{xx}}{\partial z^2} = 2\frac{\partial^2 \varepsilon_{xz}}{\partial x \partial z} \\ \frac{\partial^2 \varepsilon_{xx}}{\partial y \partial z} = -\frac{\partial^2 \varepsilon_{yz}}{\partial x^2} + \frac{\partial^2 \varepsilon_{zx}}{\partial x \partial y} + \frac{\partial^2 \varepsilon_{xy}}{\partial x \partial z} \\ \frac{\partial^2 \varepsilon_{yy}}{\partial z \partial x} = -\frac{\partial^2 \varepsilon_{zx}}{\partial y^2} + \frac{\partial^2 \varepsilon_{xy}}{\partial y \partial z} + \frac{\partial^2 \varepsilon_{yz}}{\partial y \partial x} \\ \frac{\partial^2 \varepsilon_{zz}}{\partial x \partial y} = -\frac{\partial^2 \varepsilon_{xy}}{\partial z^2} + \frac{\partial^2 \varepsilon_{yz}}{\partial z \partial x} + \frac{\partial^2 \varepsilon_{zx}}{\partial z \partial y} \end{cases} \tag{2.8.4}$$

Finally, it is perhaps worth noting that the compatibility conditions impose a kinematic constraint on the strains in a continuum where the mechanical behavior is as yet unspecified.

3. **Constitutive Equation**

Constitutive equation is mathematical model of the macro properties, also known as the constitutive relationship. Generalizing from the macroscopic experimental results, the establishment of relevant material constitutive relation is one of the important research topics in physics. The most familiar constitutive relation would be Hooke's law, Newton's law of viscosity, the ideal gas state equation, heat conduction equation, and seepage equation, etc. It is the central task of Plastic Mechanics of Geomaterial to establish a constitutive model of geomaterial in the stage of plastic deformation. It would be dissertated in detail in the following chapter in this book.

2.8.2 Boundary Condition and Initial Value

Solving of plastic mechanics problem not only requires us to know the basic equations, we also need to determine the boundary conditions and initial value.

1. **Stress boundary condition**

As shown in Fig. 2.14, the external force boundary condition of point a is (X_n, Y_n), the relation of internal stress and external force would be

$$\begin{aligned} X_n &= \sigma_x \cos\alpha + \tau_{yx} \sin\alpha \\ Y_n &= \tau_{xy} \cos\alpha + \sigma_y \sin\alpha \end{aligned} \tag{2.8.5}$$

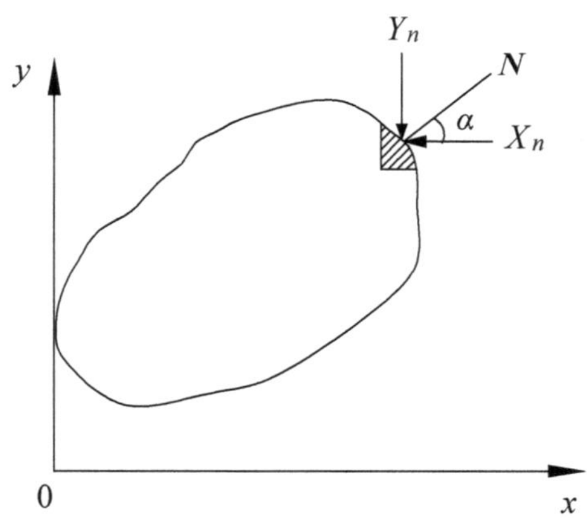

Fig. 2.14 Stress boundary condition

2. **Boundary condition of displacement**

If the displacement of boundary point i is known as $\boldsymbol{u}_0$, the displacement boundary conditions would be

$$\boldsymbol{u}_i = \boldsymbol{u}_0 \tag{2.8.6}$$

3. **Initial value**

If the displacement of point i is known at the initial moment, the initial displacement condition would be

$$\boldsymbol{u}_i|_{t=0} = \boldsymbol{u}_1 \tag{2.8.7}$$

After establishing the basic equations of Plastic Mechanics of Geomaterial, engineering problem could be solved under certain boundary conditions. It would be determined for the internal stress, strain, and displacement of materials or structures, and then the stability state could be evaluated. But now, it is rarely performed for theoretical analysis, and mainly through numerical calculation.

Questions

1. What is the number of stress state? How to understand it?
2. By calculating the eigenvalue, stress can also be expressed as the three principal values. Is it the decrease in the number of stress state?
3. What variables do we need know to get stress state of any direction of a point, and how to calculate it?
4. What are the similarities and differences between octahedral stress and stress on plane π?
5. How can we transform the two kinds of stress and strain invariants?
6. What is the concept of stress path and its realization method?
7. How can we calculate octahedral strain?
8. How can we calculate stress component on plane π?
9. What is the stress path for pure variation of Lode's Angle of stress?

References

1. Zheng YR, Kong L (2010) Geotechnical plastic mechanics. China Architecture & Building Press, Beijing
2. Davis RO, Selvadurai APS (2002) Plasticity and geomechanics. Cambridge University Presss, UK

Chapter 3
The Basic Mechanical Characteristics of the Geomaterial

The difference between Plastic Mechanics of Geomaterial and the classical plastic mechanics is determined by the distinction of mechanical characteristics for geomaterial and metal material during the plastic stage. Metal is a crystal material, but the geomaterial is the heterogeneity body which is composed of the granular materials, also being called the heterogeneity body with the friction. Thus, the geomaterial has different mechanical characteristics from the mental material.

Basic mechanical properties of geomaterial refer to the mechanical properties that have an important influence on total deformation stages of all kinds of geomaterial, and it is a sign of geomaterial different from other materials. The mechanical characteristics of geomaterial are the basis to set up its constitutive model, and the scientific evidence to evaluate a constitutive model.

3.1 Pressure-Hardening

In certain scope, the shear strength and the rigidity of geomaterial increase with the increasing of the confining stress. This characteristic is called the pressure-hardening. The shear strength of geomaterial is caused not only by the cohesion between the particles but also by the friction between these particles. This is because the geomaterial is made up of the grain stack which belongs to the frictional material, thus its shear strength is related to the angle of internal friction and the confining stress. Whereas, the metal material does not possess this characteristic, so its shear strength and rigidity have nothing to do with the confining stress.

As shown in Fig. 3.1, the phenomenon can be well described by the well-known Coulomb formula for the shear strength of geomaterial increasing with the confining pressure.

$$\tau_{\mathrm{f}} = c + \sigma \tan \varphi \tag{3.1.1}$$

Y. Liu and Y. Zheng, *Plastic Mechanics of Geomaterial*, Springer Geophysics,
https://doi.org/10.1007/978-981-13-3753-6_3

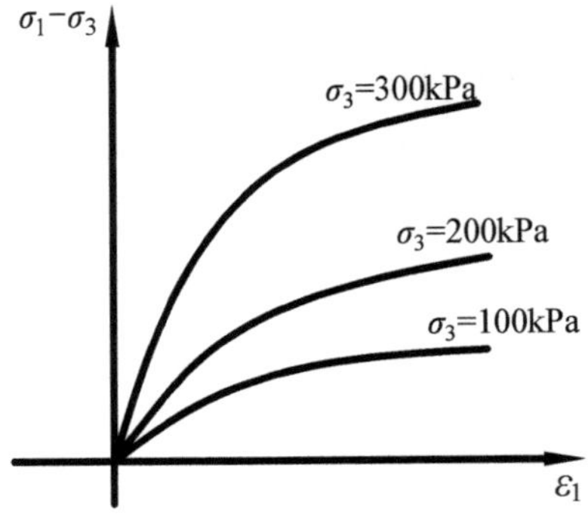

Fig. 3.1 Pressure-hardening of geomaterial

where τ, c, σ, and φ are shear strength, cohesion, normal stress, and inner friction angle, respectively.

Coulomb formula also can be expressed as the following.

$$\begin{cases} \sigma_1 = \sigma_3 \tan^2(45^\circ + \frac{\varphi}{2}) + 2c\tan(45^\circ + \frac{\varphi}{2}) \\ \sigma_3 = \sigma_1 \tan^2(45^\circ - \frac{\varphi}{2}) - 2c\tan(45^\circ - \frac{\varphi}{2}) \end{cases} \quad (3.1.2)$$

where σ_1, σ_3 are maximum principal stress and minimum principal stress, respectively.

In Fig. 3.1, it is clear that the stiffness of geomaterial increases with the increasing of confining pressure, and the initial elastic modulus can be expressed as

$$E_i = K\sigma_3^{\mathrm{n}} \quad (3.1.3)$$

where E_i, σ_3, K, and n are initial modulus, confining pressure, and two constants, respectively.

3.2 Yield Caused by Hydrostatic Pressure

The geomaterial is a porous medium. The water or gas would escape from the geomaterial after the pore collapse under the action of hydrostatic pressure. This is called the volumetric yield by hydrostatic pressure, which is significantly different from the metal. The metal just manifests the elastic deformation under large hydrostatic pressure except for the porous metal such as cast iron.

In Fig. 3.2, Loading curve of isotropic consolidation can be formulated as semilog function for geomaterial.

$$\varepsilon_{\mathrm{v}} = \lambda \ln \frac{p}{p_0} \quad (3.2.1)$$

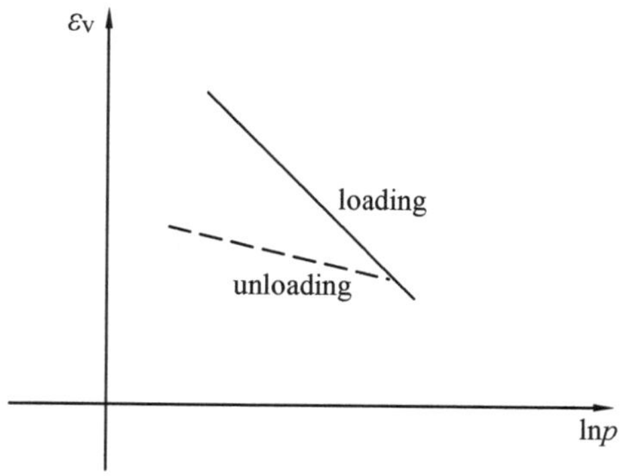

Fig. 3.2 Experiments of isotropic consolidation of soil

where ε_v, λ, p, and p_0 are volumetric strain, compression coefficient, isotropic pressure, and initial pressure, respectively.

The unloading curves of isotropic consolidation also can be expressed as semilog function for geomaterial.

$$\varepsilon_v^e = k \ln \frac{p}{p_0} \tag{3.2.2}$$

where ε_v^e, k are elastic volumetric strain, swell coefficient, respectively.

The difference between loading–unloading curves is plastic volumetric deformation caused by isotropic compression, i.e., yield of isotropic compression.

$$\varepsilon_v^p = \varepsilon_v - \varepsilon_v^e = (\lambda - k) \ln \frac{p}{p_0} \tag{3.2.3}$$

where ε_v^p is the plastic volumetric strain.

3.3 Dilatancy

The geomaterial is a heterogeneity material with the intergranular pore. Thus, the pore would collapse or expand under the action of shear stress, and then the volumetric strain appears with water or gas discharge. It is called as dilatancy for volume changes caused by shear stress. Of course, the volumetric strain of metal material is independent to the shear stress. Otherwise, the shear strain of geomaterial is also related with mean stress. These phenomena do not exist in the elasticity theory or the classical plastic mechanics. Strictly speaking, these phenomena have violated the concept of the classical continuum mechanics.

We inspect two micro units about the granular friction material (geomaterial) and the general continuous medium material (metal). Figure 3.3a is the general continuous medium material. The hydraulic stress only leads to the volumetric strain,

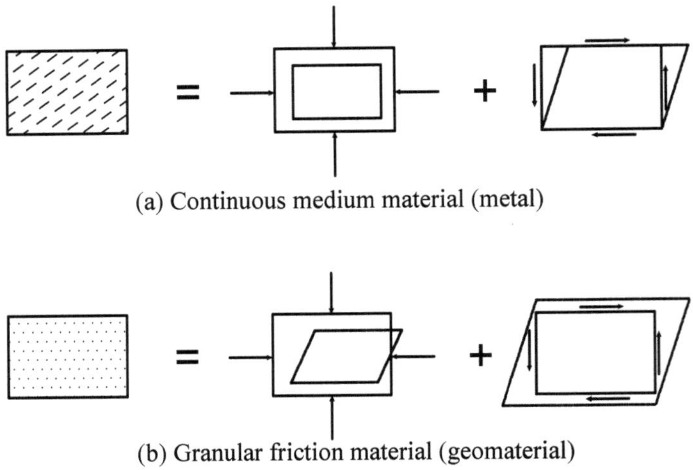

Fig. 3.3 The mechanical model of metal and geomaterial

and the deviator stress only results in the shear strain. But granular friction material infinitesimal is expressed in Fig. 3.3b, and the overlapping influence exists for hydraulic stress and deviatoric stress. Obviously, this is unreasonable in the traditional continuum mechanics. Thus, the continual material infinitesimal is not suitable for granular friction material infinitesimal (geomaterial).

It is shown in Fig. 3.4 for shear experimental results under constant hydraulic stress for soil (solid curve for overconsolidated soil, dashed curve for normal consolidation soil). In the test of stress path, the hydraulic stress is constant, and only the shear stress increases, namely constant p stress path. From Fig. 3.4 with the increase of shear stress, the volume of overconsolidated soil reduces at the beginning (negative dilatancy), and then volume expands (dilatancy). With the increase of shear stress, the volume of normal consolidation soil only shrinks without expanding, i.e., just negative dilatancy.

General continuum mechanics model can be expressed briefly as follows.

$$\begin{cases} \varepsilon_v = \dfrac{p}{K} \\ \varepsilon_s = \dfrac{q}{G} \end{cases} \tag{3.3.1}$$

where ε_v, ε_s, p, q, K, and G are the volumetric strain, generalized shear strain, hydrostatic stress, generalized shear stress, bulk modulus, and shear modulus, respectively.

General mechanical model of geomaterial should be formulated as

$$\begin{cases} \varepsilon_v = \dfrac{p}{K_1} + \dfrac{\tau}{K_2} \\ \varepsilon_s = \dfrac{p}{G_1} + \dfrac{\tau}{G_2} \end{cases} \tag{3.3.2}$$

where K_1, G_1, K_2, and G_2 are the bulk modulus, shear modulus, and two coefficients of cross influence.

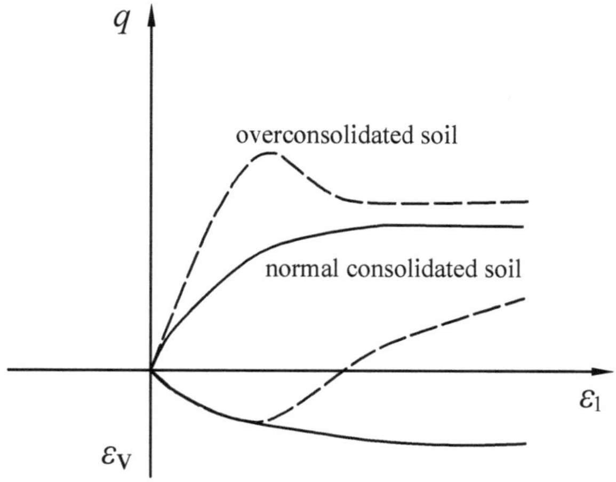

Fig. 3.4 Dilatancy of geomaterial (solid curve—overconsolidated soil, dashed curve—normal consolidated soil)

3.4 Plastic Deformation Dependent on Stress Path

The plastic deformation of geomaterial relies on the stress path. In other words, the computation parameter selecting is related to the stress path for geomaterial. For example, the direction change of the plastic strain increment would be caused by the sudden transition of stress path. In other words, the direction of plastic strain increment is related to the direction of the stress increment. Whereas, the direction of plastic strain increment is only related to the stress state, and has nothing to do with the stress increment in the classical plastic theory. Also, the plastic strain increment could not be figured out based on the classical plastic theory which is caused by the rotation of principal stress axes for the principal stress value is invariable. At present, no satisfactory solution is obtained for the question about the plastic deformation relying on the stress path which calls for the further development of the plastic mechanics.

The deformation of geomaterial depends not only on the current stress state and stress increment, but also on the stress history. For example, the influence of stress history on settlement of building a foundation: the same buildings built on the same soil with different stress history, and their deformations are completely different. The deformation is significantly smaller for overconsolidated soil, bigger for normal consolidation soil, and very larger for underconsolidated soil, and it could be significantly beyond permissible value of codes.

A lot of experiments were performed for the influence of the stress path on the mechanical characteristics of geomaterial. The experimental results provided by Anandarajah [1] are shown in Fig. 3.5 for the influence of stress path. The stress increments are put forward in Fig. 3.5a that are applied on the same stress state with equal magnitude but different direction, and the corresponding plastic strain

Fig. 3.5 The correlation of plastic strain increment and stress increment

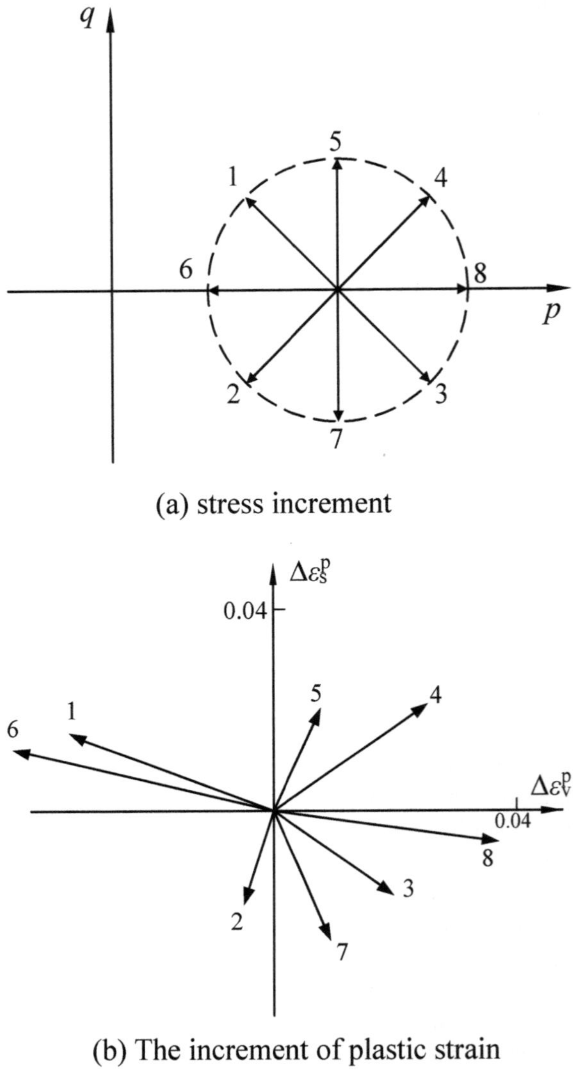

increment is shown in Fig. 3.5b. Figure 3.5 shows the magnitude of the plastic strain increment is closely related to the direction of stress increment.

Nakai [2] presented experimental results on the stress path of soil as shown in Fig. 3.6. There are four kinds of stress path in Fig. 3.6a with the same starting and end points. The corresponding plastic strain is put forward in Fig. 3.6b. Although the starting and ending points are the same for the four stress path, in the end, the magnitudes of plastic strain are different, and some disparity is very large (the plastic volumetric strain is displayed in Fig. 3.6b, and the plastic shear strain in Fig. 3.6c). It is clear for the correlation of total plastic strain and stress path.

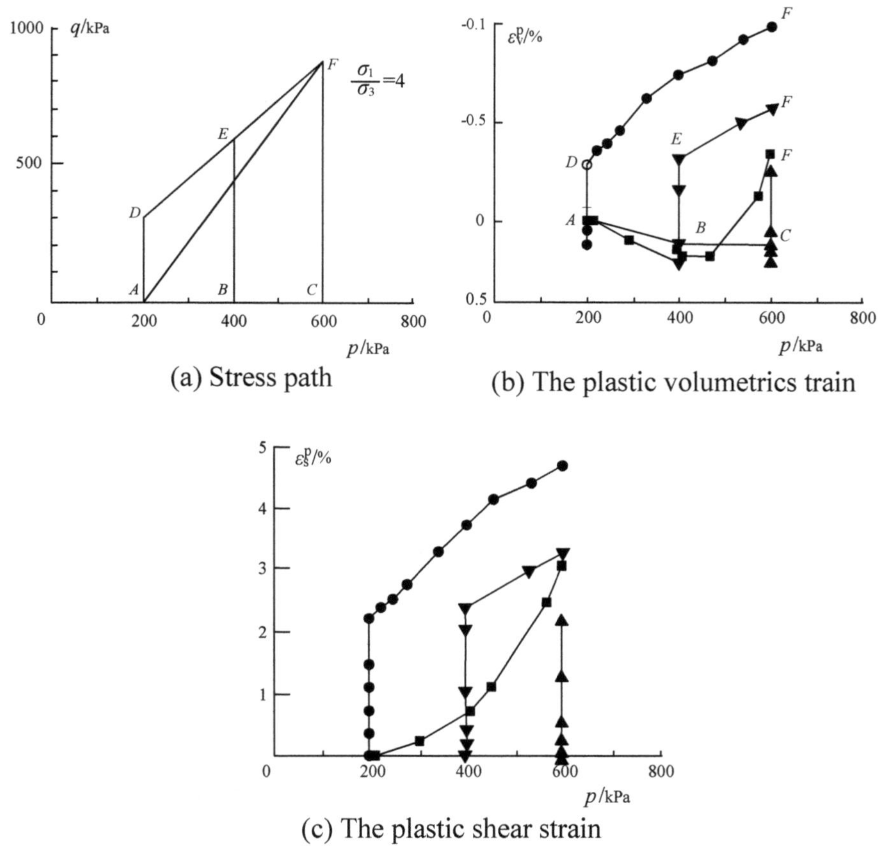

Fig. 3.6 The total plastic strain and stress path

3.5 Other Important Characteristics

The above four characteristics can be regarded as the basic characteristics for geomaterial. A reasonable constitutive model should reflect these characteristics for geomaterial. In addition, the geomaterial also possesses some characteristics different with the metal: stain softening, aeolotropy (the initial aeolotropy or the induced aeolotropy caused by the stress state) and elastoplastic coupling, etc. Being manifested at some deformation stages, these mechanical characteristics may not be embodied in all deformation process of geomaterial. It is decisive under certain special circumstances.

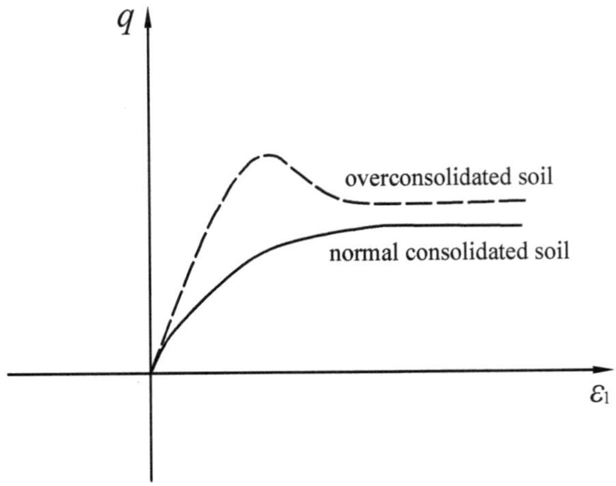

Fig. 3.7 Strain softening (dashed curve for overconsolidated soil, solid curve for normal consolidated soil)

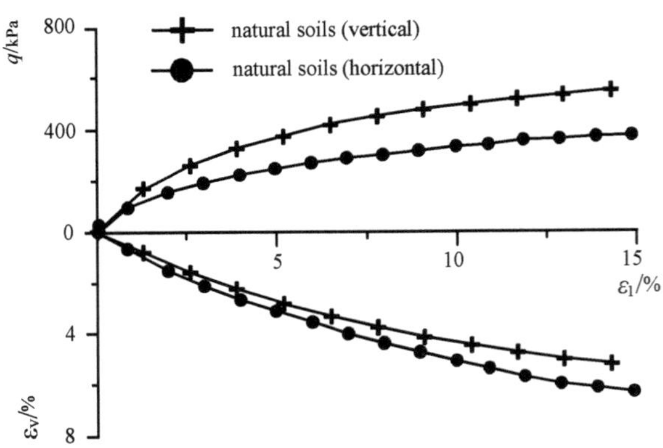

Fig. 3.8 Anisotropy of Nanjing clay

Generally, geomaterial displays strain hardening in the first stage of loading, but some also reveals the stain softening after it reaches its strength peak. Whereas, the metal material belongs to the stable material, which does not show strain softening. It is more convenient to describe the strain softening in the strain space (Fig. 3.7).

Due to the influence of sedimentary condition, the natural geomaterial has obvious native anisotropy. The experimental results of anisotropy of Nanjing clay are displayed in Fig. 3.8 [3]. The sediment direction is marked for soil sample sampled in situ. Specimens are cut along the vertical direction (sedimentary direction) and horizontal direction, respectively, for conventional triaxial test. The results are denoted as "+" for the vertical sample and the black spots for the horizontal sample. It is significant that the strength and stiffness is larger for the vertical sample (its axial direction is the same as the gravity direction).

3.6 Mechanical Characteristic at Small Strain of Geomaterial

Small strain problem in geotechnical engineering has attracted the attention of a great many research institutions in recent years. A large number of in situ tests reveal the importance of the study of small strain problem in geotechnical engineering. A large number of engineering practices shows that the deformation of rock mass is small in the general geotechnical engineering (such as tunnels and excavation), and the mechanical properties in small strain of geomaterial are difficult to be simulated by traditional constitutive model. Based on the monitor of lateral movement of soft soil caused by deep excavation of a high-rise building foundation pit in London and the settlement of large towers with soft limestone base, Burland [4] found that the strain in rock and soil was very small long time ago, usually less than 0.03%. The former foundation sedimentation measurement of a Germany top hotel on medium density sand showed that most of strain in the foundation was less than 0.1%, and the maximum value was only 0.3%. Along with the development of construction technology, requirements also have become stricter in all kinds of building codes, and strain will be smaller caused by underground engineering construction. The underground reverse construction method has been used in Tiananmen West Station of Beijing Metro Engineering, and the surface subsidence has been monitored in the tunnel construction process. The final surface settlement was small, and that of the ground was only 6 mm. The maximum strain caused by settlement was located at 9 m under the ground, and the maximum strain was only about 0.45% [5].

Currently, the analysis of stress and deformation analysis are always performed by finite element method in geotechnical engineering, and the core is the constitutive model which describes the stress–strain behavior for geomaterial. The model parameters are generally fitted on the results of the conventional triaxial experiment. The geomaterial is loaded, until they are destroyed (until the axial strain more than 15%), thus leading to the following consequences [6]:

1. The mechanical parameters are severely underestimated for the small strain stage of soil. The strain measurement is accurate in the range of 1–15% for conventional triaxial test. The secant Young's modulus of soil in the axial strain of 0.003% is 11 times more than that in the axial strain of 1%. The difference is bigger using the tangent modulus.
2. The elastic range is enlarged artificially for soil. Soil strain shows obvious non-linear when the strain is 0.01% level. With the development of the experimental method, mechanical behavior of soil is becoming more and more clear for the small strain stage. It is clear that the mechanical properties at small strain stage are difficult to be described by the traditional constitutive model.

There is significantly different mechanical behavior of the natural geomaterial from the damage state (such as the remolded soil) of the small strain region such as high rigidity, obvious nonlinearity, and aeolotropy. The experiment indicates that the natural soil shows more rigidity than the remolded soil at small strain. The strain is

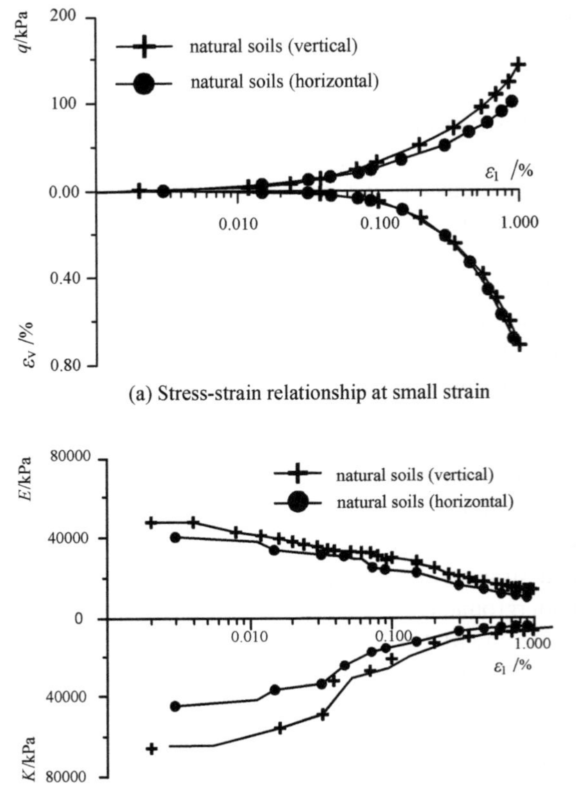

(a) Stress-strain relationship at small strain

(b) The young's modulus and the bulk modulus at small strain

Fig. 3.9 Conventional triaxial test for Nanjing undisturbed soft clay at small strain ("+" for the vertical cutting sample, "●" for horizontal cutting sample)

smaller, and the rigidity difference is bigger. Their rigidity all increase along with the strain reducing, and the natural soil is fiercer along with the strain change. Even if the strain is very small (less than 0.01%), the rigidity is not a constant. Thus, the natural soil displays nonlinearity. The aeolotropy of the natural soil is also very obvious at small strain.

In order to disclose the aeolotropy of the natural consolidated soil, the conventional triaxial tests are performed for comparison of natural soil sampling along the vertical direction and horizontal direction. The results are shown in Fig. 3.9 [3].

From Fig. 3.9, the various features can be concluded for undisturbed clay at small strain:

1. Anisotropy. As you can see, the strength and stiffness of sample cut along the vertical direction are significantly higher than the one cut horizontally for the clay. This suggests that the anisotropy cannot be ignored for natural geomaterial. The smaller the strain, the more obvious anisotropy is.
2. Nonlinear. No linear phenomenon can be seen from the graph and its stiffness and strength are changing in all the deformation stages. With the development of

testing technology, the stress–strain behavior should be nonlinear at the smaller strain region.

3. High stiffness. It is shown in Fig. 3.9b that a significant reduction of the stiffness happens with the increase of strain. That is to say, stiffness at small strain is much higher than that at the large strain for geomaterial.

3.7 Mechanical Difference for the Natural and Remolded Soil

In order to determine the mechanical parameters of geomaterial, it is needed to perform the experiments on the sample taken in situ, and some of which are just the experiments carried out with remolded sample. The following is the test for Nanjing clay, and the results reveal the mechanical difference between the natural and remolded soil.

Conventional triaxial test was carried out for the natural and remolded state of Nanjing clay. The experimental results are compared in Fig. 3.10 [3], and both of strength and stiffness of natural clay are bigger than those of the remolded clay.

The strength for the natural sample and remolded sample of Nanjing clay is shown in Fig. 3.11 [3]. From the fitting results of strength, the higher the strength of natural clay mainly displays in the larger internal friction angle.

The deformation characteristics of the natural sample and remolded one of Nanjing clay at small strain are compared in Fig. 3.12 [3]. The stiffness of natural clay is higher than the remolded one significantly at small strain. The smaller the strain is, the greater the difference of stiffness. The stiffness of both samples decreases with increasing strain, and that of natural clay decreases more obviously with larger strain.

In conclusion, stiffness and strength of undisturbed soil are obviously larger than that of the remolded soil. In order to obtain the reasonable parameters for the constitutive model, it is essential to acquire high-quality natural sample. Of course, it is best to carry out the in situ experiment.

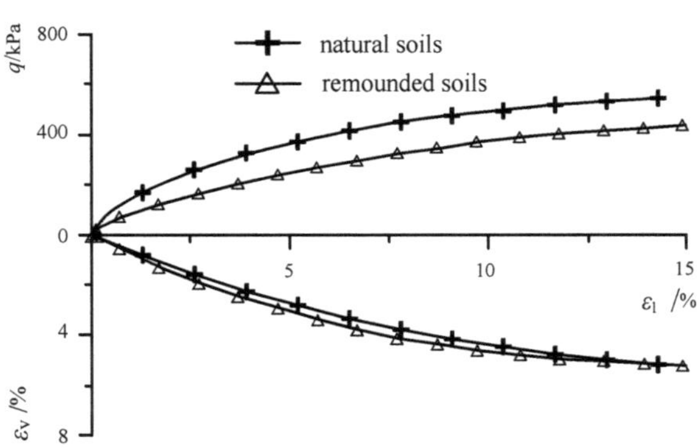

Fig. 3.10 Conventional triaxial test results for natural and remolded Nanjing clay ("+" for natural sample, "Δ" for remolded sample)

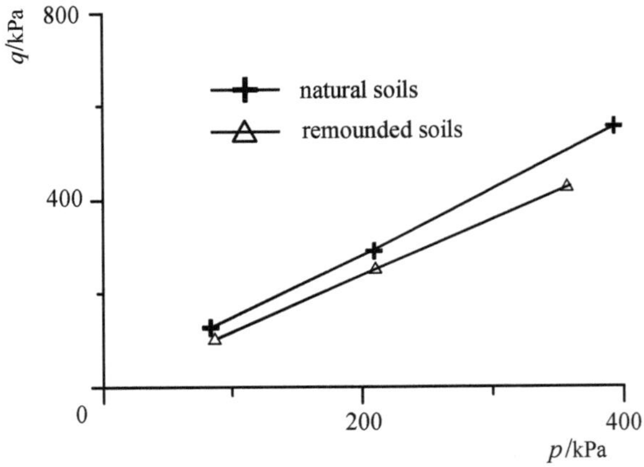

Fig. 3.11 Strength characteristics for natural and remolded clay ("+"for natural sample, "△" for remolded sample)

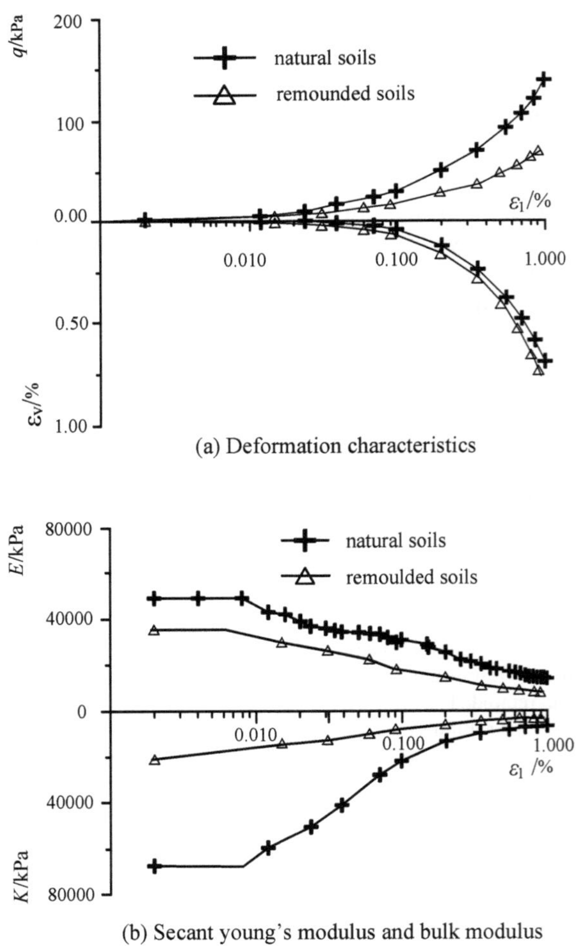

Fig. 3.12 Conventional triaxial test for natural and remolded clay at small strain ("+" for natural sample, "△" for remolded sample)

Questions

1. What are the basic mechanical characteristics of geomaterial?
2. Please design an experiment plan to verify the basic mechanical characteristics of soil by the conventional triaxial apparatus.
3. What are the mechanical characteristics of geomaterial at small strain?

References

1. Anandarajah A, Sobhan K, Kuganenthira N (1995) Incremental stress-strain behavior of granular soil. J Geotech Eng 121(1):57–68
2. Nakai T (1989) An isotropic hardening elastoplastic model considering the stress path dependency in three-dimensional stresses. Soils Found 29(1):119–137
3. Liu YX, Shi JY, Yi YF, Xu LJ (2004) Experimental study of mechanical characteristics of an incomplete consolidation silty clay. Rock Soil Mech 25(1):5–10
4. Burland JB (1989) Ninth Laurits Bjerrum memorial lectural: "small is beautiful"-the stiffness of soils at small strains. Can Geotech J 26(4):499–516
5. Luo FR, Guo B (2001) The underground reverse construction method used in Tiananmen west station of Beijing metro engineering. Chin J Geotech Eng 23(1):75–78
6. Liu YX, Shi JY, Yin GZ, Lu X (2004) A constitutive model of natural soft soils based on transformation of stress space. J Hydraul Eng 6:14–20

Chapter 4
The Elastic Model of Geomaterial

The elastic model of geomaterial is the mechanical model based on the elastic theory. There are two purposes to study the elastic model of geomaterial: one is to describe the elastic part of elastoplastic deformation, and the other is to calculate the integral calculation of geomaterial deformation.

Geomaterial is the material of nonlinear mechanical property. The development of computational models for geomaterial behavior has been strongly influenced by this property.

In certain circumstances, geomaterial (even clay) can show a substantially linear response. For instance, the large-scale field trial [1, 2] on lightly overconsolidated postglacial clay has shown that they are stiffer and more nearly linear than suggested by many models. This is confirmed by laboratory tests on specimens which have been carefully sampled and trimmed. Studies of yielding on a wide variety of natural clays have shown that they exhibit substantially linear stress–strain behavior at stresses which do not cause yielding. Stresses which pass beyond the initial yield locus produce larger strains, higher pore water pressures (undrained condition), slower excess pore water pressure dissipation, and higher creep rates.

The evidence for linear pre-yield behavior is strong and has been obtained from testing programs in many laboratories. Three examples from different clays are shown in Fig. 4.1. The initial straight sections of the curves are considered to be elastic and the break in each stress–strain curve is considered to be a yield point. The Mastemyr, Lyndhurst, and Belfast clays are quick marine silty clay, lacustrine sensitive silty clay, and organic plastic marine clay [3–5], respectively.

The purpose of this chapter is to propose mathematical techniques for describing the pre-yield or elastic part mechanical properties of geomaterial using elasticity theory, and for determining appropriate material parameters from triaxial test data.

Y. Liu and Y. Zheng, *Plastic Mechanics of Geomaterial*, Springer Geophysics,
https://doi.org/10.1007/978-981-13-3753-6_4

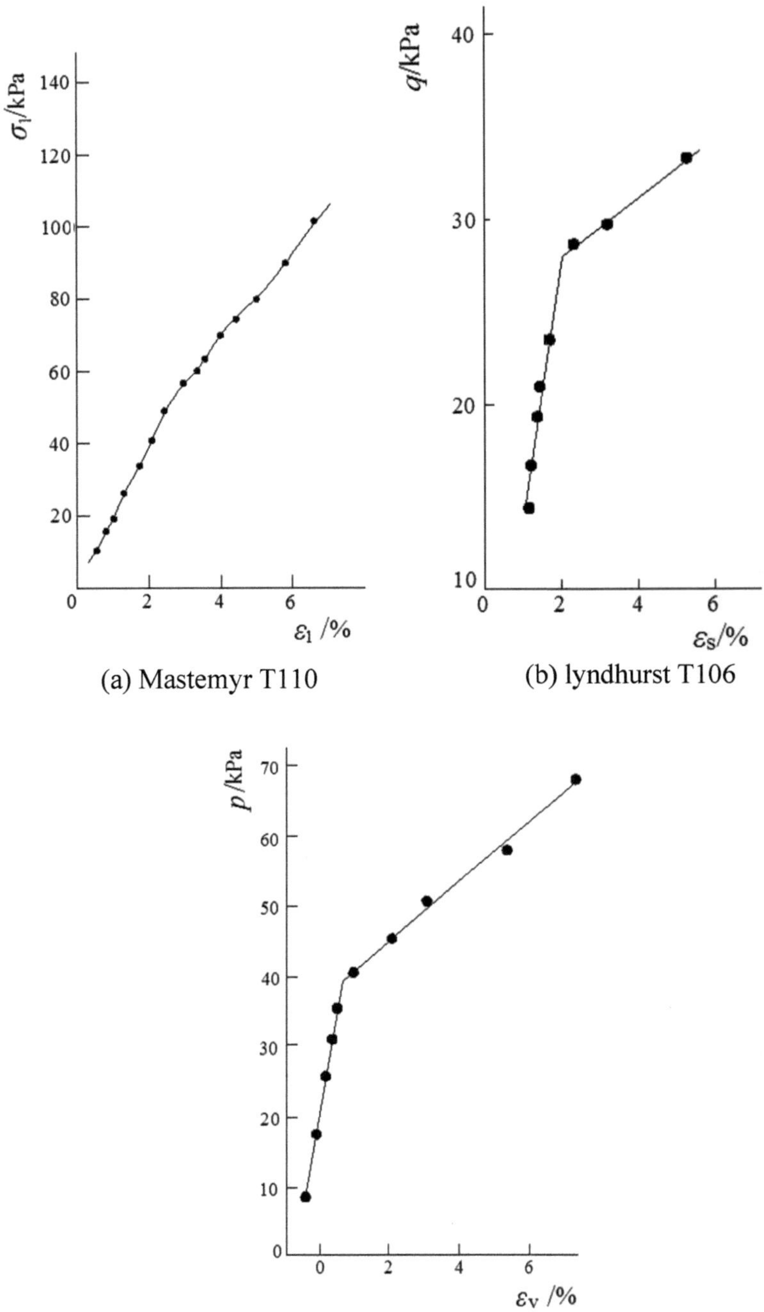

Fig. 4.1 Bilinearity and yielding in soft, lightly overconsolidated clays

4.1 Nonlinear Elastic Theory

The nonlinear elastic theory is deduced from the Hooke's law. According to the different assumptions of generalization, it is divided into three theories.

4.1.1 *Variable Elasticity Theory*

It is assumed in Variable Elastic Theory that there is a one-to-one relationship between stress and strain, not necessarily linear. The relation of stress and strain is independent of stress path, and can be described with full total strain theory, and also known as the Cauchy Model. It can be expressed as

$$\boldsymbol{\sigma} = F(\boldsymbol{\varepsilon}) \tag{4.1.1}$$

$$\text{or} \quad \boldsymbol{\varepsilon} = f(\boldsymbol{\sigma}) \tag{4.1.2}$$

There is a one-to-one relationship between stress and strain, not necessarily linear. The incremental form of Eq. (4.1.1) can be expressed as follows:

$$\mathrm{d}\boldsymbol{\sigma} = \mathrm{d}F(\boldsymbol{\varepsilon}) = \frac{\partial F(\boldsymbol{\varepsilon})}{\partial \boldsymbol{\varepsilon}}\mathrm{d}\boldsymbol{\varepsilon} = F_1(\boldsymbol{\varepsilon})\mathrm{d}\boldsymbol{\varepsilon} = F_1\left[F^{-1}(\boldsymbol{\sigma})\right]\mathrm{d}\boldsymbol{\varepsilon} = \boldsymbol{D}(\boldsymbol{\sigma})\mathrm{d}\boldsymbol{\varepsilon} \tag{4.1.3}$$

where $\boldsymbol{D}(\boldsymbol{\sigma})$ is the stiffness matrix, just a function of stress.

The incremental form of Eq. (4.1.2) can be expressed as follows:

$$\mathrm{d}\boldsymbol{\varepsilon} = \mathrm{d}f(\boldsymbol{\sigma}) = \frac{\partial f(\boldsymbol{\sigma})}{\partial \boldsymbol{\sigma}}\mathrm{d}\boldsymbol{\sigma} = f_2(\boldsymbol{\sigma})\mathrm{d}\boldsymbol{\sigma} = f_2\left[f^{-1}(\boldsymbol{\varepsilon})\right]\mathrm{d}\boldsymbol{\varepsilon} = \boldsymbol{C}(\boldsymbol{\varepsilon})\mathrm{d}\boldsymbol{\sigma} \tag{4.1.4}$$

where $\boldsymbol{C}(\boldsymbol{\sigma})$ is the flexibility matrix, only a function of strain.

The parameters of variable elastic model are just a function of stress or strain has nothing to do with the stress path.

4.1.2 *Hyperelastic Theory*

The Hyperelastic Theory has more strict restrictions on the Variable Elastic Theory, and the unique relation exists in elastic strain energy and stress or strain. It is also called Green Model. Elastic energy can be expressed as

$$W = \int \boldsymbol{\sigma}\, \mathrm{d}\boldsymbol{\varepsilon} = f(\boldsymbol{\varepsilon}) \tag{4.1.5}$$

The relation of stress and strain is as follows:

$$\boldsymbol{\sigma} = \frac{\partial W}{\partial \boldsymbol{\varepsilon}} = f'(\boldsymbol{\varepsilon}) \tag{4.1.6}$$

Incremental relationship between stress and strain is as follows:

$$\mathrm{d}\boldsymbol{\sigma} = \mathrm{d}(\frac{\partial W}{\partial \boldsymbol{\varepsilon}}) = \frac{\partial^2 W}{\partial \boldsymbol{\varepsilon} \partial \boldsymbol{\varepsilon}} \mathrm{d}\boldsymbol{\varepsilon} = \boldsymbol{D}(\boldsymbol{\varepsilon})\mathrm{d}\boldsymbol{\varepsilon} \tag{4.1.7}$$

A one-to-one relationship also exists between stress and strain, not necessarily linear, and has nothing to do with the stress path. The elastic constant is a function of stress or strain and has nothing to do with stress or strain increment.

Hyperelastic Theory looks similar to Variable Elastic Theory. Actually, the Hyperelastic Theory is more strict, requiring the existence of elastic potential W. That is to say, hyperelasticity has a concept of potential, and Laplace's equation is thus set up.

$$\Delta W = \frac{\partial^2 W}{\partial x^2} + \frac{\partial^2 W}{\partial y^2} + \frac{\partial^2 W}{\partial z^2} = 0 \tag{4.1.8}$$

4.1.3 Hypoelastic Theory

Hypoelastic Theory relaxes some restriction on the Variable Elastic Theory (Cauchy Model), assuming that there is no one-to-one relation between total stress and strain. Only elastic relationship exists in incremental sense. The description of the elasticity is not only related to the stress state, but it may also be related to the strain state and stress paths. It can be expressed as

$$\mathrm{d}\boldsymbol{\sigma} = f(\boldsymbol{\sigma}, \boldsymbol{\varepsilon}, \mathrm{d}\boldsymbol{\varepsilon}) \tag{4.1.9}$$

A simplified model can be expressed as

$$\mathrm{d}\boldsymbol{\sigma} = \boldsymbol{D}(\boldsymbol{\sigma}, \boldsymbol{\varepsilon})\mathrm{d}\boldsymbol{\varepsilon} \tag{4.1.10}$$

The stiffness matrix $\boldsymbol{D}(\boldsymbol{\sigma}, \boldsymbol{\varepsilon})$ is associated with stress and strain. In other words, it is related to the stress path, which is the difference with Variable Elastic Theory.

4.2 The Anisotropic Elastic Theory

A material in which the mechanical properties depend on the orientation of the sample is said to be anisotropic (or aeolotropic). In such a material, a linear relationship must be expressed among six independent stress increment components and six independent strain increments. This relationship is given by a 6 × 6 matrix of moduli.

$$\boldsymbol{\sigma} = [\boldsymbol{D}_\mathrm{e}]\boldsymbol{\varepsilon} \tag{4.2.1}$$

where $\boldsymbol{\sigma}$, $\boldsymbol{\varepsilon}$, and $[\boldsymbol{D}_\mathrm{e}]$ are stress, strain, and elastic stiffness matrix, respectively.

$$\boldsymbol{\sigma} = \left[\sigma_{11}\ \sigma_{22}\ \sigma_{33}\ \sigma_{12}\ \sigma_{13}\ \sigma_{23}\right]^\mathrm{T} \tag{4.2.2}$$

$$\boldsymbol{\varepsilon} = \left[\varepsilon_{11}\ \varepsilon_{22}\ \varepsilon_{33}\ \varepsilon_{12}\ \varepsilon_{13}\ \varepsilon_{23}\right]^\mathrm{T} \tag{4.2.3}$$

$$[\boldsymbol{D}_\mathrm{e}] = \begin{bmatrix} d_{11} & d_{12} & d_{13} & d_{14} & d_{15} & d_{16} \\ d_{21} & d_{22} & d_{23} & d_{24} & d_{25} & d_{26} \\ d_{31} & d_{32} & d_{33} & d_{34} & d_{35} & d_{36} \\ d_{41} & d_{42} & d_{43} & d_{44} & d_{45} & d_{46} \\ d_{51} & d_{52} & d_{53} & d_{54} & d_{55} & d_{56} \\ d_{61} & d_{62} & d_{63} & d_{64} & d_{65} & d_{66} \end{bmatrix} \tag{4.2.4}$$

This stiffness matrix must be symmetric because of the thermodynamic requirement that it must be possible to derive the stresses during recoverable elastic behavior by differentiation of an elastic strain energy potential. Thus from the 36 components of the 6 × 6 stiffness matrix, a total of 21 independent parameters are needed if the anisotropic elasticity of the material is to be fully described. The components of $[\boldsymbol{D}_\mathrm{e}]$ would obey the following equation.

$$d_{ij} = d_{ji}$$

Many geomaterials, however, show more limited forms of anisotropy. For example, a transversely isotropic material possesses an axis of symmetry in the sense that its properties are independent of rotation of a sample about the axis of symmetry. Transverse isotropy has been more commonly called cross-anisotropy in the soil mechanics literature.

The elastic properties of a geomaterial depend on its mode of deposition and its stress history. A geomaterial deposited vertically and then subjected to equal horizontal stresses will therefore be expected to exhibit a vertical axis of symmetry and be transversely isotropic. This assumption is almost invariably made for many geomaterials, and particularly for the sedimentation. However, in some cases, certain tectonic and geomorphological process such as crustal movements, tilting, moving ice sheets, erosion, and solifluction may produce geomaterial with stresses varying in different horizontal directions.

The simplest anisotropy is isotropic.

4.2.1 Isotropic Elastic Constitutive Model

General Hooke's law describes the isotropic elastic constitutive relation. It is the most commonly used elastic model for geomaterial, being widely used in deformation calculation in geotechnical engineering. At this time, only two elastic parameters are needed for expression.

In principal stress space, the elastic constants are chosen as elastic modulus E and Poisson's ratio υ, the isotropic elastic constitutive relation can be expressed as

$$\boldsymbol{\sigma} = \begin{bmatrix} \sigma_1 & \sigma_2 & \sigma_3 \end{bmatrix}^{\mathrm{T}} \tag{4.2.5}$$

$$\boldsymbol{\varepsilon} = \begin{bmatrix} \varepsilon_1 & \varepsilon_2 & \varepsilon_3 \end{bmatrix}^{\mathrm{T}} \tag{4.2.6}$$

$$[\boldsymbol{D}_{\mathrm{e}}] = \frac{E\nu}{(1+\nu)(1-2\nu)} \begin{bmatrix} \frac{1-\nu}{\nu} & 1 & 1 \\ 1 & \frac{1-\nu}{\nu} & 1 \\ 1 & 1 & \frac{1-\nu}{\nu} \end{bmatrix} \tag{4.2.7}$$

In general stress space, the isotropic elastic constitutive relation can be expressed as

$$\boldsymbol{\sigma} = \begin{bmatrix} \sigma_{11} & \sigma_{22} & \sigma_{33} & \sigma_{12} & \sigma_{13} & \sigma_{23} \end{bmatrix}^{\mathrm{T}} \tag{4.2.8}$$

$$\boldsymbol{\varepsilon} = \begin{bmatrix} \varepsilon_{11} & \varepsilon_{22} & \varepsilon_{33} & \varepsilon_{12} & \varepsilon_{13} & \varepsilon_{23} \end{bmatrix}^{\mathrm{T}} \tag{4.2.9}$$

$$[\boldsymbol{D}_{\mathrm{e}}] = \frac{E}{(1+\nu)(1-2\nu)} \begin{bmatrix} 1-\nu & \nu & \nu & 0 & 0 & 0 \\ \nu & 1-\nu & \nu & 0 & 0 & 0 \\ \nu & \nu & 1-\nu & 0 & 0 & 0 \\ 0 & 0 & 0 & 1-2\nu & 0 & 0 \\ 0 & 0 & 0 & 0 & 1-2\nu & 0 \\ 0 & 0 & 0 & 0 & 0 & 1-2\nu \end{bmatrix} \tag{4.2.10}$$

In general stress space, the isotropic elastic constitutive relation can also be expressed as

$$\left.\begin{aligned} \varepsilon_x &= \tfrac{1}{E}\left[\sigma_x - \nu\left(\sigma_y + \sigma_z\right)\right], & \gamma_{yz} &= \tfrac{\tau_{yz}}{G} \\ \varepsilon_y &= \tfrac{1}{E}\left[\sigma_y - \nu(\sigma_z + \sigma_x)\right], & \gamma_{zx} &= \tfrac{\tau_{zx}}{G} \\ \varepsilon_z &= \tfrac{1}{E}\left[\sigma_z - \nu\left(\sigma_x + \sigma_y\right)\right], & \gamma_{xy} &= \tfrac{\tau_{xy}}{G} \end{aligned}\right\} \tag{4.2.11}$$

where $G = \frac{E}{2(1+\nu)}$ is the shear modulus.

Equation (4.2.11) can be described by tensor

$$\varepsilon_{ij} = \frac{\sigma_{ij}}{2G} - \frac{3\nu}{E}\sigma_{\mathrm{m}}\delta_{ij} \tag{4.2.12}$$

where $\sigma_{\mathrm{m}} = \frac{\sigma_{ii}}{3}$ is known as the mean stress.

Add the three normal strains, the volumetric strain can be obtained.

$$\varepsilon_{\mathrm{v}} = \varepsilon_{ii} = \frac{\sigma_{ii}}{2G} - \frac{3v}{E}\sigma_{ii} = \frac{1-2v}{E}\sigma_{ii} = \frac{1}{K}3\sigma_{\mathrm{m}} \tag{4.2.13}$$

where $K = \frac{E}{3(1-2v)}$ is the bulk elastic modulus.

Using Ramy constant, the generalized Hooke's law can be deduced as

$$\left.\begin{aligned} \sigma_x &= 3\lambda\varepsilon_{\mathrm{m}} + 2\mu\varepsilon_x \\ \sigma_y &= 3\lambda\varepsilon_{\mathrm{m}} + 2\mu\varepsilon_y \\ \sigma_z &= 3\lambda\varepsilon_{\mathrm{m}} + 2\mu\varepsilon_z \\ \tau_{zx} &= \mu\gamma_{zx} \\ \tau_{xy} &= \mu\gamma_{xy} \\ \tau_{yz} &= \mu\gamma_{yz} \end{aligned}\right\} \tag{4.2.14}$$

where $\lambda = \frac{Ev}{(1+v)(1-2v)}$, $\mu = G$ is the Ramy constant.

If the tensor representation is used, the Ramy Eq. (4.2.14) can be written as

$$\sigma_{ij} = 2G\varepsilon_{ij} + 3\lambda\varepsilon_{\mathrm{m}}\delta_{ij} \tag{4.2.15}$$

The relation of five pairs of frequently used elastic constants for Hooke's law is listed in Table 4.1.

Table 4.1 The relationship of the five pairs of elastic constants

Pairs of elastic constants	E, ν	K, G	λ, μ	K, ν	K, λ
Elastic modulus $E =$	E	$\frac{9KG}{3K+G}$	$\frac{3\lambda+2\mu}{\lambda+\mu}\mu$	$3K(1-2\nu)$	$\frac{9K(K-\lambda)}{3K-\lambda}$
Poisson's ratio $\nu =$	ν	$\frac{3K-2G}{2(3K+G)}$	$\frac{\lambda}{2(\lambda+\mu)}$	ν	$\frac{\lambda}{3K-\lambda}$
Bulk modulus $K =$	$\frac{E}{3(1-2\nu)}$	K	$\lambda+\frac{2}{3}\mu$	K	K
Shear modulus $G =$	$\frac{E}{2(1+\nu)}$	G	μ	$\frac{3K(1-2\nu)}{2(1+\nu)}$	$\frac{3}{2}(K-\lambda)$
Ramy constant $\lambda =$	$\frac{E\nu}{(1+\nu)(1-2\nu)}$	$K-\frac{2}{3}G$	λ	$\frac{3K\nu}{1+\nu}$	λ
Ramy constant $\mu =$	$\frac{E}{2(1+\nu)}$	G	μ	$\frac{3K(1-2\nu)}{2(1+\nu)}$	$\frac{3}{2}(K-\lambda)$

Other expressions for elastic stiffness matrix are

$$[\boldsymbol{D}_{\mathrm{e}}] = \begin{bmatrix} K + \frac{4}{3}G & K - \frac{2}{3}G & K - \frac{2}{3}G & 0 & 0 & 0 \\ K - \frac{2}{3}G & K + \frac{4}{3}G & K - \frac{2}{3}G & 0 & 0 & 0 \\ K - \frac{2}{3}G & K - \frac{2}{3}G & K + \frac{4}{3}G & 0 & 0 & 0 \\ 0 & 0 & 0 & 2G & 0 & 0 \\ 0 & 0 & 0 & 0 & 2G & 0 \\ 0 & 0 & 0 & 0 & 0 & 2G \end{bmatrix} \tag{4.2.16}$$

$$[\boldsymbol{D}_{\mathrm{e}}] = \begin{bmatrix} M & M - 2G & M - 2G & 0 & 0 & 0 \\ M - 2G & M & M - 2G & 0 & 0 & 0 \\ M - 2G & M - 2G & M & 0 & 0 & 0 \\ 0 & 0 & 0 & 2G & 0 & 0 \\ 0 & 0 & 0 & 0 & 2G & 0 \\ 0 & 0 & 0 & 0 & 0 & 2G \end{bmatrix} \tag{4.2.17}$$

where M is the confined deformation modulus.

$$M = \frac{E(1 - \nu)}{(1 + \nu)(1 - 2\nu)}$$

It can also be expressed by the inverse form of stress–strain.

$$\boldsymbol{\varepsilon} = [\boldsymbol{C}_{\mathrm{e}}]\boldsymbol{\sigma} \tag{4.2.18}$$

where $[\boldsymbol{C}_{\mathrm{e}}]$ is the elastic flexibility matrix.

$$[\boldsymbol{C}_{\mathrm{e}}] = \frac{1}{E}\begin{bmatrix} 1 & -\nu & -\nu & 0 & 0 & 0 \\ -\nu & 1 & -\nu & 0 & 0 & 0 \\ -\nu & -\nu & 1 & 0 & 0 & 0 \\ 0 & 0 & 0 & 1 + \nu & 0 & 0 \\ 0 & 0 & 0 & 0 & 1 + \nu & 0 \\ 0 & 0 & 0 & 0 & 0 & 1 + \nu \end{bmatrix} \tag{4.2.19}$$

Under the condition of plane strain,

$$\varepsilon_{yz} = 0,\ \varepsilon_{zx} = 0,\ \varepsilon_{z} = 0$$

Then the relationship between the stress and strain can be expressed as

$$\boldsymbol{\sigma}^{\mathrm{T}} = \begin{bmatrix} \sigma_x & \sigma_y & \sigma_{xy} \end{bmatrix} \tag{4.2.20}$$

$$\boldsymbol{\varepsilon}^{\mathrm{T}} = \begin{bmatrix} \varepsilon_x & \varepsilon_y & \varepsilon_{xy} \end{bmatrix} \tag{4.2.21}$$

$$[\boldsymbol{D}_e] = \frac{E}{(1+\nu)(1-2\nu)}\begin{bmatrix} 1-\nu & \nu & 0 \\ \nu & 1-\nu & 0 \\ 0 & 0 & 1-2\nu \end{bmatrix} \tag{4.2.22}$$

The corresponding elastic flexibility matrix is

$$[\boldsymbol{C}_e] = \frac{1}{E}\begin{bmatrix} 1 & -\nu & 0 \\ -\nu & 1 & 0 \\ 0 & 0 & 1+\nu \end{bmatrix} \tag{4.2.23}$$

4.2.2 The Elastic Constitutive Model with Cross-Anisotropy

A cross-anisotropy material possesses an axis of symmetry. Geomaterial always exhibits a vertical axis of symmetry. For a transverse isotropic geomaterial, its mechanical properties are the same in the horizontal direction, but it is different between the vertical and horizontal directions. There are five independent parameters for the determination of stiffness matrix $[\boldsymbol{D}_e]$ and its inverse matrix, the flexibility matrix $[\boldsymbol{C}_e]$. The elastic constitutive model with transverse isotropy can be described as

$$\boldsymbol{\varepsilon} = [\boldsymbol{C}_e]\boldsymbol{\sigma} \tag{4.2.24}$$

$$[\boldsymbol{C}_e] = \begin{bmatrix} \frac{1}{E_h} & -\frac{\nu_{hh}}{E_h} & -\frac{\nu_{hv}}{E_h} & 0 & 0 & 0 \\ -\frac{\nu_{hh}}{E_h} & \frac{1}{E_h} & -\frac{\nu_{hv}}{E_v} & 0 & 0 & 0 \\ -\frac{\nu_{hv}}{E_v} & -\frac{\nu_{hv}}{E_v} & \frac{1}{E_v} & 0 & 0 & 0 \\ 0 & 0 & 0 & \frac{1}{2G_{vh}} & 0 & 0 \\ 0 & 0 & 0 & 0 & \frac{1}{2G_{vh}} & 0 \\ 0 & 0 & 0 & 0 & 0 & \frac{1+\nu_{hh}}{E_h} \end{bmatrix} \tag{4.2.25}$$

where E_h, ν_{hh}, E_ν, $\nu_{h\nu}$, and $G_{\nu h}$ are elastic modulus, Poisson's ratio in horizontal direction, elastic modulus, Poisson's ratio, and shear modulus in vertical direction, respectively.

4.3 The Isotropic Nonlinear Elastic Model of Geomaterial

An isotropic elastic material is one in which the elastic properties are independent of the coordinate axes to which the properties are referred. Thus, a sample cut from such a material displays the same elastic properties irrespective of the orientation of the sample. Such a material may be described by two elastic constants, e.g., Young's modulus E and Poisson's ratio ν, or the bulk modulus K and shear modulus G. Any two of these properties are sufficient to define the elastic behavior, since simple function relationships exist between them which arise from their definitions.

The application in engineering is very wide for isotropic elastic model of geomaterial. The simplest isotropic elastic model is the linear elastic model, i.e., generalized Hooke's law, and two elastic constants are enough. The mechanical properties would not be so simple for geomaterial in natural engineering, and the isotropic nonlinear elastic theory would be adopted. Currently and especially in China, Duncan-Chang model [6] is the most widely used nonlinear elastic model, and a simple introduction is presented below.

4.3.1 The Basic Principle of Duncan-Chang Model

Duncan-Chang model is a popular geomaterial model. To simulate correlation of stress path for geomaterial, the Variable Elastic Theory and the Hyperelastic Theory cannot be used, and only the Hypoelastic Theory can be used, namely, the mechanical behavior of geomaterial just meets the incremental elastic relationship. To facilitate the convenient application, the isotropy hypothesis is used. The stress–strain relationship meets the incremental generalized Hooke's law, and the model can be expressed as

$$\mathrm{d}\boldsymbol{\sigma} = \boldsymbol{D}(\boldsymbol{\sigma})\mathrm{d}\boldsymbol{\varepsilon} \tag{4.3.1}$$

where $\boldsymbol{D}(\boldsymbol{\sigma})$ is the stiffness matrix, a function of stress state.

The stiffness matrix $\boldsymbol{D}(\boldsymbol{\sigma})$ satisfies the general Hooke's law.

$$\boldsymbol{D}(\boldsymbol{\sigma}) = \frac{\boldsymbol{E}(\boldsymbol{\sigma})}{[1+\boldsymbol{\nu}(\boldsymbol{\sigma})][1-2\boldsymbol{\nu}(\boldsymbol{\sigma})]}\begin{bmatrix} 1-\nu(\sigma) & \nu(\sigma) & \nu(\sigma) & 0 & 0 & 0 \\ \nu(\sigma) & 1-\nu(\sigma) & \nu(\sigma) & 0 & 0 & 0 \\ \nu(\sigma) & \nu(\sigma) & 1-\nu(\sigma) & 0 & 0 & 0 \\ 0 & 0 & 0 & 1-2\nu(\sigma) & 0 & 0 \\ 0 & 0 & 0 & 0 & 1-2\nu(\sigma) & 0 \\ 0 & 0 & 0 & 0 & 0 & 1-2\nu(\sigma) \end{bmatrix} \tag{4.3.2}$$

where $\boldsymbol{E}(\boldsymbol{\sigma})$, $\boldsymbol{\nu}(\boldsymbol{\sigma})$ are the incremental elastic parameters, as well as a function of stress, rather than a constant.

The core of the isotropic hyperelastic model is to determine the two elastic parameters changing with stress state.

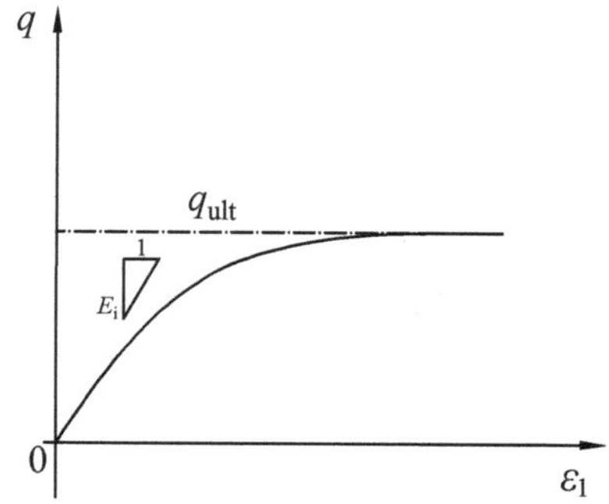

Fig. 4.2 Experimental curve for conventional triaxial compression of soil

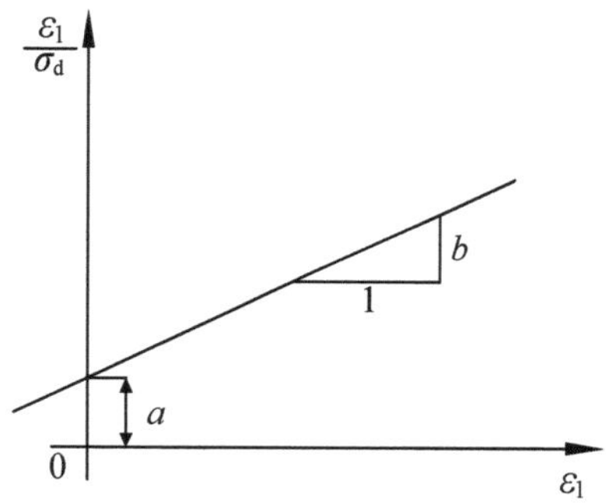

Fig. 4.3 Experimental results of conventional triaxial test

4.3.2 *Two Elastic Function of Duncan-Chang Model*

1. **The function of elastic modulus**

The parameters of Duncan-Chang model are determined by the results of conventional triaxial test of soil, which are shown in Fig. 4.2.

It can be seen from Fig. 4.2, the nonlinear stress–strain relationship is more obvious with the increase of axial stress, and the shear stress-axial strain can be described as hyperbolic.

$$\sigma_d = q = \sigma_1 - \sigma_3 = \frac{\varepsilon_1}{a + b\varepsilon_1} \tag{4.3.3}$$

where a, b are constants, and both of which are the function of confining pressure σ_3.

Equation (4.3.3) can be rewritten as

$$\frac{\varepsilon_1}{\sigma_d} = a + b\varepsilon_1 \tag{4.3.4}$$

This line is shown in Fig. 4.3, which enables us to determine easily the parameters a, b under different confining pressures.

According to Eq. (4.3.3), when the strain is smaller enough, we may obtain the initial elastic modulus E_i.

$$E_i = \left(\frac{\sigma_d}{\varepsilon_1}\right)_{\varepsilon_1 \to 0} = \left.\frac{\varepsilon_1}{a + b\varepsilon_1}\right|_{\varepsilon_1 \to 0} = \frac{1}{a} \tag{4.3.5}$$

where a is the reciprocal of the initial elastic modulus.

In the different experiments, confining pressure σ_3 may change, and the experimental curve also changes (pressure-hardening). But all of these curves can be expressed by Eq. (4.3.3). E_i is changing with σ_3, and the following formula is suggested:

$$E_i = K P_a \left(\frac{\sigma_3}{P_a}\right)^n \tag{4.3.6}$$

where K, n are constants. For different soils, K may be less than 100, and also greater than 3500, the value of n can span between 0.2 and 1.0.

Then a can be expressed as

$$a = \frac{1}{E_i} = \frac{1}{K\sigma_3^n} = \frac{1}{K}\sigma_3^{-n} \tag{4.3.7}$$

According to Eq. (4.3.3), when the strain is infinitely large, the limit of the deviate stress under lateral confining compression can be achieved.

$$(\sigma_d)_{ult} = (\sigma_d)_{\varepsilon_1 \to \infty} = \left.\frac{\varepsilon_1}{a + b\varepsilon_1}\right|_{\varepsilon_1 \to \infty} = \frac{1}{b} \tag{4.3.8}$$

Equation (4.3.8) is the asymptotic line of the strain–stress curve. When the sample is broken, the actual strength $(\sigma_d)_f$ is reached, but it is always less than the limit value $(\sigma_d)_{ult}$. And the ratio of $(\sigma_d)_f$ to $(\sigma_d)_{ult}$ is called the destructive ratio R_f.

$$R_f = \frac{(\sigma_d)_f}{(\sigma_d)_{ult}} = \frac{(\sigma_d)_f}{\frac{1}{b}} = b(\sigma_d)_f \tag{4.3.9}$$

b is achieved as

$$b = \frac{R_f}{(\sigma_d)_f} \tag{4.3.10}$$

The value R_f is selected from 0.75 to 1.0. Generally, it is considered independency with confining pressure σ_3. But from many domestic experiment results, R_f is not a constant, it changes with σ_3.

According to the Mohr–Coulomb destructive criterion, we have

$$(\sigma_d)_f = (\sigma_1 - \sigma_3)_f = \frac{2c\cos\varphi + 2\sigma_3\sin\varphi}{1 - \sin\varphi} \tag{4.3.11}$$

Then b is determined.

$$b = \frac{R_f}{(\sigma_d)_f} = \frac{R_f}{\frac{2(c\cos\varphi + \sigma_3\sin\varphi)}{1-\sin\varphi}} = \frac{R_f(1 - \sin\varphi)}{2(c\cos\varphi + \sigma_3\sin\varphi)} \tag{4.3.12}$$

After determining parameters a, b, we can get the elastic modulus from the tangent slope of the curve of Eq. (4.3.3).

$$E = \frac{d\sigma_1}{d\varepsilon_1} = \frac{d(\sigma_1 - \sigma_3)|_{\sigma_3=C}}{d\varepsilon_1} = \frac{d\sigma_d}{d\varepsilon_1} = \frac{a}{(a + b\varepsilon_1)^2} \tag{4.3.13}$$

The strain in Eq. (4.3.13) needs to be converted to the function of stress for convenience of calculation.

Equation (4.3.3) can be rewritten as

$$\varepsilon_1 = \frac{a}{\frac{1}{\sigma_d} - b}$$

Substituting it to Eq. (4.3.13), we can obtain the function of elastic modulus.

$$E(\boldsymbol{\sigma}) = \frac{(1 - b\sigma_d)^2}{a}$$

Substituting a, b, and $q = \sigma_d$, then

$$\begin{aligned} E(\boldsymbol{\sigma}) &= \left[1 - \frac{R_f(1 - \sin\varphi)q}{2c\cos\varphi + 2\sigma_3\sin\varphi}\right]^2 E_i \\ &= \left[1 - \frac{R_f(1 - \sin\varphi)q}{2c\cos\varphi + 2\sigma_3\sin\varphi}\right]^2 K P_a \left(\frac{\sigma_3}{P_a}\right)^n \end{aligned} \tag{4.3.14}$$

So an elastic parameter is set up and its values change with stress state. The function is involving five parameters c, φ, R_f, K, and n.

2. **The function of Poisson's Ratio $\nu(\sigma)$**

The initial method of setting up function of Poisson's ratio $\nu(\boldsymbol{\sigma})$ is the same as that of elastic modulus in Duncan-Chang model. The results of the conventional triaxial test of normally consolidated soil are illustrated in Fig. 4.4. Hyperbolic function is used to describe the relationship of lateral strain ε_3 and axial strain ε_1.

$$\varepsilon_3 = \frac{h\varepsilon_1}{1 - d\varepsilon_1} = f_2(\sigma_d, \varepsilon_3) \tag{4.3.15}$$

$$h = G - F\log\left(\frac{\sigma_3}{P_a}\right) \tag{4.3.16}$$

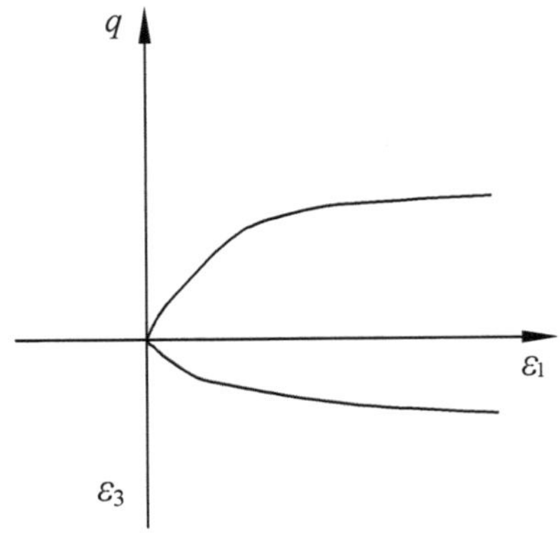

Fig. 4.4 The conventional triaxial test of normally consolidated soils

where G, F, d are experimental constants. The value of F is among 0.1–0.2 generally, so we can set up the equation $h \approx G$.

According to Eq. (4.3.13), another elastic function can be established.

$$\nu(\boldsymbol{\sigma}) = \frac{\mathrm{d}\varepsilon_3}{\mathrm{d}\varepsilon_1} = \frac{G - F\log\left(\frac{\sigma_3}{P_\mathrm{a}}\right)}{(1 - \mathrm{d}\varepsilon_1)^2} = \frac{G - F\log\left(\frac{\sigma_3}{P_\mathrm{a}}\right)}{\left\{1 - \frac{\mathrm{d}q}{KP_\mathrm{a}\left(\frac{\sigma_3}{P_\mathrm{a}}\right)^n\left[1 - \frac{R_\mathrm{f}q(1-\sin\varphi)}{2c\cos\varphi} + 2\sigma_3\sin\varphi\right]}\right\}^2} \tag{4.3.17}$$

If the calculated value of ν_t is equal or greater than 0.5, we may assume $\nu_\mathrm{t} = 0.49$.

The model was called Duncan-Chang model or E–ν model. It is applied widely in the world as a practical model of geomaterial. When it is used in real engineering, the weakness is recognized. Because the calculated value ν is always greater than the actual value. Duncan and Wong changed ν to the bulk modulus K, and named it as E–K model.

Duncan adopted the bulk modulus K as the calculated parameter instead of Poisson's ratio.

$$K(\boldsymbol{\sigma}) = \frac{\mathrm{d}p}{\mathrm{d}\varepsilon_\mathrm{v}} = k_\mathrm{b}P_\mathrm{a}\left(\frac{\sigma_3}{P_\mathrm{a}}\right)^m \tag{4.3.18}$$

where K_b and m are experimental constants. To a majority of soils, the value of m is between 0.0 and 1.0.

When the soil is unloading or reloading, the bulk modulus value is K_ur.

$$K_\mathrm{ur} = k_\mathrm{ur}P_\mathrm{a}\left(\frac{\sigma_3}{P_\mathrm{a}}\right)^m \tag{4.3.19}$$

Table 4.2 Parameters for Duncan-Chang model

	Great soil group	Soft clay	Hard clay	Sand	Pebble	Stone
Shearing parameter	c/Pa	0–0.1	0.1–0.5	0	0	0
	φ	20–30°	20–30°	30–40°	30–40°	40–50°
	R_f	0.7–0.9	0.7–0.9	0.6–0.85	0.6–0.85	0.6–1.0
	K	20–200	200–500	300–1000	500–2000	300–1000
	K_{ur}	3.0 K	1.50–2.0 K			
	n	0.5–0.8	0.3–0.6	0.3–0.6	0.4–0.7	0.1–0.5
Duncan	K_b	20–100	100–500	50–1000	100–2000	50–1000
	m	0.4–0.7	0.2–0.5	0–0.5	0–0.5	−0.2–0.4

where k_{ur} is defined by the experiment, and generally $k_{ur} > k_b$.

With E, K, $\nu(\sigma)$ can be calculated.

$$\nu(\boldsymbol{\sigma}) = \frac{6K - E}{6K} \tag{4.3.20}$$

After determining the two elastic parameters $E(\boldsymbol{\sigma})$, $\nu(\boldsymbol{\sigma})$, the stress increment can be calculated by Eqs. (4.3.1)–(4.3.2) for any strain increment under arbitrary stress state.

Duncan-Chang model is a widely used one in the world, with which rich experiences of each kind of soil have been accumulated, and the parameter scopes are given in Table 4.2 and provide reference for preliminary calculation.

4.3.3 *Review of Duncan-Chang Model*

1. **Advantages**

The model is relatively simple and the mechanical parameters can be determined by the conventional triaxial experiment. Experiences have been accumulated through the domestically wide application.

2. **Disadvantages**

Duncan-Chang model is based on the theory of isotropic elastic theory. It cannot reflect the mechanical characteristics of dilatancy and anisotropy. The results of the conventional triaxial experimental are generalized to the general stress state, giving no consideration of the influence of stress path. So it is suitable for stability analysis of soil, and engineering problems in which confining pressure is near constant.

It is suitable for both the cohesion soil, and the sand, but not suitable for highly overconsolidated soil, such as the dense sands and hard clay. In addition, according to Duncan' assumption when $\sigma_3 = 0$, the value of E and K are zero, which is not true with the fact.

It is advised that the previous consolidation pressure is chosen as σ_3 when the confining pressure is less than the previous consolidation pressure, while the current confining pressure is adopted when the confining pressure is greater than the previous consolidation pressure.

4.4 The Elastic Model with Transverse Isotropy

The behavior of a transversely isotropic material may be described by five constants which may be shown as Eqs. (4.2.24)–(4.2.25).

The following two elastic models with transverse isotropy are put forward by Gong [7] and Graham [8].

4.4.1 Xiao-nan Gong Model

Based on the concept of transverse isotropy elastic theory, a transverse isotropic elastic constitutive model is built by Gong [7] for Jinshan clay in Zhejiang province, China. The model framework is the same as that of Eqs. (4.2.24)–(4.2.25). His core work is to determine the model parameters by experiments.

The anisotropy is detected by the unconfined compression test performed by Xiao-nan Gong on Jinshan clay. The measured stress–strain relationship is shown in Fig. 4.5 and the samples are derived along the vertical, horizontal, and inclined 45° direction, respectively.

The horizontal elastic modulus E_{h}, Poisson's ratio $\nu_{\mathrm{h}\upsilon}$ are obtained by the conventional triaxial test results on soil samples cutting along the horizontal direction. The vertical elastic modulus E_{v}, Poisson's ratio ν_{hh} are obtained for that along the vertical direction.

According to the coordinate transformation, the elastic modulus for the direction with an angle θ to the horizontal is computed in the following for transverse isotropic body.

$$\frac{1}{E_{\theta}} = \frac{\cos^4\theta}{E_{\mathrm{h}}} + \frac{\sin^4\theta}{E_{\mathrm{v}}} + \left(\frac{1}{G_{\mathrm{v}}} - \frac{2\nu_{\mathrm{hh}}}{E_{\mathrm{v}}}\right)\cos^2\theta\sin^2\theta \tag{4.4.1}$$

The corresponding elastic modulus E_{45} can be got by the test on sample cutting along the direction tilted 45°.

Then the vertical shear modulus can be set.

$$G_{\mathrm{vh}} = \frac{E_{\mathrm{v}}}{A + 2(1 + \nu_{\mathrm{vh}})} \tag{4.4.2}$$

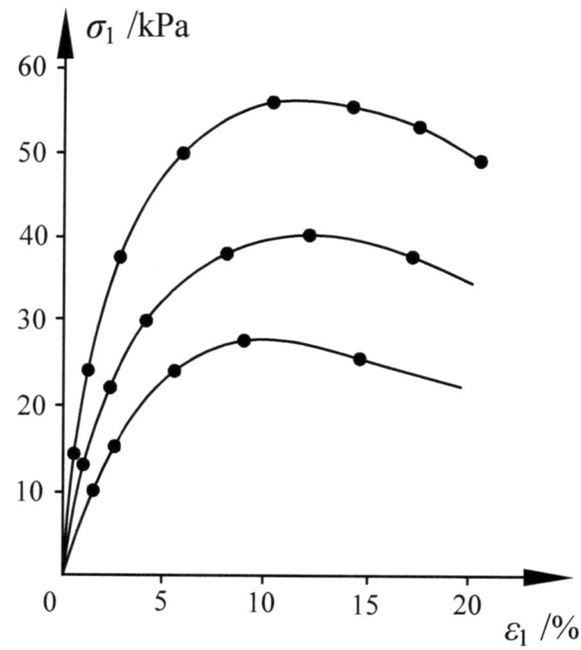

Fig. 4.5 The stress–strain relationship for soil sample cutting along different directions at the unconfined compression test

where A, n, and n_{45} are expressed in the following equation:

$$A = \frac{4n - n_{45} - 3nn_{45}}{nn_{45}}, \quad n = \frac{E_h}{E_v}, \quad n_{45} = \frac{E_{45}}{E_v} \tag{4.4.3}$$

Thus five parameters are determined for the transverse isotropy, and this model can be used in engineering calculation.

4.4.2 Graham Model

A lightly overconsolidated natural clay exhibits cross-anisotropy in Lake Agassiz, Winnipeg, Canada. Graham and Houlsby [8] proposed a simplified cross-anisotropy elastic model for it.

1. **Theoretic basis**

The behavior of a transversely isotropic material can be described with five constants.

$$\begin{bmatrix} d\sigma_{11} \\ d\sigma_{22} \\ d\sigma_{33} \\ d\sigma_{23} \\ d\sigma_{31} \\ d\sigma_{12} \end{bmatrix} = \begin{bmatrix} A & B & B & 0 & 0 & 0 \\ B & C & D & 0 & 0 & 0 \\ B & D & C & 0 & 0 & 0 \\ 0 & 0 & 0 & C-D & 0 & 0 \\ 0 & 0 & 0 & 0 & F & 0 \\ 0 & 0 & 0 & 0 & 0 & F \end{bmatrix} \begin{bmatrix} d\varepsilon_{11} \\ d\varepsilon_{22} \\ d\varepsilon_{33} \\ d\varepsilon_{23} \\ d\varepsilon_{31} \\ d\varepsilon_{12} \end{bmatrix} \tag{4.4.4}$$

If a triaxial test is carried out on a transversely isotropic material, with the axis of symmetry of the test corresponding to the vertical direction (a vertical cut specimen), then no shear stresses are applied to the specimen nor are shear strains measured. So, only the top left-hand corner (3 × 3) of the (6 × 6) matrix shown in Eq. (4.4.4) may be investigated.

$$\begin{bmatrix} d\sigma_{11} \\ d\sigma_{22} \\ d\sigma_{33} \end{bmatrix} = \begin{bmatrix} A & B & B \\ B & C & D \\ B & D & C \end{bmatrix} \begin{bmatrix} d\varepsilon_{11} \\ d\varepsilon_{22} \\ d\varepsilon_{33} \end{bmatrix} \tag{4.4.5}$$

Furthermore, all horizontal stresses and strains are equal in a good quality triaxial test. The matrix may be further reduced to

$$\begin{bmatrix} d\sigma_{11} \\ d\sigma_{22} \end{bmatrix} = \begin{bmatrix} A & 2B \\ B & C+D \end{bmatrix} \begin{bmatrix} d\varepsilon_{11} \\ d\varepsilon_{22} \end{bmatrix} \tag{4.4.6}$$

The last matrix is unsymmetric. Let $C+D=H$, only three elastic constants would be found from triaxial tests on vertically cut samples of transversely isotropic material, compared with the five parameters needed to define its properties fully. In this case, the calculation for general stress states requires assumptions to be made regarding the relationships between the five elastic parameters in Eq. (4.4.5). In order to do this, consider an isotropic material first, for which the stiffness matrix may be expressed as

$$\begin{bmatrix} d\sigma_{11} \\ d\sigma_{22} \\ d\sigma_{33} \end{bmatrix} = \begin{bmatrix} A & B & B \\ B & A & B \\ B & B & A \end{bmatrix} \begin{bmatrix} d\varepsilon_{11} \\ d\varepsilon_{22} \\ d\varepsilon_{33} \end{bmatrix} \tag{4.4.7}$$

Suppose now that the anisotropy is expressed by multiplying the stiffness coefficients to increase the stiffness in a horizontal direction by an anisotropy factor α. This is achieved by multiplying the terms in the second and third rows by α to give a matrix.

$$\begin{bmatrix} A^* & B^* & B^* \\ \alpha B^* & \alpha A^* & \alpha B^* \\ \alpha B^* & \alpha B^* & \alpha A^* \end{bmatrix}$$

A and B have been replaced by A^* and B^* to emphasize the change in their use. In order to preserve the symmetry of the matrix, some further adjustment is necessary, and this may be achieved, for instance, by multiplying the second and third columns by α to give the form.

$$\begin{bmatrix} A^* & \alpha B^* & \alpha B^* \\ \alpha B^* & \alpha^2 A^* & \alpha^2 B^* \\ \alpha B^* & \alpha^2 B^* & \alpha^2 A^* \end{bmatrix} \tag{4.4.8}$$

In Eq. (4.4.8), when $\alpha = 1$, the material is isotropic. When $\alpha < 1$, the material is stiffer vertically than horizontally. While $\alpha > 1$, the material is stiffer horizontally than vertically. The factor α^2 is the ratio of the direct stiffness in the horizontal and vertical directions, (i.e., the ratio of the second and first terms on the leading diagonal of the stiffness matrix in Eq. (4.4.5)). The factor α may therefore be seen as a rational measure of anisotropy with an easily understood significance.

A simple rule for deriving the terms in the stiffness matrix from the matrix for the isotropic material is that for every occurrence of 2 or 3 in the labels identifying the row and column of the matrix entry, that entry is multiplied by $\sqrt{\alpha}$. From this, it may be verified that the full 6×6 stiffness matrix for a transversely isotropic material under these assumptions may be expressed as Eq. (4.4.9). In this equation, the isotropic E and ν values have been replaced by modified parameters E^* and ν^* for anisotropic soil.

$$\begin{bmatrix} d\sigma_{11} \\ d\sigma_{22} \\ d\sigma_{33} \\ d\sigma_{23} \\ d\sigma_{31} \\ d\sigma_{12} \end{bmatrix} = \frac{E^*}{(1+\nu^*)(1-2\nu^*)} \begin{bmatrix} 1-\nu^* & \alpha\nu^* & \alpha^2\nu^* & & & \\ \alpha\nu^* & \alpha^2(1-\nu^*) & \alpha\nu^* & & & \\ \alpha\nu^* & \alpha^2\nu^* & \alpha^2(1-\nu^*) & & & \\ & & & \alpha^2(1-2\nu^*) & & \\ & & & & \alpha^2(1-2\nu^*) & \\ & & & & & \alpha^2(1-2\nu^*) \end{bmatrix} \begin{bmatrix} d\varepsilon_{11} \\ d\varepsilon_{22} \\ d\varepsilon_{33} \\ d\varepsilon_{23} \\ d\varepsilon_{31} \\ d\varepsilon_{12} \end{bmatrix} \tag{4.4.9}$$

This matrix can be inverted to give the compliance matrix Eq. (4.4.10).

$$\begin{bmatrix} d\varepsilon_{11} \\ d\varepsilon_{22} \\ d\varepsilon_{33} \\ d\varepsilon_{23} \\ d\varepsilon_{31} \\ d\varepsilon_{12} \end{bmatrix} = \frac{1}{E^*} \begin{bmatrix} 1 & -\frac{\nu^*}{\alpha} & -\frac{\nu^*}{\alpha} & & & \\ -\frac{\nu^*}{\alpha} & \frac{1}{\alpha^2} & -\frac{\nu^*}{\alpha^2} & & & \\ -\frac{\nu^*}{\alpha} & -\frac{\nu^*}{\alpha^2} & \frac{1}{\alpha^2} & & & \\ & & & \frac{1+\nu^*}{\alpha^2} & & \\ & & & & \frac{1+\nu^*}{\alpha} & \\ & & & & & \frac{1+\nu^*}{\alpha} \end{bmatrix} \begin{bmatrix} d\sigma_{11} \\ d\sigma_{22} \\ d\sigma_{33} \\ d\sigma_{23} \\ d\sigma_{31} \\ d\sigma_{12} \end{bmatrix} \tag{4.4.10}$$

Comparison Eq. (4.4.4) with Eq. (4.4.9), the assumptions about the relationships between A, B, C, D, and F are

$$\left(\frac{D}{B}\right)^2 = \frac{C}{A} = \left(\frac{C-D}{F}\right)^2 \tag{4.4.11}$$

It is of interest to compare the parameters used above with the five parameters more commonly used to describe a transversely isotropic material, which are often chosen as E_{v}, E_{h}, ν_{vv}, ν_{vh}, and G_{vh} in Eq. (4.4.12).

$$\begin{bmatrix} d\varepsilon_{11} \\ d\varepsilon_{22} \\ d\varepsilon_{33} \\ d\varepsilon_{23} \\ d\varepsilon_{31} \\ d\varepsilon_{12} \end{bmatrix} = \begin{bmatrix} \frac{1}{E_\nu} & -\frac{\nu_{\nu\nu}}{E_\nu} & -\frac{\nu_{\nu\nu}}{E_\nu} & & & \\ -\frac{\nu_{\nu\nu}}{E_\nu} & \frac{1}{E_{\text{h}}} & -\frac{\nu_{\nu\text{h}}}{E_{\text{h}}} & & & \\ -\frac{\nu_{\nu\nu}}{E_\nu} & -\frac{\nu_{\nu\text{h}}}{E_{\text{h}}} & \frac{1}{E_{\text{h}}} & & & \\ & & & \frac{1+\nu_{\nu\text{h}}}{E_{\text{h}}} & & \\ & & & & \frac{1}{2G_{\nu\text{h}}} & \\ & & & & & \frac{1}{2G_{\nu\text{h}}} \end{bmatrix} \begin{bmatrix} d\sigma_{11} \\ d\sigma_{22} \\ d\sigma_{33} \\ d\sigma_{23} \\ d\sigma_{31} \\ d\sigma_{12} \end{bmatrix} \tag{4.4.12}$$

Compare the above Eq. (4.4.12) with Eq. (4.4.10) and the following Eq. (4.4.13) can be established.

$$E_\nu = E^*,\ \ E_{\text{h}} = \alpha^2 E^*,\ \ \nu_{\nu\nu} = \frac{\nu^*}{\alpha},\ \ \nu_{\nu\text{h}} = \nu^* \text{ and } 2G_{\nu\text{h}} = \frac{\alpha E^*}{1+\nu^*} \tag{4.4.13}$$

Advantages are obtained by investigating the elasticity of a soil, not through its resistance to direct stresses expressed by E and v parameters, but rather in terms of its change tendency of volume and shape. The stress–strain behavior in triaxial tests is more conveniently described using the parameters p, q, ε_{v}, and ε_{s}. In the case of an isotropic elastic material, it is expressed by the following matrix:

$$\begin{bmatrix} dp \\ dq \end{bmatrix} = \begin{bmatrix} K & 0 \\ 0 & G \end{bmatrix} \begin{bmatrix} d\varepsilon_{\text{v}} \\ d\varepsilon_{\text{s}} \end{bmatrix} \tag{4.4.14}$$

In the case of transversely isotropic material, this matrix may be modified to

$$\begin{bmatrix} dp \\ dq \end{bmatrix} = \begin{bmatrix} K^* & J \\ J & G^* \end{bmatrix} \begin{bmatrix} d\varepsilon_{\text{v}} \\ d\varepsilon_{\text{s}} \end{bmatrix} \tag{4.4.15}$$

K^* and G^* have been introduced to emphasize the fact that the material is no longer isotropic, and J is a new parameter expressing the cross dependence of the shear strain on the mean pressure p, and of volumetric strain on the shear stress q. The parameters K^*, G^* and J provide the three parameters required for describing a transversely isotropic material in triaxial test (Eq. (4.4.4)). The symmetry of the matrix in Eq. (4.4.15) is required for thermodynamic considerations. With the use of the definitions of the p, q, ε_v, and ε_s parameters, the stiffness coefficients may be related to A^*, B^* and α in Eq. (4.4.15) by

$$K^* = \frac{A^* + 4\alpha B^* + 2\alpha^2(A^* + B^*)}{9} \tag{4.4.16}$$

$$G^* = \frac{A^* - 2\alpha B^* + \frac{\alpha^2(A^*+B^*)}{2}}{3} \tag{4.4.17}$$

$$J^* = \frac{A^* + \alpha B^* - 2\alpha^2(A^* + B^*)}{3} \tag{4.4.18}$$

It is also convenient to invert Eq. (4.4.15) to give the compliance matrix.

$$\begin{bmatrix} \mathrm{d}\varepsilon_v \\ \mathrm{d}\varepsilon_s \end{bmatrix} = \begin{bmatrix} C_1 & C_2 \\ C_2 & C_3 \end{bmatrix} \begin{bmatrix} \mathrm{d}p \\ \mathrm{d}q \end{bmatrix} \tag{4.4.19}$$

where $C_1 = \frac{G^*}{Det}$, $C_2 = -\frac{J}{Det}$, $C_3 = \frac{K^*}{Det}$ and $Det = (3\,K^* G^* - J^2)$, the determinant of the stiffness matrix.

2. **Solution for parameters**

In any single triaxial test, measurements will be made of $\mathrm{d}\sigma_{11}$, $\mathrm{d}\sigma_{33}$, $\mathrm{d}\varepsilon_{11}$, and $\mathrm{d}\varepsilon_{33}$, from which $\mathrm{d}p$, $\mathrm{d}q$, $\mathrm{d}\varepsilon_v$, and $\mathrm{d}\varepsilon_s$ may be calculated. Substitute them into Eq. (4.4.19) and give two equations in three variables. This is insufficient to solve for all three variables. At least two tests with different $\left(\frac{\mathrm{d}q}{\mathrm{d}p}\right)$ ratios are required to solve for the anisotropic properties. However, if two tests are available, there are then four equations in three variables. More than two tests are usually performed in most testing programs, and they produce a considerable amount of mathematically redundant information about the anisotropic elasticity of the clay. Because of the variation of the behavior of real clays from the idealization assumed in Eq. (4.4.19), the natural differences which occur between similar samples, the presence of small experimental errors in measurement, and the equations obtained from a test program will be mutually inconsistent to some extent. A commonly accepted way of treating such a set of equations is to solve for the required parameters (K^*, G^*, and J) using the least squares procedure to minimize random errors.

A least squares solution may be obtained using stresses, strains, or a combination of both as independent variables. In the following analysis, the stresses are treated as independent variables, and the strains are considered dependent. If measured values of $\mathrm{d}p$ and $\mathrm{d}q$ are obtained, then the corresponding calculated volumetric strain $\mathrm{d}\varepsilon_v$ predicted by Eq. (4.4.19) is given by

$$\mathrm{d}\varepsilon_v = C_1 \mathrm{d}p + C_2 \mathrm{d}q \tag{4.4.20}$$

However, volumetric strains $d\varepsilon_v$ corresponding to the stress increments dp and dq have actually been measured. The error in volumetric strain $d\varepsilon_{ve}$ is then given by the difference between the calculated and measured values.

$$d\varepsilon_{ve} = d\varepsilon_v - d\varepsilon_{vc} = C_1 dp + C_2 dq - d\varepsilon_{vc} \quad (4.4.21)$$

Similarly, the error in the calculated shear strain $d\varepsilon_s$ from the measured value of $d\varepsilon_{sc}$ is given by $d\varepsilon_{se}$.

$$d\varepsilon_{se} = d\varepsilon_s - d\varepsilon_{sc} = C_2 dp + C_3 dq - d\varepsilon_{sc} \quad (4.4.22)$$

The sum of the squares of the errors in the strains for all the tests available is given by

$$e = \sum_{Tests} (C_1 dp + C_2 dq - d\varepsilon_{vc})^2 + (C_2 dp + C_3 dq - d\varepsilon_{sc})^2 \quad (4.4.23)$$

The least squares solution for the parameters C_1, C_2 and C_3 from the set of redundant equations is found by setting the differentials of the error measure e with respect to each of the parameters C_1, C_2 and C_3 in turn to zero.

$$\frac{\partial e}{\partial C_1} = \sum 2(C_1 dp + C_2 dq - d\varepsilon_{vc})\, dp = 0 \quad (4.4.24)$$

$$\frac{\partial e}{\partial C_2} = \sum 2(C_1 dp + C_2 dq - d\varepsilon_{vc})\, dq + 2(C_2 dp + C_3 dq - d\varepsilon_{sc}) dp = 0 \quad (4.4.25)$$

$$\frac{\partial e}{\partial C_3} = 2(C_2 dp + C_3 dq - d\varepsilon_{sc}) dp = 0 \quad (4.4.26)$$

This may be reduced to the solution for C_1, C_2 and C_3 of the matrix

$$\begin{bmatrix} \sum d\varepsilon_{vc} dp \\ \sum d\varepsilon_{vc} dq + d\varepsilon_{sc} dp \\ \sum d\varepsilon_{sc} dq \end{bmatrix} = \begin{bmatrix} \sum dp^2 & \sum dq dp & 0 \\ \sum dp dq & \sum dq^2 + dp^2 & \sum dq dp \\ 0 & \sum dp dq & \sum dq^2 \end{bmatrix} \begin{bmatrix} C_1 \\ C_2 \\ C_3 \end{bmatrix} \quad (4.4.27)$$

or more general Eq. (4.4.28), where w_1 and w_2 are weightings applied to the measurements of $d\varepsilon_{vc}$ and $d\varepsilon_{se}$ for each test. The weightings simply reflect the relative confidence in each measurement, and may take account of equipment or procedural problems that arise during testing.

$$\begin{bmatrix} \sum w_1 d\varepsilon_{vc} dp \\ \sum w_1 d\varepsilon_{vc} dq + w_2 d\varepsilon_{sc} dp \\ \sum w_2 d\varepsilon_{sc} dq \end{bmatrix} = \begin{bmatrix} \sum w_1 dp^2 & \sum w_1 dq dp & 0 \\ \sum w_1 dp dq & \sum w_1 dq^2 + w_2 dp^2 & \sum w_2 dq dp \\ 0 & \sum w_2 dp dq & \sum w_2 dq^2 \end{bmatrix} \begin{bmatrix} C_1 \\ C_2 \\ C_3 \end{bmatrix} \quad (4.4.28)$$

After seeking solutions for C_1, C_2 and C_3 by inverting the 3×3 matrix in Eq. (4.4.28), the values of K^*, G^* and J in Eq. (4.4.15) are given by

$$K^* = \frac{C_3}{C_1 C_3 - C_2^2} \tag{4.4.29}$$

$$G^* = \frac{C_1}{C_1 C_3 - C_2^2} \tag{4.4.30}$$

$$J = -\frac{C_2}{C_1 C_3 - C_2^2} \tag{4.4.31}$$

In order to be compared with more common engineering parameters, these may then be converted algebraically to the A^*, B^* and α parameters in Eq. (4.4.8) to get

$$A^* = K^* + \frac{4}{3}G^* + \frac{4}{3}J \tag{4.4.32}$$

$$\alpha = \frac{\sqrt{9\left(K^* - \frac{2}{3}G^* + \frac{1}{3}J\right)^2 + 8\left(3K^*G^* - J^2\right)} - \left(K^* - \frac{2}{3}G^* + \frac{1}{3}J\right)}{2A^*} \tag{4.4.33}$$

$$B^* = \frac{K^* - \frac{2}{3}G^* + \frac{1}{3}J}{\alpha} \tag{4.4.34}$$

Questions

1. Please describe the concept of the elasticity combining examples in daily life.
2. Please describe the anisotropy with examples in daily life.
3. Please analyze the theoretic basis, qualitative, advantages, disadvantages, and improving methods of Duncan-Chang model.
4. What is the elastic theory with transverse isotropy?

References

1. Heog K, Andersland OB, Rolfsen EN (1969) Undrained behavior of quick clay under load tests at Asrum. Geotechnique 19(1):101–115
2. Wood DM (1980) Yielding in soft clay at Backebol. Sweden. Geotechnique 30(1):49–65
3. Graham J (1969) Results of direct shear, oedometer and triaxial tests from Mastemyr. Internal report F. 372-3. Norwegian Geotechnical Institute, Oslo
4. Graham J (1974) Laboratory testing of sensitive clay from Lyndhurst, Ontario. Research report CE74-2. Royal Military College of Canada, Kingston
5. Crooks JHA, Graham J (1976) Geotechnical properties of Belfast estuarine deposits. Geotechnique 26(2):293–315
6. Duncan JM, Chang CY (1970) Nonlinear analysis of stress and strain in soils. J Soil Mech Found Div 96(5):1629–1653
7. Gong XN (1986) The preliminary discussion on the anisotropy of soft clay foundation. J Zhejiang Univ (Eng Sci) 30(4):103–115
8. Graham J, Houlsby GT (1983) Anisotropic elasticity of a natural clay. Geotechnique 33(2):165–180

Chapter 5
Classical Plastic Theory

5.1 Potential Function and Thermodynamics

In elastic theory, the relationship is defined as follows for stress, elastic strain, and elastic potential:

$$\boldsymbol{\sigma} = \frac{\partial \phi}{\partial \boldsymbol{\varepsilon}^{\mathrm{e}}}$$

where $\boldsymbol{\sigma}$, $\boldsymbol{\varepsilon}^{\mathrm{e}}$, ϕ are stress, elastic strain and elastic potential function, respectively.

Elastic potential, i.e., elastic potential energy and strain energy. The elastic force is obtained by partial derivative of elastic potential with respect to strain. Its variation of displacement is also the elastic force. This concept is the basis for a wide range of applications, as among which the generating of original immersed boundary method is just a clever use of this concept in the coupling of fluid structure, and the finite element is also taking advantages from this idea in solid mechanics. The list also includes a lot of new conception of mechanics development, such as damage potential, the dissipation potential, flow potential, etc.

In 1928, based on the concept of elastic potential of elastic mechanics, the concept of plastic potential and flow rule were put forward by Mises, and the potential function was needed to create the constitutive relation of material in the Classical Plastic Theory and the plastic theory on account of thermodynamics. In this section, the relation of the various potential function and thermodynamic will be briefly explored from the perspective of thermodynamics.

5.1.1 First Law of Thermodynamics

All the substances in the nature possess energy. Although the energy has various different forms and can transit from one form to another and from one object to

Y. Liu and Y. Zheng, *Plastic Mechanics of Geomaterial*, Springer Geophysics,
https://doi.org/10.1007/978-981-13-3753-6_5

another. Energy conversion should be satisfied in these processes. The thermodynamics framework can be obtained in interrelated literature [1, 2]. Generally, in any nonlinear dynamic system, there will be an internal energy (U) and a kinetic energy (K), which is as follows:

$$U = \int_v \rho u \mathrm{d}v,\ K = \frac{1}{2}\int_v \rho \mathbf{v}\cdot\mathbf{v}\mathrm{d}v \tag{5.1.1}$$

where ρ is the mass density, u is the internal energy density, $\mathbf{v}$ is velocity vector, ν is the body occupied spatial domain.

Similarly, $\mathrm{d}W$ is mechanical work increment and $\mathrm{d}Q$ is thermal increment, and can be written as

$$\mathrm{d}W = \int_v \mathbf{fv}\mathrm{d}v + \oint_{\partial v} \mathbf{pv}\mathrm{d}A \tag{5.1.2}$$

$$\mathrm{d}Q = \int_v \rho \boldsymbol{r}\mathrm{d}v - \oint_{\partial v} \mathbf{q}\boldsymbol{n}\mathrm{d}A \tag{5.1.3}$$

where $\boldsymbol{f}$ is a body force vector, $\boldsymbol{p}$ is surface force vector, $\partial\nu$ is the boundary, $\mathrm{d}A$ is infinitesimal area, $\boldsymbol{r}$ is a body thermal vector, $\boldsymbol{q}$ is surface thermal vector, and $\boldsymbol{n}$ is the unit vector of outward normal.

The first law of thermodynamics (energy conservation) in incremental form can be written as

$$\mathrm{d}K + \mathrm{d}U = \mathrm{d}W + \mathrm{d}Q \tag{5.1.4}$$

ν is arbitrary and according to the local form of the Energy Conservation Law, we get

$$\rho \mathrm{d}u - \rho \boldsymbol{r} + \mathrm{div}\boldsymbol{q} - \boldsymbol{\sigma} : \mathrm{d}\boldsymbol{\varepsilon} = 0 \tag{5.1.5}$$

5.1.2 Second Law of Thermodynamics

The Second Law of Thermodynamics, which determines the direction of energy transition, states that the entropy of a thermodynamic system cannot decrease. It can be stated as

$$\mathrm{d}S - Q_{\mathrm{T}} \ge 0,\ Q_{\mathrm{T}} = \int_v \frac{\rho \boldsymbol{r}}{T}\mathrm{d}v - \oint_{\partial v} \frac{\boldsymbol{q}}{T}\cdot \boldsymbol{n}\mathrm{d}A \tag{5.1.6}$$

where $S = \int_v \rho s \mathrm{d}v$ is entropy, and T is absolute temperature.

Applying the divergence theorem to the heat flux across the boundary of a unit volume,

$$\oint_{\partial v} \frac{\boldsymbol{q}}{T} n \mathrm{d}A = \int_v \mathrm{div}(\frac{\boldsymbol{q}}{T}) \mathrm{d}v = \int_v (\frac{1}{T} \mathrm{div}\boldsymbol{q} - \frac{\boldsymbol{q}}{T^2} \mathrm{grad}T) \mathrm{d}v \geq 0 \tag{5.1.7}$$

where grad is gradient operator.

The local form for the Second Law of Thermodynamics is

$$\mathrm{d}S - \frac{\rho \boldsymbol{r} - \mathrm{div}\boldsymbol{q} + \frac{\boldsymbol{q}}{T}\mathrm{grad}T}{T} \geq 0 \tag{5.1.8}$$

Apply Eq. (5.1.6) into Eq. (5.1.8), and the local form is obtained for the Clausius–Duhem inequality.

$$\boldsymbol{\sigma} \mathrm{d}\boldsymbol{\varepsilon} - \rho(\mathrm{d}u - T\mathrm{d}S) - \frac{\boldsymbol{q}}{T}\mathrm{grad}T \geq 0 \tag{5.1.9}$$

The factor ρ simply appears as a multiplier throughout the analysis. If the extensive quantities are all converted to a per unit volume, rather than per unit mass, then this factor disappears [3].

$$\boldsymbol{\sigma} \mathrm{d}\boldsymbol{\varepsilon} - \mathrm{d}u + T\mathrm{d}S - \frac{\boldsymbol{q}}{T}\mathrm{grad}T \geq 0 \tag{5.1.10}$$

5.1.3 Thermodynamics Potential and Dissipative Inequality

Thermodynamic potential is a function of a set of independent state variables. This function is only associated with the value of an independent state variable, and has nothing to do with the process of change. Independent state variables include plastic strain ε^{p} and damage variable D, and those which associated with an independent state are the corresponding thermodynamic forces. Generally, the thermodynamic potential can be expressed as

$$\Pi = \Pi(\chi_0, \chi_1, \ldots)$$

Its differential would be

$$\mathrm{d}\Pi = \frac{\partial \Pi}{\partial \chi_0}\mathrm{d}\chi_0 + \frac{\partial \Pi}{\partial \chi_1}\mathrm{d}\chi_1 + \cdots \tag{5.1.11}$$

Equation (5.1.11) is a common expression of thermodynamic potential, within which $\frac{\partial \Pi}{\partial \chi_i}$ is the thermodynamic force corresponding to χ_i.

In particular, the free energy ψ would be

$$\psi = u - TS \tag{5.1.12}$$

The incremental change in free energy can be written as

$$\mathrm{d}\psi = \mathrm{d}u - T\mathrm{d}S - \mathrm{d}TS \tag{5.1.13}$$

If Eq. (5.1.13) is applied into Eq. (5.1.10), the following relation is derived:

$$\sigma \mathrm{d}\varepsilon - (\mathrm{d}\psi + S\mathrm{d}T) - \frac{1}{T}\boldsymbol{q}\,\mathrm{grad}T = 0 \tag{5.1.14}$$

On the other hand, ψ can be written as

$$\psi = \psi(\boldsymbol{\varepsilon}^{\mathrm{e}}, k, D, T) \tag{5.1.15}$$

where $\boldsymbol{\varepsilon}^{\mathrm{e}}$, k, D are elastic strain, plastic internal variable, and damage variable, respectively.

Then the chain rule yields

$$\mathrm{d}\psi = \frac{\partial\psi}{\partial\boldsymbol{\varepsilon}^{\mathrm{e}}}\mathrm{d}\boldsymbol{\varepsilon}^{\mathrm{e}} + \frac{\partial\psi}{\partial k}\mathrm{d}k + \frac{\partial\psi}{\partial D}\mathrm{d}D + \frac{\partial\psi}{\partial T}\mathrm{d}T \tag{5.1.16}$$

So Eq. (5.1.14) can take the following form:

$$(\boldsymbol{\sigma} - \frac{\partial\psi}{\partial\boldsymbol{\varepsilon}^{e}})\mathrm{d}\boldsymbol{\varepsilon}^{e} + \boldsymbol{\sigma}\mathrm{d}\boldsymbol{\varepsilon}^{p} - \frac{\partial\psi}{\partial k}\mathrm{d}k - \frac{\partial\psi}{\partial D}\mathrm{d}D - (S + \frac{\partial\psi}{\partial T})\mathrm{d}T + qg \geq 0 \tag{5.1.17}$$

where $\mathrm{d}\boldsymbol{\varepsilon}^{\mathrm{p}} = \mathrm{d}\boldsymbol{\varepsilon} - \mathrm{d}\boldsymbol{\varepsilon}^{\mathrm{e}}$ is plastic strain increment, $g = \frac{1}{T}\mathrm{grad}T$.

$\mathrm{d}\boldsymbol{\varepsilon}^{\mathrm{e}}$ and $\mathrm{d}T$ are arbitrary, there will be

$$\boldsymbol{\sigma} = \frac{\partial\psi}{\partial\boldsymbol{\varepsilon}^{\mathrm{e}}},\ S = -\frac{\partial\psi}{\partial T} \tag{5.1.18}$$

Define thermodynamics force K, Y as

$$K = -\frac{\partial\psi}{\partial k},\ Y = -\frac{\partial\psi}{\partial D} \tag{5.1.19}$$

Using Eq. (5.1.17), the energy equation and local entropy production inequality take the following forms:

$$\boldsymbol{\sigma}\mathrm{d}\boldsymbol{\varepsilon}^{\mathrm{p}} + K\mathrm{d}k + Y\mathrm{d}D + qg \geq 0 \tag{5.1.20}$$

It is assumed that there is no local heat source, which means the thermal dissipation is zero. Then Eq. (5.1.20) takes the following form:

$$\boldsymbol{\sigma}\mathrm{d}\boldsymbol{\varepsilon}^{\mathrm{p}} + K\mathrm{d}k + Y\mathrm{d}D \geq 0 \tag{5.1.21}$$

Assumes that there is a dissipation potential

$$\Omega = \Omega(\varepsilon_{\mathrm{v}}^{\mathrm{p}}, \varepsilon_{\mathrm{s}}^{\mathrm{p}}, \theta_{\varepsilon^{\mathrm{p}}}, k, D) \tag{5.1.22}$$

where $\varepsilon_{\mathrm{v}}^{\mathrm{p}}, \varepsilon_{\mathrm{s}}^{\mathrm{p}}, \theta_{\varepsilon^{\mathrm{p}}}$ are plastic volumetric strain, generalized plastic shear strain, and Lode's angle of plastic strain, respectively.

The dual dissipation potential Ω^* is given via the Legendre transform of Ω.

$$\Omega^* = \Omega^*(\boldsymbol{\sigma}, K, Y, \boldsymbol{\varepsilon}^{\mathrm{e}}, \boldsymbol{\varepsilon}^{\mathrm{p}}, k, D) \tag{5.1.23}$$

Then, the total relation would be yielded for constitutive model.

$$\mathrm{d}\boldsymbol{\varepsilon}^{\mathrm{p}} = \mathrm{d}\lambda\frac{\partial\Omega^*}{\partial\boldsymbol{\sigma}},\quad \mathrm{d}k = \mathrm{d}\lambda\frac{\partial\Omega^*}{\partial K},\quad \mathrm{d}D = \mathrm{d}\lambda\frac{\partial\Omega^*}{\partial Y} \tag{5.1.24}$$

Assume that the plastic and damage dissipation are mutually uncorrelated, namely decoupling, then the following can be further obtained.

$$\Omega(\varepsilon_{\mathrm{v}}^{\mathrm{p}}, \varepsilon_{\mathrm{s}}^{\mathrm{p}}, \theta_{\varepsilon^{\mathrm{p}}}, k, D) = \Omega_{\mathrm{p}}(\varepsilon_{\mathrm{v}}^{\mathrm{p}}, \varepsilon_{\mathrm{s}}^{\mathrm{p}}, \theta_{\varepsilon^{\mathrm{p}}}, k) + \Omega_{\mathrm{d}}(D) \tag{5.1.25}$$

where $\Omega = \Omega(\varepsilon_{\mathrm{v}}^{\mathrm{p}}, \varepsilon_{\mathrm{s}}^{\mathrm{p}}, \theta_{\varepsilon^{\mathrm{p}}}, k, D)$ is total dissipation potential; $\Omega_{\mathrm{p}} = \Omega_{\mathrm{p}}(\varepsilon_{\mathrm{v}}^{\mathrm{p}}, \varepsilon_{\mathrm{s}}^{\mathrm{p}}, \theta_{\varepsilon^{\mathrm{p}}}, k)$ is plastic dissipative potential; $\Omega_{\mathrm{d}} = \Omega_{\mathrm{d}}(D)$ is damage dissipative potential.

Then

$$\mathrm{d}\boldsymbol{\varepsilon}^{\mathrm{p}} = \mathrm{d}\lambda\frac{\partial\Omega_{\mathrm{p}}^*}{\partial\boldsymbol{\sigma}},\quad \mathrm{d}k = \mathrm{d}\lambda\frac{\partial\Omega_{\mathrm{p}}^*}{\partial K},\quad \mathrm{d}D = \mathrm{d}\lambda\frac{\partial\Omega_{\mathrm{p}}^*}{\partial Y} \tag{5.1.26}$$

This is the continuum thermodynamics basis for classical plastic theory and damage theory expressed similar to the classical plastic theory.

The above reasoning shows there are two bases for this expression:

1. There is a dissipation potential function (Eq. (5.1.23));
2. Plastic and damage dissipation are unrelated, i.e., decoupling.

Based on the basic mechanical properties of geomaterial, the nonexistence of the dissipative potential is verified, and the non-decoupling of the dissipation potential has also been proved [4, 5]. Similarly, the nonexistence and non-decoupling are verified for the flow potential [6]. These show that the classical plastic theory is not based on the theory of thermodynamics.

5.2 Plastic Postulate

5.2.1 Drucker's Stability Postulate

The notations of normality and convexity of yield surface outlined earlier are just mathematical ideas. In an attempt to provide a missing link between material behavior and these mathematical ideas, Drucker [7–9] introduced a fundamental stability postulate. In essence, Drucker's Stability Postulate is a generalization of simple facts which are valid for certain classes of materials, and is not a statement of any thermodynamic principle [10], as it is often presented.

Figure 5.1 shows two types of typical stress–strain behavior observed in experiments on real engineering materials. In the case of Fig. 5.1a, the stress increases with increasing strain and the material is actually hardening from the beginning to the end. In other words, an additional loading (i.e., $\Delta\sigma_{ij} > 0$) gives rise to an additional strain (i.e., $\Delta\varepsilon_{ij} > 0$), with the product $\Delta\sigma_{ij}\varepsilon_{ij} > 0$. The additional stress $\Delta\sigma_{ij}$, therefore, does positive work as represented by the shaded triangle in Fig. 5.1a. Behavior of this kind is called stable.

In the case of Fig. 5.1b, the deformation curve has a descending branch which follows a strain-softening section. In the descending section, the strain increases with decreasing stress. In other words, the additional stress does negative work (i.e., $\Delta\sigma_{ij}\varepsilon_{ij} < 0$). Behavior of this kind is called unstable.

In the light of this basic fact, Drucker [7–9] introduced the idea of a stable plastic material. This postulate, when applied to an element of elastoplastic materials in equilibrium under the action of surface loads and body forces, may be stated as follows.

Assume an element initially in some state of stress, to which an additional set of stresses is slowly applied and slowly removed by an external agency. Then, during the application of the added stresses and in a cycle of application and removal of the added stresses (Fig. 5.2), the work done by the external agency is nonnegative. It can be illustrated as

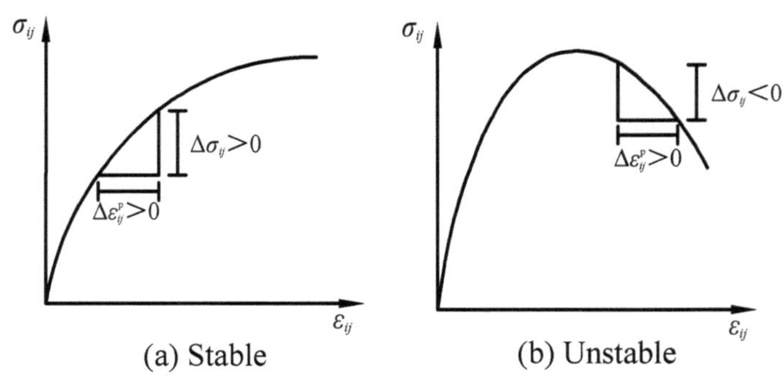

Fig. 5.1 Drucker's stability postulate

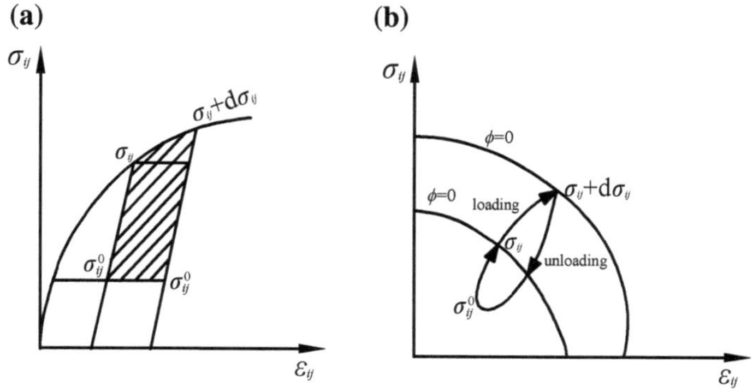

Fig. 5.2 Additional stress cycle

$$W_{\mathrm{D}} = \oint_{\sigma} (\boldsymbol{\sigma} - \boldsymbol{\sigma}_0)\mathrm{d}\boldsymbol{\varepsilon} \geq 0 \tag{5.2.1}$$

Elastic strain is reversible in stress cycle.

$$\oint_{\sigma} (\boldsymbol{\sigma} - \boldsymbol{\sigma}_0)\mathrm{d}\boldsymbol{\varepsilon}^{\mathrm{e}} = 0$$
$$W_{\mathrm{D}} = W^{\mathrm{p}} = \oint_{\sigma} (\boldsymbol{\sigma} - \boldsymbol{\sigma}_0)\mathrm{d}\boldsymbol{\varepsilon}^{\mathrm{p}} \geq 0$$

Within the whole stress cycle, the plastic deformation can be only produced at the loading stage of $\sigma_{ij} \rightarrow \sigma_{ij} + \mathrm{d}\sigma_{ij}$.

$$W^{\mathrm{p}} = (\boldsymbol{\sigma} - \boldsymbol{\sigma}_0 + a\mathrm{d}\boldsymbol{\sigma})\mathrm{d}\boldsymbol{\varepsilon}^{\mathrm{p}} \geq 0, \quad 1 \geq a \geq 0.5 \tag{5.2.2}$$

When $\boldsymbol{\sigma} \neq \boldsymbol{\sigma}_0$,

$$W^{\mathrm{p}} = (\boldsymbol{\sigma} - \boldsymbol{\sigma}_0)\mathrm{d}\boldsymbol{\varepsilon}^{\mathrm{p}} \geq 0 \tag{5.2.3}$$

When $\boldsymbol{\sigma} = \boldsymbol{\sigma}_0$,

$$\mathrm{d}\boldsymbol{\sigma}\mathrm{d}\boldsymbol{\varepsilon}^{\mathrm{p}} \geq 0 \tag{5.2.4}$$

5.2.2 Inference of Drucker's Postulate

1. **Corollary 1**

The yield surface must be convex.

Yield surface: The surface formed by the stress point at which plastic deformation begins to appear in the stress space.

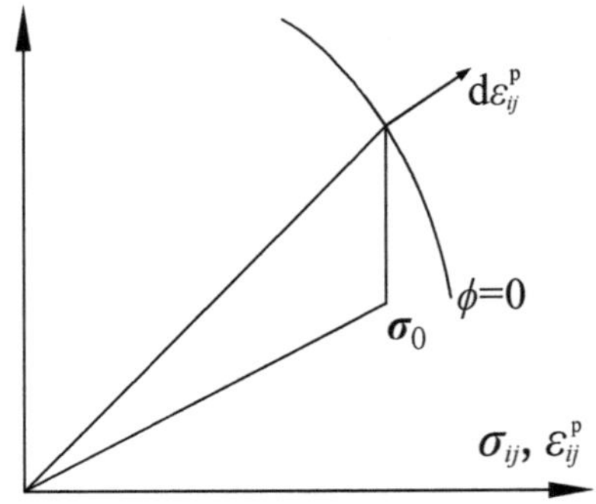

Fig. 5.3 Relation of stress increment direction and plastic strain direction

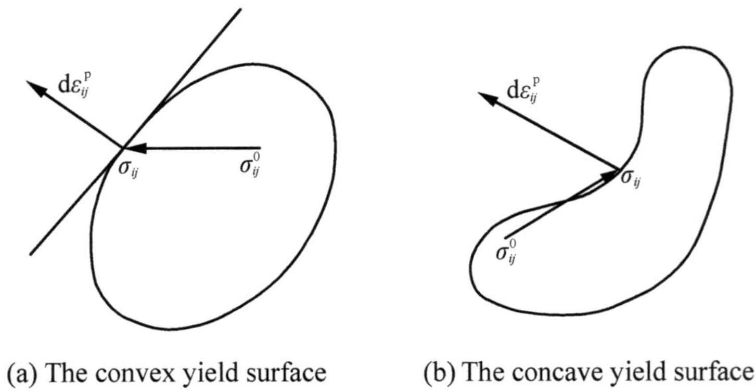

Fig. 5.4 Illustration of convex of yield surface

Equation (5.2.3) shows that the angle is not greater than 90° between stress increment direction and plastic strain direction (Fig. 5.3).

As shown in Fig. 5.4a, a tangent plane for the yield surface is drawn at a yield point, and all possible initial stress state is in the right side of the tangent plane. Then the yield surface must be convex, and the direction of plastic strain increment should be the same as outer normal direction of yield surface. Otherwise, as shown in Fig. 5.4b, there is an initial stress state within the concave yield surface, the angle is larger than 90° between the loading stress incremental and plastic strain increment, and it would be contrary to Eq. (5.2.3).

2. **Corollary 2**

The direction of plastic strain increment must be the same as the normal of the yield surface (i.e., with an associated flow rule).

Equation (5.2.4) shows that the angle should be less than 90° between the stress increment and the induced plastic strain increment. The yield surface must be convex, and only when the normal of the yield surface is the same as plastic strain increment, it can be ensured that the angle is less than 90° between that all the load stress increment and the induced plastic strain increment (as shown in Fig. 5.5).

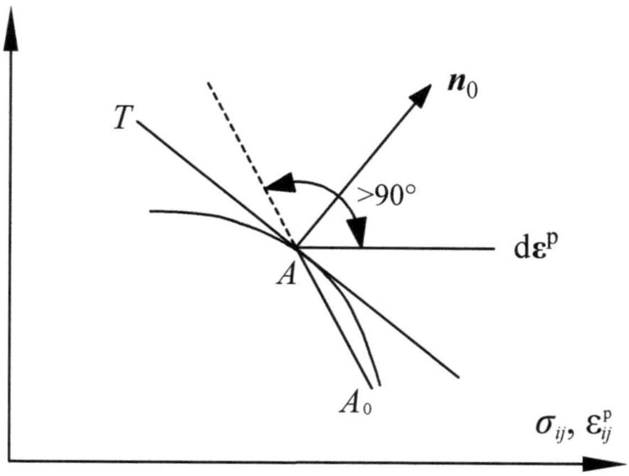

Fig. 5.5 The direction of plastic strain increment orthogonal to the yield surface

So the corollary is achieved that the plastic strain increment is in the same direction with the normal of the yield surface.

It can be concluded that

$$d\boldsymbol{\varepsilon}^p = d\lambda \frac{\partial F}{\partial \boldsymbol{\sigma}} \tag{5.2.5}$$

where F is the yield surface; $d\lambda$ is plastic factor which is the magnitude of plastic strain increment. They are calculated according to the law of hardening.

Equation (5.2.5) means that the yield surface F (to determine whether it enters into the state of loading) is the same as plastic potential surface Q (its outward normal is in the same direction as the plastic strain increment), also known as the associated flow. It is verified the direction of plastic strain increment depends only on the stress state.

3. **Corollary 3**

Loading–unloading criterion.

The loading–unloading criterion is a rule to judge whether the stress increment would induce the plastic deformation.

$$d\boldsymbol{\sigma} d\boldsymbol{\varepsilon}^p \geq 0$$

Then

$$d\boldsymbol{\sigma}\, \boldsymbol{n} \geq 0$$

where $\boldsymbol{n}$ is the unit normal vector for the yield surface.

This equation shows that the stress increment can lead to plastic deformation only when the angle is less than 90° between the stress increment and yield surface normal.

When the yield surface is adopted, the loading–unloading criterion can be expressed as

$$\begin{cases} \frac{\partial F}{\partial \boldsymbol{\sigma}} \mathrm{d}\boldsymbol{\sigma} > 0 & \text{loading} \\ \frac{\partial F}{\partial \boldsymbol{\sigma}} \mathrm{d}\boldsymbol{\sigma} = 0 & \text{neutral loading} \\ \frac{\partial F}{\partial \boldsymbol{\sigma}} \mathrm{d}\boldsymbol{\sigma} < 0 & \text{unloading} \end{cases} \tag{5.2.6}$$

Iliushin proposed the plastic postulate in strain space. Similarly, based on Iliushin Plastic Postulate, the classic plastic mechanics system can be created in the strain space.

5.3 The Constitutive Model Based on the Classic Plastic Theory

5.3.1 The Framework of the Classic Plastic Theory

The Classic Plastic Theory mainly includes the yield surface, associated flow rule and the loading–unloading criterion.

1. **The yield surface**

The yield surface is the core of the Classical Plastic Theory. It is used to determine the mechanical response of material. Within the yield surface, it is the elastic area, and the mechanical response of the material is elastic. When reaching the yield surface, the judgment of loading–unloading is needed to perform. Loading will produce plastic deformation, while unloading still brings about the elastic response. Both the plastic flow rule and loading–unloading criterion of the Classical Plastic Theory are related to the yield surface.

2. **The associated flow rule**

The flow rule is used to determine direction of plastic flow under loading stress increment. The Classical Plastic Theory adopts associated flow rule. The plastic potential function (Q) is the same as the yield surface (F), and the increment of plastic strain ($\mathrm{d}\boldsymbol{\varepsilon}^{\mathrm{p}}$) can be expressed as

$$\mathrm{d}\boldsymbol{\varepsilon}^{\mathrm{p}} = \mathrm{d}\lambda \frac{\partial Q}{\partial \boldsymbol{\sigma}} = \mathrm{d}\lambda \frac{\partial F}{\partial \boldsymbol{\sigma}}$$

3. **Loading–unloading criterion**

The loading–unloading criterion is used to determine whether the mechanical response of material is the elastic behavior, or plastic deformation. The performing of the judgments of loading–unloading by the Classical Plastic Theory is based on the yield surface.

$$\begin{cases} \frac{\partial F}{\partial \boldsymbol{\sigma}} \mathrm{d}\boldsymbol{\sigma} > 0 & \text{loading} \\ \frac{\partial F}{\partial \boldsymbol{\sigma}} \mathrm{d}\boldsymbol{\sigma} = 0 & \text{neutral loading} \\ \frac{\partial F}{\partial \boldsymbol{\sigma}} \mathrm{d}\boldsymbol{\sigma} < 0 & \text{unloading} \end{cases}$$

4. **Plastic factor**

Plastic factor $\mathrm{d}\lambda$ is used to characterize the magnitude of the plastic deformation, and the calculating process according to the law of hardening will be discussed in detail in the next chapter.

5.3.2 Commonly Used Models

For the Classic Plastic Theory, the two most widely used plastic models are those proposed by Tresca [11] and Mises [12] initially for metals. Experience suggests that under undrained conditions, fully saturated cohesive soils (i.e., clay) can be modeled accurately by either Tresca or von Mises plastic theory.

1. **Tresca model**

After a series of experiments on metals, Tresca [11] concluded that the yielding occurred when the maximum shear stress reached a certain value. In proposing this, Hill suggested that Tresca was probably influenced by a more general law for the failure of soils, proposed many years earlier by Coulomb. Tresca's yield criterion is

$$f = \sigma_1 - \sigma_3 - 2S_\mathrm{u} = 0 \tag{5.3.1}$$

where $\sigma_1 \geq \sigma_2 \geq \sigma_3$, and S_u is the undrained shear strength. From a computational point of view, it is more useful to write the Eq. (5.3.1) in terms of the second invariant of deviatoric stress J_2 and Lode's angle θ_σ as follows:

$$f = \sqrt{J_2}\cos\theta_\sigma - S_\mathrm{u} = 0 \tag{5.3.2}$$

As shown in Fig. 5.6, the Tresca yield surface is a regular hexagon on a deviatoric plane.

When a saturated clay is loaded under undrained condition, the volume remains constant. As a result, it is suitable to adopt an associated plastic flow rule by treating the yield function Eq. (5.3.2) as the plastic potential as well. Therefore,

$$g = \sqrt{J_2}\cos\theta_\sigma - S_\mathrm{u} = 0 \tag{5.3.3}$$

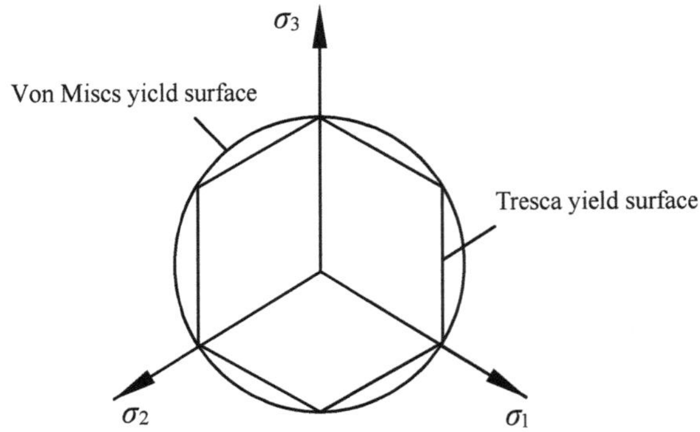

Fig. 5.6 Tresca and Von Mises yield criteria on a deviatoric plane

As described before, the complete relation between a stress rate and a strain rate would be put forward for an elastic—perfectly plastic solid. To determine the complete stress–strain relation for Tresca materials, we need to determine $\frac{\partial f}{\partial \boldsymbol{\sigma}}$ and $\frac{\partial g}{\partial \boldsymbol{\sigma}}$, which can be obtained using the chain rule.

$$\frac{\partial f}{\partial \boldsymbol{\sigma}} = \frac{\partial f}{\partial J_2}\frac{\partial J_2}{\partial \boldsymbol{\sigma}} + \frac{\partial f}{\partial \theta_\sigma}\frac{\partial \theta_\sigma}{\partial \boldsymbol{\sigma}} = \frac{\cos\theta_\sigma}{2\sqrt{J_2}}\frac{\partial J_2}{\partial \boldsymbol{\sigma}} - \sqrt{J_2}\sin\theta_\sigma\frac{\partial \theta_\sigma}{\partial \boldsymbol{\sigma}} \tag{5.3.4}$$

$$\frac{\partial g}{\partial \boldsymbol{\sigma}} = \frac{\partial g}{\partial J_2}\frac{\partial J_2}{\partial \boldsymbol{\sigma}} + \frac{\partial g}{\partial \theta_\sigma}\frac{\partial \theta_\sigma}{\partial \boldsymbol{\sigma}} = \frac{\cos\theta_\sigma}{2\sqrt{J_2}}\frac{\partial J_2}{\partial \boldsymbol{\sigma}} - \sqrt{J_2}\sin\theta_\sigma\frac{\partial \theta_\sigma}{\partial \boldsymbol{\sigma}} \tag{5.3.5}$$

where $\frac{\partial J_2}{\partial \boldsymbol{\sigma}}$ and $\frac{\partial \theta_\sigma}{\partial \boldsymbol{\sigma}}$ are independent of the form of yield functions and plastic potentials as they only depend on the definitions of the second invariant of deviatoric stress and Lode's Angle.

However, it is noted that Tresca's yield function and plastic potential are not differentiable at certain corner points from Fig. 5.6. These singularities deserve special treatment as they are important for some practical problems. Several approaches exist for dealing with these singularities. One of the classical approaches has been developed by Nayak and Zienkienicz, and it consists of using only one yield function in combination with a rounding off procedure for points at which two planes of the yield function meet (the so-called corner point). Sloan and Booker have adopted a modified surface to round off the corners so that a smooth yield surface may be obtained. Although these approaches have proven to be effective to some extent, they are mathematically inconvenient and somewhat physically artificial. A rigorous method has been proposed by Yu [13] for stress states sited on the corners.

2. **Mises model**

A slightly better alternative to the Tresca yield criterion is the criterion proposed by Mises [12]. Mises suggested that yielding occurred when the second invariant of deviatoric stress reached a critical value, Mises' yield criterion is expressed as follows:

$$f = \sqrt{J_2} - k = 0 \tag{5.3.6}$$

or

$$f = (\sigma_1 - \sigma_2)^2 + (\sigma_2 - \sigma_3)^2 + (\sigma_3 - \sigma_1)^2 - 6k^2 = 0 \tag{5.3.7}$$

where k is the undrained shear strength of the soil in pure shear.

As shown in Fig. 5.6, the Mises yield surface is a circle on a deviatoric plane. Like the Tresca yield surface, the Von Mises yield criterion does not depend on the mean stress. A physical interpretation of the Mises yield criterion is that Eq. (5.3.7) implies that yielding begins when the elastic energy of distortion reaches a critical value.

By choosing the appropriate value of the strength parameter k in Eq. (5.3.6), we can make the Mises circle pass through the corners of the Tresca hexagon (as shown in Fig. 5.6), which happens when

$$k = \frac{S_\mathrm{u}}{\cos\theta_\sigma} = \frac{2}{\sqrt{3}} S_\mathrm{u} \tag{5.3.8}$$

By comparing Tresca's and Mises' yield criteria, it is obvious that the Mises yield criterion generally implies slightly higher undrained shear strength. The difference depends on Lode's Angle that indicates the direction of shear stress.

For undrained loading, the plastic volumetric strain is zero so that an associated flow rule is adequate, namely

$$g = \sqrt{J_2} - k = 0 \tag{5.3.9}$$

which, together with Eq. (5.3.6), leads to

$$\frac{\partial f}{\partial \boldsymbol{\sigma}} = \frac{\partial f}{\partial J_2}\frac{\partial J_2}{\partial \boldsymbol{\sigma}} = \frac{1}{2\sqrt{J_2}}\frac{\partial J_2}{\partial \boldsymbol{\sigma}} \tag{5.3.10}$$

$$\frac{\partial g}{\partial \boldsymbol{\sigma}} = \frac{\partial g}{\partial J_2}\frac{\partial J_2}{\partial \boldsymbol{\sigma}} = \frac{1}{2\sqrt{J_2}}\frac{\partial J_2}{\partial \boldsymbol{\sigma}} \tag{5.3.11}$$

Questions

1. In 1928, Mises put forward the plastic potential concept and flow rule based on the concept of elastic potential in elastic mechanics. In 1957, Drucker improved the theoretic basis by proposing Drucker's postulate. Is the foundation strong for the theoretic basis of the Classic Plastic Theory?
2. What is the framework of the Classical Plastic Theory?
3. According to the Classical Plastic Theory, how can you establish the constitutive model for geomaterial.
4. Is the Classical Plastic Theory perfect?

References

1. Huang ZP (2003) Fundamentals of continuum mechanics. Higher Education Press, Beijing
2. Voyiadjis GZ, Shojaei A, Li GQ (2011) A thermodynamic consistent damage and healing model for self healing materials. Int J Plast 27(7):1025–1044
3. Houlsby GT, Puzrin AM (2000) A thermomechanical framework for constitutive models for rate-independent dissipative materials. Int J Plast 16:1017–1047
4. Liu YX, Zhang Y, Wu RZ, Zhou JW, Zheng YR (2015) Nonexistence and non-decoupling of the dissipative potential for geo-materials. Geomech Eng 9(5):569–583
5. Zhou JW, Liu YX, Lu X, Zheng YR (2011) Inexistence and un-decoupling of the dissipation potential for geomaterials. Chin J Geotech Eng 33(4):607–617
6. Zhou JW, Liu YX, Lu X, Zheng YR (2012) Condition of existence and decoupling for the flow potential of geomaterial. Rock Soil Mech 33(2):375–375
7. Ducker DC, Prager W (1952) Soil mechanics and plastic analysis or limit design. Q Appl Math 10(2):157–165
8. Drucker DC (1953) Limit analysis of two and three dimensional soil mechanics problems. J Mech Phys Solids 1(4):217–226
9. Drucker DC, Gibson RE, Henkel DH (1957) Soil mechanics and work-hardening theories of plasticity. Trans. ASCE 122:94–112
10. Huang SJ (1988) The thermodynamics principle of stability postulate in plastic mechanics. J Solid Mech 9(2):95–101
11. Tresca H (1864) Comptes rendus. Academic Science, Paris 59:754
12. Mises V (1913) Gottinger nachrichten. Math Phys Kl 582
13. Yu HS (2006) Plasticity and Geotechnics. Springer Science Business Media, USA

Chapter 6
The Development of the Plastic Theory of Geomaterial

6.1 Study of Several Basic Problems in Plastic Theory of Geomaterial

The former constitutive models of geomaterial are always based on classical theory of plasticity, such as the famous Cam-Clay model. However, a large number of geotechnical tests show that the following basic mechanical characteristics of geomaterial cannot be reflected by the models based on the Classical Plastic Theory:

1. According to the Classical Plastic Theory, the direction of plastic strain increment is only determined by stress states. However, a number of experimental results [1, 2] show that the direction of plastic strain increment is influenced by stress increment significantly, that is, geomaterial does not satisfy the uniqueness assumption between the direction of plastic strain increment and stress. Dependency of sand's plastic strain increment direction on stress increment and the variation of such dependency with stress state are provided in details by Anandarajah.
2. The rotation of principal stress axes will cause plastic strain and it cannot be calculated by the Classical Plastic Theory [3–6].
3. The elastoplastic model with single-yield surface based on the Classical Plastic Theory cannot describe the dilatancy (deviatoric stress will bring about volumetric deformation) of geomaterial reasonably [7].

To overcome the above disadvantages, a lot of works have been done by scholars in the world. Some constitutive models were presented which disobey the Classical Plastic Theory, such as double-yield surface model [8–10] and multi-yield surfaces models [11]. Nonassociated flow rule was used to remedy excessive dilatancy [10, 12]. The Classical Plastic Theory is based on plastic postulate. Some important works have been done on the study of plastic postulate. Huang [13] found that Drucker Postulate and Illyushin Postulate are independent of the laws of thermodynamics, so these two postulates would not be satisfied by some materials.

Y. Liu and Y. Zheng, *Plastic Mechanics of Geomaterial*, Springer Geophysics,
https://doi.org/10.1007/978-981-13-3753-6_6

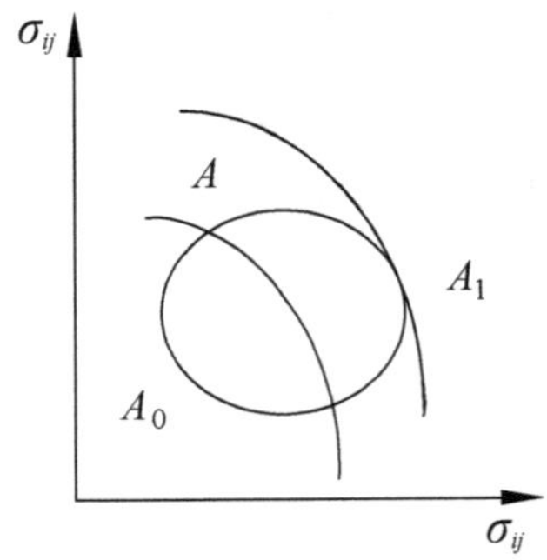

Fig. 6.1 The illustration of additional stress circulation

Based on the basic mechanical properties of geomaterial, it was proved that Drucker [14] Postulate and the principle of Classical Plastic Theory are not suitable for geomaterial. Several basic problems are discussed for the Plastic Theory of Geomaterial.

6.1.1 Proving that Drucker Postulate Is Unsuitable for Geomaterial

Drucker Postulate can be expressed as the work of arbitrary additional stress circulation will not be negative, and it is illustrated in Fig. 6.1.

$$W_D = \oint_{\sigma} (\sigma - \sigma_0) d\varepsilon \geq 0$$

An additional stress circulation is illustrated in Fig. 6.1. During an additional stress circulation, the work caused by elastic deformation is zero, and the plastic deformation will not be generalized in the stage of elastic loading $A_0 \longrightarrow A$ and the stage of unloading $A_1 \longrightarrow A_0$. Then

$$\begin{aligned} W_D &= \int_A^{A_1} (\boldsymbol{\sigma} - \boldsymbol{\sigma}_0) d\boldsymbol{\varepsilon}^p \\ &= (\boldsymbol{\sigma} - \boldsymbol{\sigma}_0) d\boldsymbol{\varepsilon}^p + \frac{1}{2} d\boldsymbol{\sigma} \cdot d\boldsymbol{\varepsilon}^p \end{aligned} \tag{6.1.1}$$

When the relation of point $A_0(\boldsymbol{\sigma}_0)$ and point $A(\boldsymbol{\sigma})$ is not superposing, the second term in the right of Eq. (6.1.1) can be ignored, then

$$W_D = (\boldsymbol{\sigma} - \boldsymbol{\sigma}_0) d\boldsymbol{\varepsilon}^p \tag{6.1.2}$$

The elastoplastic models of geomaterial are always formulated in the plane of p(mean stress)-q(deviatoric stress), which can be written as

Fig. 6.2 The illustration of additional stress circulation in plane *p*-*q*

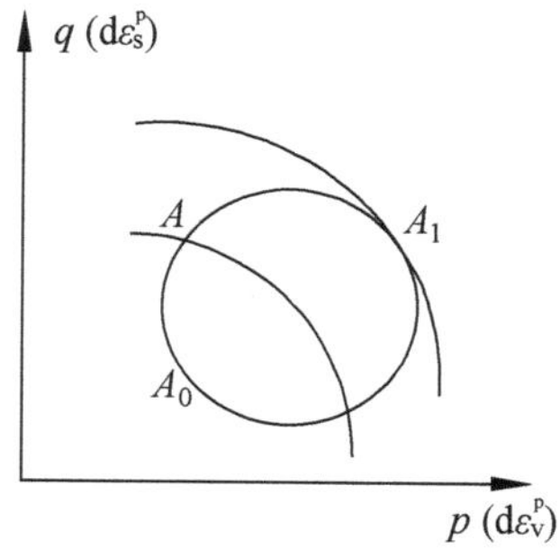

$$\begin{cases} \mathrm{d}\varepsilon_v^p = E\mathrm{d}p + B\mathrm{d}q \\ \mathrm{d}\varepsilon_s^p = C\mathrm{d}p + D\mathrm{d}q \end{cases} \tag{6.1.3}$$

where $\mathrm{d}\varepsilon_v^p$, $\mathrm{d}\varepsilon_s^p$ are the increments of plastic volumetric strain and plastic deviatoric strain, respectively.

An additional stress circulation in *p*-*q* plane is illustrated in Fig. 6.2. The coordinates are $A_0(p_0, q_0)$, $A(p, q)$, and $A_1(p+\mathrm{d}p, q+\mathrm{d}q)$. Then, Eq. (6.1.2) can be rewritten as

$$W_D = (\boldsymbol{\sigma} - \boldsymbol{\sigma}_0)\mathrm{d}\boldsymbol{\varepsilon}^p = (p - p_0)\mathrm{d}\boldsymbol{\varepsilon}_v^p + (q - q_0)\mathrm{d}\boldsymbol{\varepsilon}_s^p \tag{6.1.4}$$

The dilatancy of geomaterial is recognized by geomechanics circle. Generalized dilatancy phenomenon includes dilatancy (volumetric swell is caused by deviatoric stress, and parameter *B* in Eq. (6.1.3) will satisfy $B<0$) and negative dilatancy (volumetric contraction is caused by deviatoric stress, and parameter *B* in Eq. (6.1.3) will satisfy $B>0$). Grounded on dilatancy, Eq. (6.1.4) will be discussed further.

1. **Proving that Drucker Postulate is unsuitable for geomaterial in condition of dilatancy**

A special stress circulation in *p*-*q* plane is illustrated in Fig. 6.3. Corresponding to Fig. 6.2, the coordinates in Fig. 6.3 will be $A_0(p_0, q_0)$, $A(p, q_0)$, and $A_1(p, q_0+\mathrm{d}q)$. Thus

$$\begin{aligned} W_D &= (\boldsymbol{\sigma} - \boldsymbol{\sigma}_0)\mathrm{d}\boldsymbol{\varepsilon}^p \\ &= (p - p_0)\mathrm{d}\boldsymbol{\varepsilon}_v^p + 0 \cdot \mathrm{d}\boldsymbol{\varepsilon}_s^p \\ &= (p - p_0)\mathrm{d}\boldsymbol{\varepsilon}_v^p \\ &= (p - p_0)(E \cdot 0 + B \cdot \mathrm{d}q) \\ &= B(p - p_0)\mathrm{d}q \end{aligned} \tag{6.1.5}$$

In the loading stage $A \rightarrow A_1$ of Fig. 6.3, $\mathrm{d}q>0$, $(p-p_0)>0$, $B<0$ (in condition of dilatancy), then Eq. (6.1.5) will satisfy

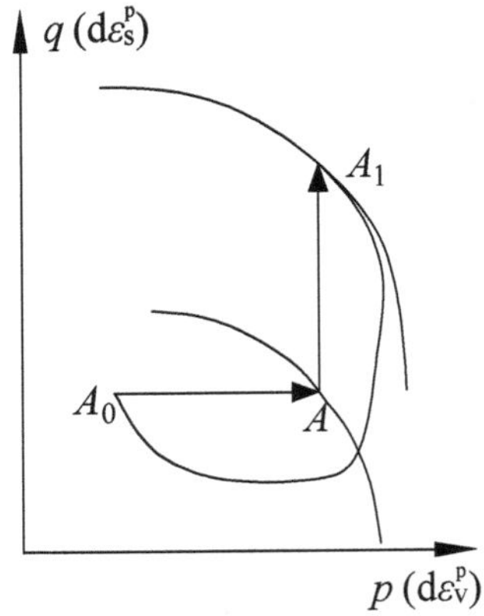

Fig. 6.3 The illustration of a special loading path in p-q plane

$$W_D = B(p - p_0)\mathrm{d}q < 0 \tag{6.1.6}$$

Equation (6.1.6) indicates that Drucker Postulate will not be satisfied in condition of dilatancy. That is, Eq. (6.1.7) will not be satisfied in condition of dilatancy.

$$W_D = (\boldsymbol{\sigma} - \boldsymbol{\sigma}_0)\mathrm{d}\boldsymbol{\varepsilon}^p > 0 \tag{6.1.7}$$

2. **Proving that Drucker Postulate is unsuitable for geomaterial in condition of negative dilatancy**

Equation (6.1.3) can be rewritten as

$$\begin{bmatrix} \mathrm{d}\varepsilon_v^p \\ \mathrm{d}\varepsilon_s^p \end{bmatrix} = \begin{bmatrix} E & B \\ C & D \end{bmatrix} \begin{bmatrix} \mathrm{d}p \\ \mathrm{d}q \end{bmatrix} \tag{6.1.8}$$

In condition of negative dilatancy, $B>0$, and other geomaterial parameters will satisfy $E>0$, $D>0$, $C<0$. So

$$0 > \frac{E}{C} \neq \frac{B}{D} > 0$$

The last equation shows that elements in each line of the plastic coefficients matrix in Eq. (6.1.8) are not in proportion to each other. The plastic strain increment should be expressed by two linearly independent base vectors.

$$\begin{bmatrix} \mathrm{d}\varepsilon_v^p \\ \mathrm{d}\varepsilon_s^p \end{bmatrix} = k_1\boldsymbol{\xi}_1 + k_2\boldsymbol{\xi}_2 \tag{6.1.9}$$

where $\boldsymbol{\xi}_1$, $\boldsymbol{\xi}_2$ are two linearly independent base vectors; k_1, k_2 are the corresponding constants.

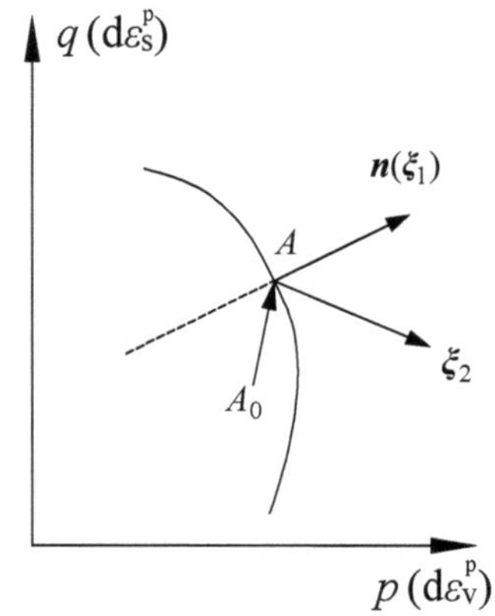

Fig. 6.4 The illustration of strain increment direction in *p-q* plane

The last equation shows that the possible directions of plastic increment are unlimited and the uniqueness relation is nonexistent between the directions of plastic strain increment of geomaterial and stress state. The direction of plastic strain increment is illustrated in Fig. 6.4. Let the exterior normal direction of yield surface $\boldsymbol{n}$ coincide with the base vector $\boldsymbol{\xi}_1$. As another base vector $\boldsymbol{\xi}_2$ is linearly independent of $\boldsymbol{\xi}_1$, $\boldsymbol{\xi}_2$ must not coincide with $\boldsymbol{n}(\boldsymbol{\xi})$, and it is illustrated in Fig. 6.4. If the initial stress states A_0 and $\boldsymbol{\xi}_2$ are located on the same side of the exterior normal of yield surface $\boldsymbol{n}$, then the following will be obtained:

$$W_D = (\boldsymbol{\sigma} - \boldsymbol{\sigma}_0)\mathrm{d}\boldsymbol{\varepsilon}^p = \mathbf{k}_3 \cdot \mathbf{A}_0\mathbf{A} \cdot \boldsymbol{\xi}_2 < 0$$

So Drucker Postulate will not be satisfied in condition of negative dilatancy, too.

Dilatancy is one basic mechanic characteristic of geomaterial. Drucker Postulate will not be satisfied in either dilatancy or negative dilatancy, so Drucker Postulate cannot be applied to geomaterial.

Similarly, Ilyushin Postulate can be discussed in the strain space. It will be justified that Ilyushin Postulate cannot be applied to geomaterial, too.

6.1.2 Proving that the Classic Plastic Theory Is Unsuitable for Geomaterial

The core idea of the Classic Plastic Theory is the uniqueness relation between the direction of plastic strain increment and stress state, namely, unique plastic potential function Q exists in the stress space. Plastic strain increment can be expressed as

$$\mathrm{d}\boldsymbol{\varepsilon}^p = \mathrm{d}\lambda \frac{\partial Q}{\partial \boldsymbol{\sigma}} \tag{6.1.10}$$

where $d\lambda$ is plastic coefficient, signifying the magnitude of plastic strain increment.

Based on the basic mechanic characteristics of geomaterial, the unique relation between the direction of plastic strain increment and stress state will be proved to be nonexistent to geomaterial, namely, the Classic Plastic Theory cannot be applied to geomaterial.

1. **Proving that the Classic Plastic Theory is unsuitable for geomaterial in condition of negative dilatancy**

According to Sect. 6.1.1, Eq. (6.1.9) shows that elements in each line of the plastic coefficients matrix in Eq. (6.1.8) are not in proportion to each other in condition of negative dilatancy. The plastic strain increment should be expressed by two linearly independent base vectors. It means that the possible directions of plastic increment are unlimited and the uniqueness relation cannot be satisfied between the direction of plastic strain increment of geomaterial and stress state, namely, the plastic strain increment cannot be expressed in the form of Eq. (6.1.10) by unique plastic potential surface. So the Classic Plastic Theory cannot be applied to geomaterial in condition of negative dilatancy.

2. **Proving that the Classic Plastic Theory is unsuitable for geomaterial in condition of negative dilatancy**

   In condition of dilatancy, the plastic coefficients in Eq. (6.1.8) should satisfy $B<0, E>0, D>0, C<0$.
   If we assume $\frac{E}{C}=\frac{B}{D}=I<0$, two stress increments in arbitrary stress state are chosen.

   (1) $\mathrm{d}p=0$, $\mathrm{d}q=k$, the corresponding plastic strain increment will be

$$\mathrm{d}\boldsymbol{\varepsilon}_1^{\mathrm{p}}=\begin{bmatrix}\mathrm{d}\varepsilon_{\mathrm{v}1}^{\mathrm{p}}\\ \mathrm{d}\varepsilon_{\mathrm{s}1}^{\mathrm{p}}\end{bmatrix}=\begin{bmatrix}E & B\\ C & D\end{bmatrix}\begin{bmatrix}\mathrm{d}p\\ \mathrm{d}q\end{bmatrix}=\begin{bmatrix}B\mathrm{d}q\\ D\mathrm{d}q\end{bmatrix}=\begin{bmatrix}Bk\\ Dk\end{bmatrix}=\begin{bmatrix}I\\ 1\end{bmatrix}Dk \tag{6.1.11}$$

   (2) $\mathrm{d}p=k$, $\mathrm{d}q=0$, the corresponding plastic strain increment will be

$$\mathrm{d}\boldsymbol{\varepsilon}_2^{\mathrm{p}}=\begin{bmatrix}\mathrm{d}\varepsilon_{\mathrm{v}2}^{\mathrm{p}}\\ \mathrm{d}\varepsilon_{\mathrm{s}2}^{\mathrm{p}}\end{bmatrix}=\begin{bmatrix}E & B\\ C & D\end{bmatrix}\begin{bmatrix}\mathrm{d}p\\ \mathrm{d}q\end{bmatrix}=\begin{bmatrix}E\mathrm{d}p\\ C\mathrm{d}p\end{bmatrix}=\begin{bmatrix}Ek\\ Ck\end{bmatrix}=Ck\begin{bmatrix}I\\ 1\end{bmatrix}=\frac{C}{D}\mathrm{d}\boldsymbol{\varepsilon}_1^{\mathrm{p}} \tag{6.1.12}$$

   where $\frac{C}{D}<0$.

Equation (6.1.12) means that the two plastic strain increment directions in the same stress state are inverse, namely, the uniqueness relation is nonexistent between the direction of plastic strain increment of geomaterial and stress state.

In contrary, if $\frac{E}{C}=\frac{B}{D}=I<0$ is not satisfied,namely, elements in each line of the plastic coefficients matrix in Eq. (6.1.8) are not in proportion to each other.

According to Sects. 6.1.1 and 6.1.2, it will be verified that the potential directions of plastic increment are unlimited and the uniqueness relation is nonexistent between the direction of plastic strain increment of geomaterial and stress state, namely, the plastic strain increment cannot be expressed in the form of Eq. (6.1.10) by unique plastic potential surface. So the Classic Plastic Theory cannot be applied to geomaterial in condition of dilatancy.

Dilatancy is one basic mechanic characteristic of geomaterial. The Classic Plastic Theory cannot be applied to geomaterial either in dilatancy or negative dilatancy, so the Classic Plastic Theory cannot be applied to geomaterial.

Similarly, the Classic Plastic Theory can be discussed in the strain space. It will be proved that the Classic Plastic Theory cannot be applied to geomaterial in the strain space, too.

6.1.3 Study of Several Key Problems in the Plastic Theory of Geomaterial

1. **Review of current modeling theory of geomaterial**

In Sect. 6.1.2, it has been proved that the Classic Plastic Theory cannot be applied to geomaterial. Obviously, the single-yield surface model with associated flow rule, which is grounded on the Classic Plastic Theory, cannot reflect the basic mechanical characteristics of geomaterial and therefore be unreasonable to geomaterial. When the single-yield surface model with nonassociated flow rule is adopted, the excessive large dilatancy phenomenon will be improved. Nevertheless, the single-yield surface model with nonassociated flow rule is still based on the uniqueness relation between the direction of plastic strain increment and stress state, and the direction of plastic strain increment is not normal to the yield surface. This uniqueness relation has been proved to be unsuitable to geomaterial in the last chapter, so the single-yield surface model with nonassociated flow rule cannot be applied to geomaterial too. Even as that pointed by Shen [15], these models do not cast off the restriction of the Classic Plastic Theory.

The merit of multi-yield surface model involving double-yield surface model is that the uniqueness relation will not be satisfied between the direction of plastic stain increment and stress state while the materials complete yield. But the uniqueness relation between the direction of plastic stain increment and stress state will definitely exist in the partial yield zone similar to the single-yield model. Obviously, this is not reasonable, because no experiment has proved the existence of a zone where the uniqueness relation exists between the directions of plastic stain increment and stress state in geomaterial. Other important problems of current multi-yield surface model which needs to be solved are the determination of number of yield surface and whether the associated flow rule can be adopted or not.

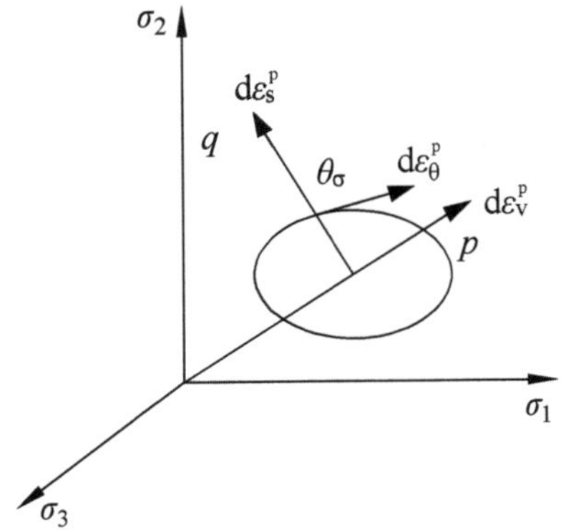

Fig. 6.5 Decomposition of plastic strain increment in principal stress space

2. **Determination of the number of yield surface**

According to the Classic Plastic Theory, the plastic potential function is the same as yield function. Let the plastic potential function be Q, then plastic deformation can be described as

$$d\boldsymbol{\varepsilon}^{\mathrm{p}} = d\lambda \frac{\partial Q}{\partial \boldsymbol{\sigma}}$$

where $d\lambda$ is the plastic coefficient.

If Mises yield surface is chosen, according to the classic theory of plasticity, plastic flow just appears along the normal direction of function $Q=q$, namely, the direction of generalized deviatoric stress (q) in π-plane. In such case, only one plastic potential function $Q=q$ is needed to describe the plastic deformation. In general case, even if principal stress axes rotation can be ignored, plastic volumetric strain increment $d\varepsilon_v^p$ in the direction of p, plastic shear strain increment $d\gamma_q^p$, $d\gamma_\theta^p$ in the direction of q and θ_σ will exist in geomaterial, as illustrated in Fig. 6.5. As a matter of fact, the values of these strain increment components depend on stress increment and are not in proportion to each other. So total plastic strain increment direction is relevant to stress increment, namely, the unique plastic potential surface does not exist corresponding to stress state.

While principal stress axes rotation can be ignored, the principal axes of plastic strain increment will coincide with that of stress, that is, plastic flow can be described in the principal stress space. Under such condition, arbitrary plastic strain increment can be uniquely decomposed into the components in three linearly independent directions. When the directions of three principal stress axes are chosen, three plastic strain increment components will be $d\boldsymbol{\varepsilon}_1^p$, $d\boldsymbol{\varepsilon}_2^p$, and $d\boldsymbol{\varepsilon}_3^p$, namely, plastic potential functions are $Q_1 = \sigma_1$, $Q_2 = \sigma_2$, $Q_3 = \sigma_3$, the corresponding plastic coefficients must be $d\lambda_1 = d\boldsymbol{\varepsilon}_1^p$, $d\lambda_2 = d\boldsymbol{\varepsilon}_2^p$, $d\lambda_3 = d\boldsymbol{\varepsilon}_3^p$. When plastic potential functions are chosen as $Q_1 = p$, $Q_2 = q$, $Q_3 = \theta_\sigma$, the corresponding plastic coefficients will be $d\lambda_1 = d\boldsymbol{\varepsilon}_v^p$, $d\lambda_2 = d\boldsymbol{\varepsilon}_q^p$, $d\lambda_3 = d\boldsymbol{\varepsilon}_\theta^p$ (Fig. 6.5). So in general case, there are three linearly independent plastic potential functions. So Eq. (6.1.10) should be generalized to be

$$d\boldsymbol{\varepsilon}^{p} = \sum_{i=1}^{3} d\lambda_i \frac{\partial Q_i}{\partial \boldsymbol{\sigma}} \tag{6.1.13}$$

where Q_k (k = 1, 2, 3) are three linearly independent plastic potential functions, and $d\lambda_k$ (k = 1, 2, 3) are three corresponding plastic coefficients, respectively.

When principal stress axes rotation cannot be ignored, the principal axes of plastic strain increment will not coincide with that of stress; obviously, it is unsuitable to describe plastic deformation in the principal stress space under such condition. In general case, plastic strain increment can be decomposed into six components in six linearly independent directions, respectively. Thus, in general stress space, plastic strain increment of geomaterial can be expressed as [16]

$$d\boldsymbol{\varepsilon}^{p} = \sum_{i=1}^{6} d\lambda_i \frac{\partial Q_i}{\partial \boldsymbol{\sigma}} \tag{6.1.14}$$

where Q_k (k = 1, 2, …, 6) are six linearly independent plastic potential functions, and $d\lambda_k$ (k = 1, 2, …, 6) are six corresponding plastic coefficients, respectively.

Based on the above analysis, the number of plastic potential surface should be the same as that of the freedoms of plastic strain increment. The maximum number of plastic potential surface should be six. Then, the number of plastic potential surface should be three in principal stress space. It will be two in p-q plane. The number of yield surfaces should be equal to the number of plastic potential surfaces and be the same as that of the freedoms of plastic strain increment, too.

3. Negation of associated flow rule to geomaterial

One disputed problem in multi-yield surface model of geomaterial is whether the associated flow rule can be applied or not. In some multi-yield surface models, either the associated flow rule or the nonassociated flow rule was applied to all yield surfaces [9–11], or they are applied simultaneously and partially [10]. So it is necessary to study the rationality of the associated flow rule to geomaterial. First, it should be pointed out that the associated flow rule is deduced by Drucker Postulate. Drucker Postulate has been proved unsuitable to geomaterial, so there is no scientific basis for the associated flow rule applied to geomaterial.

The characteristics of the elastoplastic matrix of multi-yield surface model with the associated flow rule have been deeply studied by Lu [17] and Yin [18]. The characteristics of the elastoplastic matrix of multi-yield surface model with the associated flow rule will determine whether the associated flow rule is reasonable or not. It is illustrated under the condition of two yield surfaces with the associated flow rule. Let the two yield surfaces are

$$\begin{cases} f_1(p, q) = 0 \\ f_2(p, q) = 0 \end{cases}$$

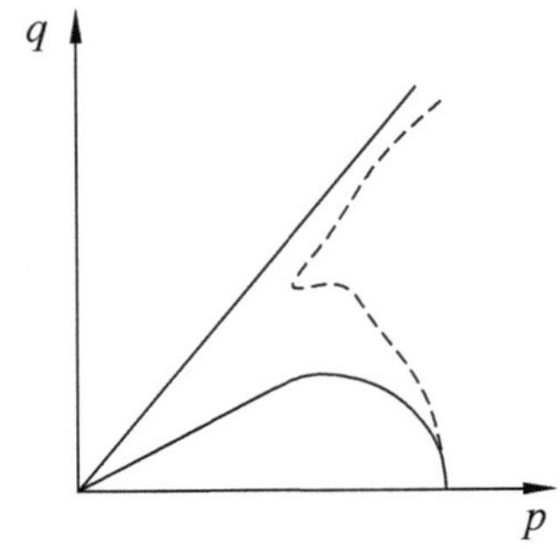

Fig. 6.6 Two undrained effective stress paths in *p*-*q* plane

The plastic coefficients *B* and *C* in Eq. (6.1.3) will be

$$B = C = \frac{1}{A_1}\frac{\partial f_1}{\partial p}\frac{\partial f_1}{\partial q} + \frac{1}{A_2}\frac{\partial f_2}{\partial p}\frac{\partial f_2}{\partial q}$$

where A_1, A_2 are plastic modules.

The plastic coefficient *C* should satisfy $C<0$, and *B* will satisfy $B<0$ in condition of dilatancy, and $B>0$ in condition of negative dilatancy. So geomaterial should not satisfy $B=C$ in general case. $B=C$ must be satisfied if the associated flow rule is adopted in multi-yield surface model, so the associated flow rule is reasonless to geomaterial and without scientific basis.

4. **Discussion about the yield surface of geomaterial**

Grounded on Drucker Postulate, the characteristics of uniqueness, convex of yield surface in stress space will be deduced. In the former part, Drucker Postulate has been proved unsuitable to geomaterial, and then yield surface of geomaterial may not satisfy these characteristics.

Under undrained experiments, the effective stress path in *p*-*q* plane can be regarded as the volumetric yield surface of geomaterial. Two effective stress paths in *p*-*q* plane under undrained test are illustrated in Fig. 6.6. The undrained effective stress path will be convex if negative dilatancy appears in the phase transformation point. But the undrained effective stress path will not be convex [19] if dilatancy appears in the phase transformation point (illustrated in Fig. 6.6 by broken curve).

The number of yield surface should be the same as that of the freedom of plastic strain increment, so the yield surface will not be unique and may not be convex. A key problem of yield surface of geomaterial is how to describe the influence of stress path.

6.2 Development of the Yield Surface for Geomaterial

6.2.1 Significance of Yield Surface

Drucker Postulate is the cornerstone of the Classical Plastic Theory. In the stress space, the yield surface is supposed as

$$f(\sigma_{ij}) = 0$$

According to Drucker Postulate, we know

$$\mathrm{d}\boldsymbol{\varepsilon}^{\mathrm{p}} = \mathrm{d}\lambda \frac{\partial f}{\partial \boldsymbol{\sigma}}$$

Knowing the yield surface for geomaterial, we can obtain plastic flow rule and loading–unloading criterion for geomaterial. That is to say, the yield surface is the core problem of the Classical Plastic Theory.

6.2.2 Yield of Geomaterial

For overconsolidated soil, structural soil, or intact rock, the plastic deformation is not obvious when stress is within a certain range. Beyond the range, the loading stress increment would produce significant plastic deformation. The border of this scope is called as the yield surface (Fig. 6.7).

Note: for geomaterial, there is no absolute yield surface; plastic deformation is not so obvious when stress changes within the elastic zone; like grading students, there is no absolute standard between excellent and good.

Yield surface usually is a function of stress, strain, time, and temperature.

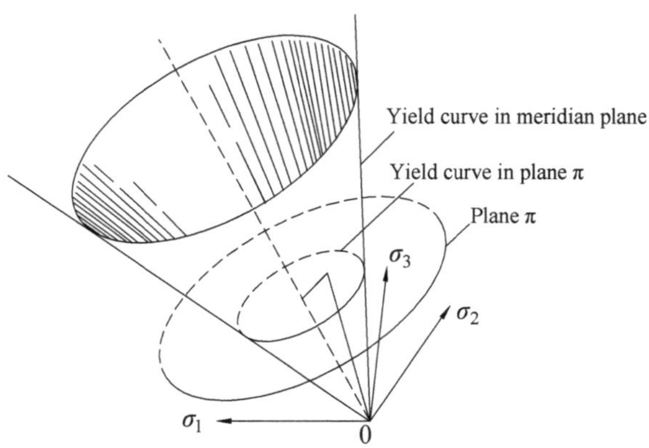

Fig. 6.7 Yield surface of geomaterial

$$F(\boldsymbol{\sigma}, \boldsymbol{\varepsilon}, t, T) = 0 \tag{6.2.1}$$

where $\boldsymbol{\sigma}, \boldsymbol{\varepsilon}, t, T$ are stress, strain, time, and temperature.

Commonly, abbreviate Eq. (6.2.1) as a function of stress.

$$F(\boldsymbol{\sigma}) = 0 \tag{6.2.2}$$

Subsequent yield: After yield of geomaterial, the size, center, or the shape of yield surface will change; the changing yield surface is known as the subsequent yield surface.

The yield function will change with one parameter.

$$F(\boldsymbol{\sigma}, H_{\alpha}) = 0 \tag{6.2.3}$$

where H_{α} is the hardening parameter, which shows the influence of plastic deformation on the subsequent yield surface.

Overconsolidated soils of which yield surface expands with the increase of previous consolidation pressure.

$$F(\boldsymbol{\sigma}, p_{c}) = 0 \tag{6.2.4}$$

6.2.3 *The Shear Yield Surface*

Initial focus of mechanical characteristics is on the strength and destruction of geomaterial, and then the failure criteria have set up, such as Coulomb formula. Thus, yield of geomaterial is analogical to shear failure, and it is called the shear yield condition for geomaterial.

1. **Mohr–Coulomb Model**

The yield criterion proposed by Coulomb is in terms of shear stress τ and normal stress σ_{n} acting on a plane. It suggests that the yielding begins as long as the shear stress and the normal stress satisfy the following equation:

$$\tau = c + \sigma_{n} \tan \varphi$$

where c and φ are the cohesion and internal angle of friction for the soil.

In terms of the principal stresses, Coulomb yield criterion can be expressed as (Figs. 6.8 and 6.9)

$$f = \sigma_1 - \sigma_3 - (\sigma_1 + \sigma_3)\sin\varphi - 2c\cos\varphi = 0$$

for $\sigma_1 \geq \sigma_2 \geq \sigma_3$.

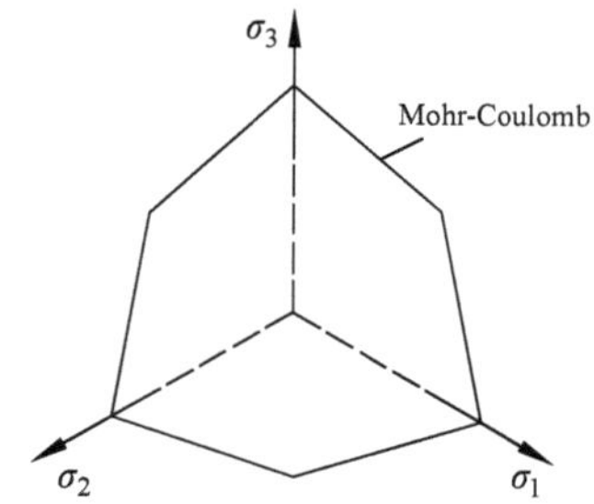

Fig. 6.8 Mohr–Coulomb yield surface on a deviatoric plane

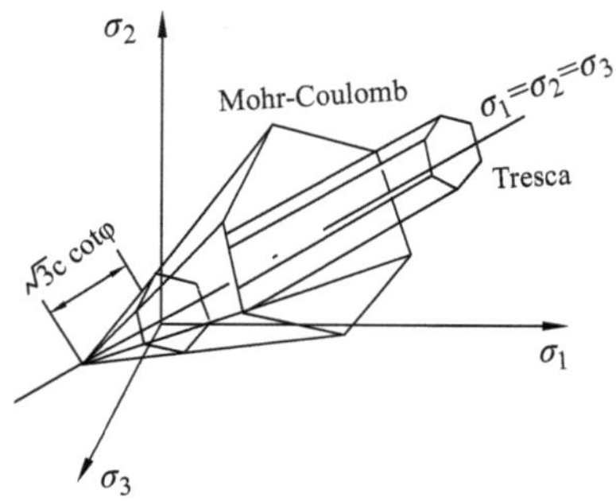

Fig. 6.9 Mohr–Coulomb yield surface in principal stress space

In terms of the stress invariants and Lode's Angle, the Mohr–Coulomb yield criterion can be written as

$$f = \sqrt{J_2} - \frac{m(\theta_\sigma, \varphi)\sin\varphi}{3} I_1 - m(\theta_\sigma, \varphi)c\cos\varphi = 0$$

where

$$m(\theta_\sigma, \varphi) = \frac{\sqrt{3}}{(\sqrt{3}\cos\theta_\sigma + \sin\theta_\sigma\sin\varphi)}$$

Alternatively, the equation can also be expressed in terms of the generalized shear stress q and the mean stress p as follows:

$$f = q - \sqrt{3}pm(\theta_\sigma, \varphi)\sin\varphi - \sqrt{3}m(\theta_\sigma, \varphi)c\cos\varphi = 0 \tag{6.2.5}$$

by noting $q = \sqrt{3J_2}$ and $p = \frac{I_1}{3}$.

The last yield surface is the limit of yield surface—the failure surface.

Obviously, before shear failure of geomaterial, shear yield must be generated. Then what does the shear yield surface look like? A natural idea would be that the yield surface is similar to the shear failure surface, like Coulomb failure condition. Generally, it is referred as like Coulomb failure surface and its subsequent yield surface is expressed as

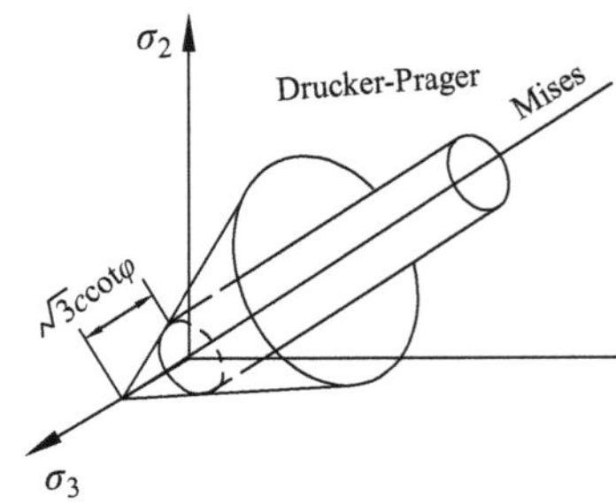

Fig. 6.10 Generalized Mises yield condition and Mises yield condition

$$p \sin \phi + \frac{1}{\sqrt{3}}(\cos \theta_\sigma - \frac{1}{\sqrt{3}} \sin \theta_\sigma \sin \phi) q - c \cos \phi = f(H_\alpha) \tag{6.2.6}$$

2. **Drucker–Prager Model**

The problem of adopting Coulomb yield condition it is a hexagon in plane π is where there are singularities at the angular point if the associated flow rule is used. Thus, Mises yield condition is selected instead of the metal.

$$F = k(H_\alpha) + \sqrt{J_2} = 0 \tag{6.2.7}$$

The Mises yield criterion is not suitable for modeling the yielding of frictional material as it does not include the effect of mean stress as observed in experiments. To overcome this limitation of the Mises yield function, Drucker and Prager [20] proposed the following revised function for frictional soils:

$$f = \sqrt{J_2} - \alpha I_1 - k = 0 \tag{6.2.8}$$

where α and k are material constants.

On a deviatoric plane, Eq. (6.2.8) plots as a circle (Fig. 6.10) for the Mises yield surface. However, in principal stress space, the Drucker–Prager yield surface is a cone while the Mises yield surface is an infinitely long cylinder.

Because of its simplicity, the Drucker–Prager yield criterion has been used quite widely in analysis of geotechnical Engineering. However, experimental research suggests that its circular shape does not agree well with experimental data on the deviatoric plane. For this reason, care is needed when the Drucker–Prager plastic model is used in analysis of geotechnical Engineering.

In plane π, the relations of circles and Coulomb inequilateral hexagon are divided into inscribed tangent circle, inscribed circle, circumscribed circle, and equivalent area circle (Fig. 6.11). The default yield surface is circumscribed circle in ANSYS, and its computation results will add to the risk of danger.

To select the material constants α and k for use in analysis, the Drucker–Prager yield surface is often matched with the Mohr–Coulomb yield surface using a certain

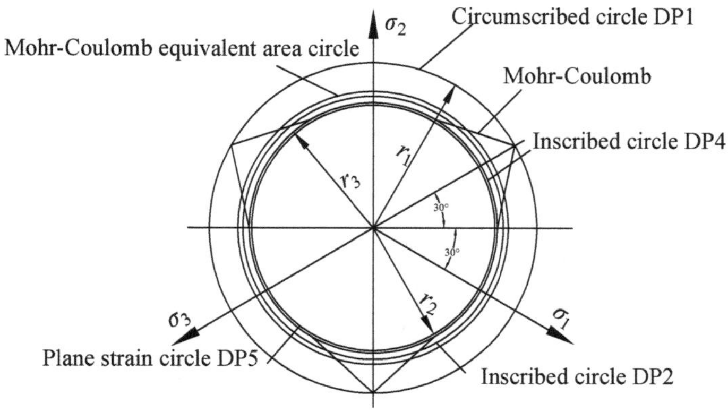

Fig. 6.11 Different D-P yield conditions in plane π

Fig. 6.12 Circumscribed circle for Coulomb yield surface on a deviatoric plane

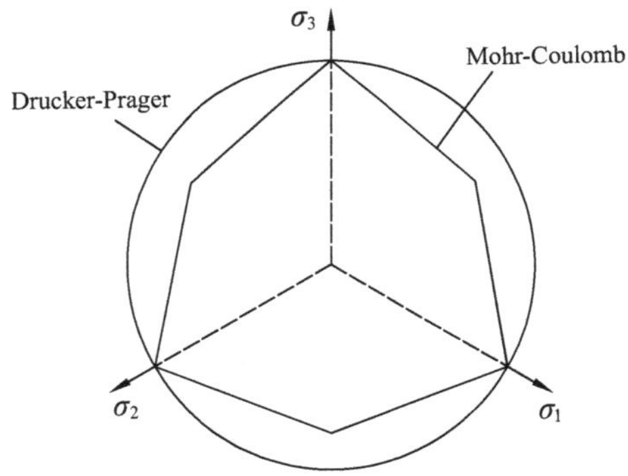

criterion. Figure 6.12 shows such a match at major vertices (circumscribed circle). Mathematically, this condition demands the following relations:

$$\alpha = \frac{2\sin\varphi}{\sqrt{3}(3-\sin\varphi)}, \quad k = \frac{6c\cos\varphi}{\sqrt{3}(3-\sin\varphi)}$$

As another example, if the Drucker Prager and Mohr–Coulomb criteria are made to give an identical limit load for a plane strain problem, then the following relationships must be set up:

$$\alpha = \frac{\tan\varphi}{\sqrt{9+12\tan^2\varphi}}, \quad k = \frac{3c}{\sqrt{9+12\tan^2\varphi}}$$

Inscribed tangent circle for Coulomb inequilateral hexagon in plane π is too safe, and so it is too wasteful.

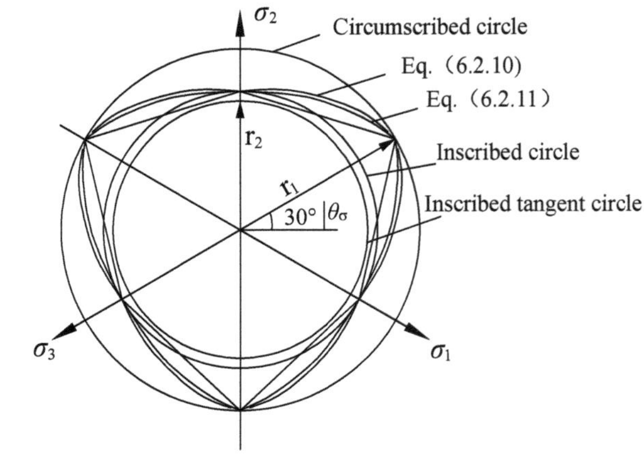

Fig. 6.13 The fitting effect of different correction functions in plane π

$$\alpha = \frac{\sin\varphi}{\sqrt{3}\sqrt{3+\sin^2\varphi}}, \quad k = \frac{\sqrt{3}c\cos\varphi}{\sqrt{3+\sin^2\varphi}}$$

Inscribed circle for Coulomb inequilateral hexagon in plane π is safe biased too, and so it is wasteful too.

$$\alpha = \frac{2\sin\phi}{\sqrt{3}(3+\sin\phi)}, \quad k = \frac{6c\cos\phi}{\sqrt{3}(3+\sin\phi)}$$

The equivalent area circle has high precision and is relatively simple to calculate compared with the coulomb inequilateral hexagons. Why is it?

$$\alpha = \frac{\sin\varphi}{\sqrt{3}(\sqrt{3}\cos\theta_\sigma - \sin\theta_\sigma\sin\varphi)}, \quad k = \frac{\sqrt{3}c\cos\varphi}{\sqrt{3}\cos\theta_\sigma - \sin\theta_\sigma\sin\varphi} \tag{6.2.9}$$

3. **Zienkweiz-Pande Model**

Coulomb condition is an inequilateral hexagonal cone in plane π. Singularity will appear if the associated flow rule is applied. Zienkweiz thinks it is reasonable overall and recommends a smooth curve to simulate the unequal cone. This smooth curve is supposed to be

$$f = f(p) + h(\frac{\sqrt{J_2}}{g(\theta_\sigma)}) = 0$$

where $g(\theta_\sigma)$ is the shape function of the yield surface in plane π (Fig. 6.13).

The main contribution of the fitting function for Coulomb inequilateral hexagonal cone:

(1) Formula of Williams-Warnke (Eq. (6.2.10))

$$g(\theta_\sigma) = \frac{(1-K^2)(\sqrt{3}\cos\theta_\sigma - \sin\theta_\sigma) + (2K-1)\sqrt{(2+\cos 2\theta_\sigma - \sqrt{3}\sin\theta_\sigma)(1-K^2) - 5K^2 - 4K}}{(1-K^2)(2+\cos 2\theta_\sigma - \sqrt{3}\sin 2\theta_\sigma) + 1 - 2K^2} \tag{6.2.10}$$

where K is the radius ratio of triaxial tensile and triaxial compression for coulomb failure conditions in plane π.

$$K = \frac{3 - \sin\phi}{3 + \sin\phi}$$

The formula solves the singularity problem of curve in plane π, but it is too complex.

(2) Formula of Gudehus–Argyris (Eq. (6.2.11))

$$g(\theta_\sigma) = \frac{2k}{(1+k) - (1-k)\sin(3\theta_\sigma)} \tag{6.2.11}$$

The formula solves the singularity problem and is comparatively simple. It was an important improvement for the calculation conditions at that time.

In order to guarantee convex, it should satisfy that $K > \frac{7}{9}$ or $\varphi < 22°$

The improvement of Eq. (6.2.10) is provided by Yingren Zheng.

$$g(\theta_\sigma) = \frac{2k}{(1+k) - (1-k)\sin(3\theta_\sigma) + \alpha\cos^2 3\theta_\sigma}$$

This equation was also convex under certain conditions. Guanghua Yang revised it again later.

4. **Yield condition of double shear**

Yield condition of double shear is proposed by Yu [21] in Xi'an Jiaotong University. The sum is zero for the three principal shear stresses of material (in Fig. 6.14). Yield of material should be caused by the sum of the two larger principal shear stresses of material reaching certain conditions.

Under $\sigma_{12} > \sigma_{23}$, the yield condition of double shear is expressed as

$$F_1 = \sigma_{13} + \sigma_{12} = \sigma_1 - \frac{\sigma_2 + \sigma_3}{2} = (\frac{3}{2}\cos\theta_\sigma - \frac{\sqrt{3}}{2}\sin\theta_\sigma)\sqrt{J_2} - k = 0, \quad \theta_\sigma \le 0 \tag{6.2.12}$$

Under $\sigma_{23} > \sigma_{12}$, the yield condition of double shear is expressed as

$$F_2 = \sigma_{13} + \sigma_{23} = \frac{\sigma_1 + \sigma_2}{2} - \sigma_3 = (\frac{3}{2}\cos\theta_\sigma + \frac{\sqrt{3}}{2}\sin\theta_\sigma)\sqrt{J_2} - k = 0, \quad \theta_\sigma \ge 0 \tag{6.2.13}$$

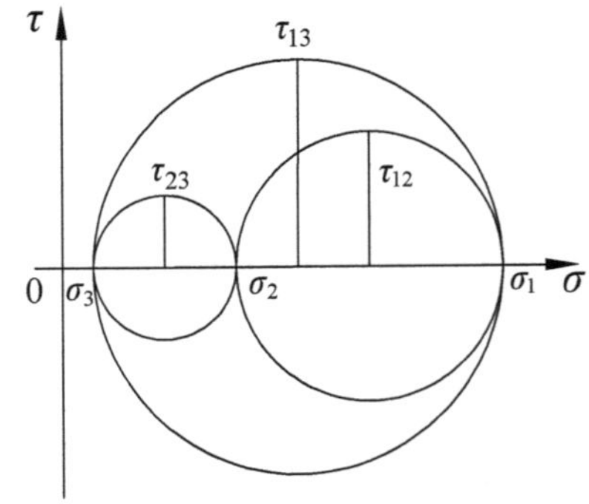

Fig. 6.14 Three principal shear stresses

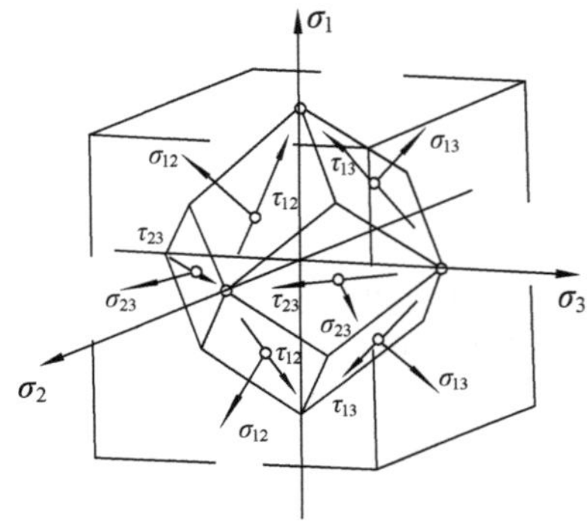

Fig. 6.15 The yield condition of double shear in 3-D space-dodecahedron stress

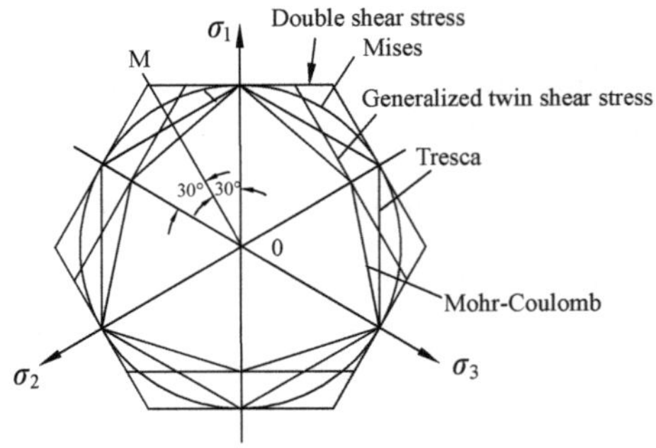

Fig. 6.16 Yield condition of double shear compared with other yield conditions

The yield condition in the 3-D space is illustrated in Fig. 6.15.

Yield condition of double shear is an original contribution in the field of mechanics. It takes into account the effect of intermediate principal stress without considering the effect of hydraulic stress, and then it has been extended to the generalized yield condition of double shear.

It can be seen from Fig. 6.16 that its area is much bigger than Coulomb yield condition, and the calculated safety factor is much larger than the actual value, thus making the engineering design tend to be dangerous. It needs to be revised through matching the Coulomb yield condition.

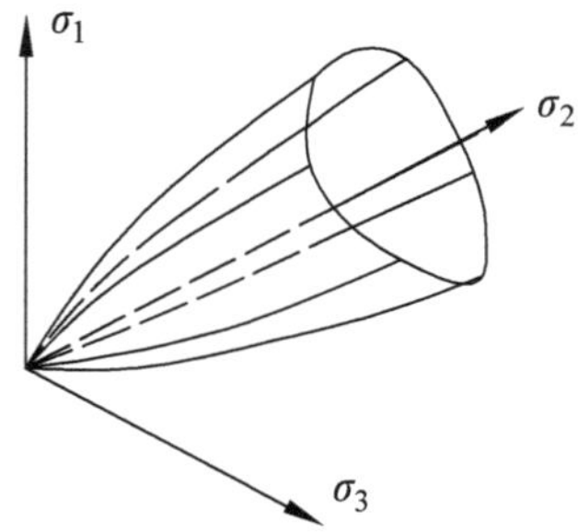

Fig. 6.17 Yield criterion of Hoek–Brown

5. **Hoek–Brown Model**

In addition to the linear Mohr–Coulomb yield criterion, a number of researchers have also used nonlinear yield criteria to analyze rock mechanics problems. The most popular development has been the empirical nonlinear criterion proposed by Hoek and Brown [22–24] to describe the yield and failure behavior of rock masses.

The yield of rock is assumed to be governed by the Hoek and Brown criterion (Fig. 6.17) in the following form:

$$f = \sigma_1 - \sigma_3 - \sqrt{mY\sigma_3 + sY^2} = 0 \tag{6.2.14}$$

where σ_1 and σ_3 are the major and minor principal stresses; Y is the uniaxial compressive strength of the intact rock material; m and s are constants depending on the nature of the rock mass and the extent to which it is broken prior to being subjected to the principal stresses σ_1 and σ_3.

The Hoek–Brown criterion offers some advantages over other approaches in determining the overall strength of in situ rock masses because it is based on one simple material property Y, and rock mass quality data that may be systematically collected and evaluated during site investigation.

For the convenience of numerical applications, the Hoek–Brown criterion may also be expressed in terms of stress invariants by the following equation:

$$f = 4J_2\cos^2\theta_\sigma + g(\theta_\sigma)\sqrt{J_2} - \alpha I_1 - k = 0 \tag{6.2.15}$$

where

$$g(\theta_\sigma) = mY(\cos\theta_\sigma + \frac{\sin\theta_\sigma}{\sqrt{3}}),\ k = sY^2$$

The Hoek–Brown criterion takes into account the influence of the nature of the rock and confining pressure on strength, which is more suitable for rock than Coulomb criterion. But the effect is not considered for intermediate principal stress and it is based mainly on the experimental results. The theoretical basis is not enough.

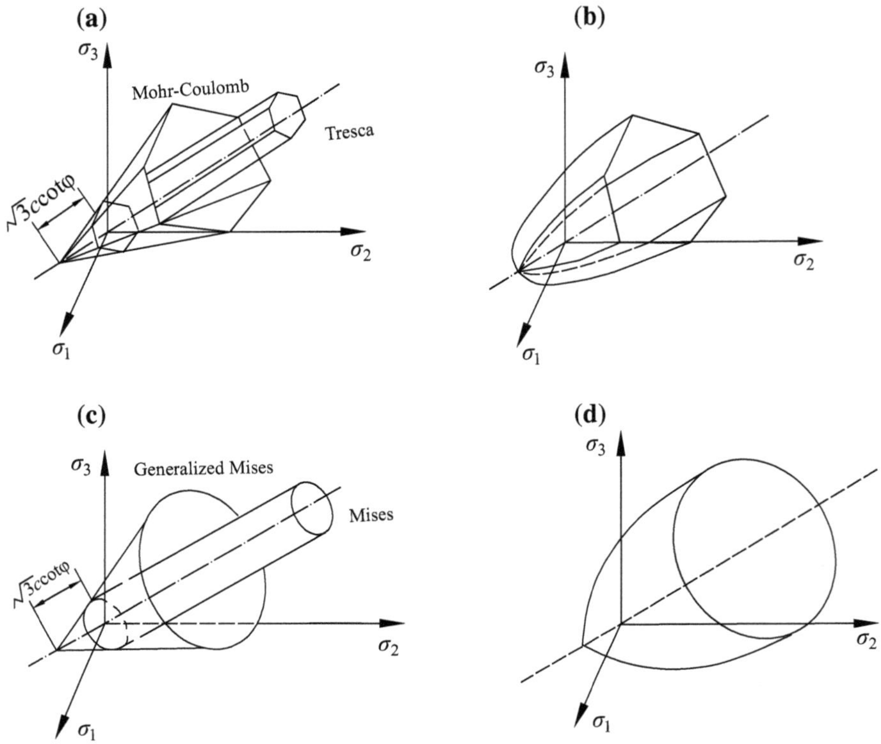

Fig. 6.18 The 3-D shear yield surfaces for geomaterial

6. **General expressions for shear yield criterion**

General shear yield surface can be expressed as

$$F = \beta p^2 + \alpha p - k(H_\alpha) + \bar{\sigma}_+^n = 0 \tag{6.2.16}$$

where $\bar{\sigma}_+ = \frac{\sqrt{J_2}}{g(\theta_\sigma)}$ is the shear stress, and the denominator is the shape function of yield surface in plane π.

The shear yield surfaces with different parameters are shown in Figs. 6.18 and 6.19.

6.2.4 Volumetric Yield Criterion

We have had a more comprehensive understanding of the yield and shear yield criterion for geomaterial in the previous sections. Let us have a look at the effect

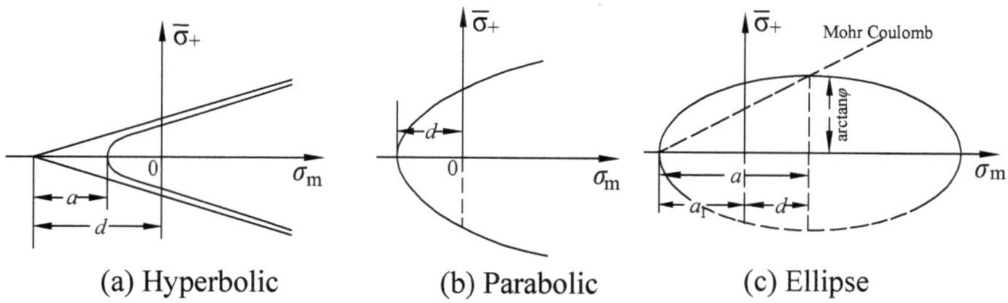

Fig. 6.19 The shear yield surfaces of geomaterial in meridian plane

Fig. 6.20 The shear yield surface of geomaterial

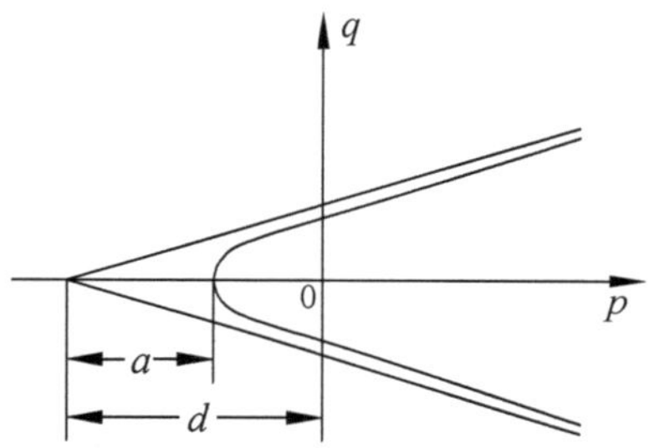

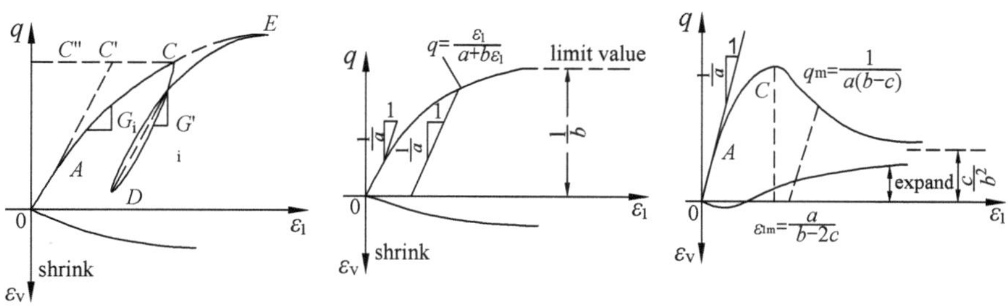

Fig. 6.21 The triaxial stress–strain curve for soil

of the application of the shear yield criterion of geomaterial in the Classical Plastic Theory.

The shear yield surface is shown in Fig. 6.20 in plane p-q. What will happen if the associated flow rule is adopted? There is a very prominent dilatancy phenomenon, and the significant dilatancy appears as long as it is yield.

Let us take a look at the basic mechanical properties of geomaterial.

The stress–strain curve is illustrated in Fig. 6.21 for soil. Only the volumetric contraction exists for normally consolidated soils (there is no dilatancy) in the whole stage of triaxial shear. Overconsolidated soil also displays the volumetric contraction at the beginning of triaxial shear, and the dilatancy will emerge afterward.

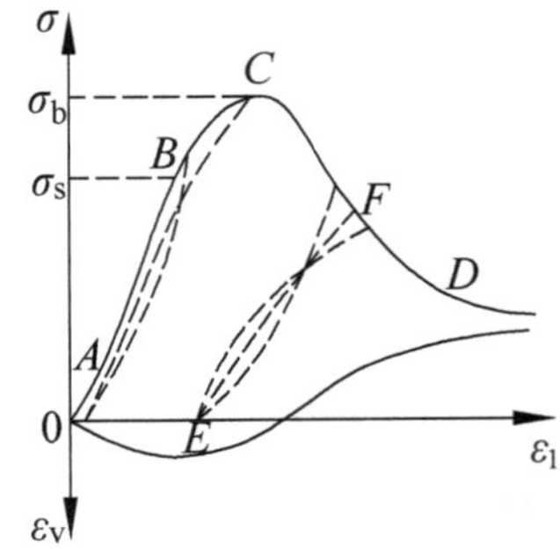

Fig. 6.22 Stress–strain curve of rock

The stress–strain curve is demonstrated in Fig. 6.22 for the whole test process of rock. Similar to overconsolidated soil, rock reveals volumetric shrinkage, and then volumetric expands (dilatancy).

Therefore, using the Classical Plastic Theory and the shear yield criterion is not reasonable for geomaterial. For numerical calculation, dilatancy will appear at the beginning, which is not in conformity with the basic mechanical characteristics of geomaterial because volumetric contraction is exhibited at the initial deformation stage for any geomaterial. The elaboration of another yield surface, volumetric yield surface, is as follows.

Volumetric yield surface is a yield surface using plastic volumetric strain as the hardening parameter.

$$F(\sigma, \varepsilon_v^p) = 0 \tag{6.2.17}$$

Shear yield surface comes from imitating the shear failure criterion. It appears like the colorful flowers and bamboo shoots in spring. But volumetric yield surface is lonesome, adopting just Cam-Clay yield surface, or its variations. The volumetric yield surface was put forward at first by Roscoe [25, 26] of the University of Cambridge, based on triaxial compression experiments of Cam-Clay.

1. **The concept of critical state of soil**

   Roscoe carried out triaxial compression experiment on Cam-Clay (normal consolidated soft clay) under drainage and undrained condition. The experimental results were formulated in space of *p-q-v* (specific volume, equivalent to void ratio or the volumetric strain), then the critical state concept was put forward.

   (1) Undrained test
   Specimens consolidate under different hydraulic stresses at first. Then, under the undrained condition, axial pressure keeps increasing until the sample fails under the condition of constant confining pressure. The results are revealed in Fig. 6.23.
   The stress paths in plane *q-p* are shown in Fig. 6.24 for these undrained experiments. As can be seen from the diagram, the shape of stress path

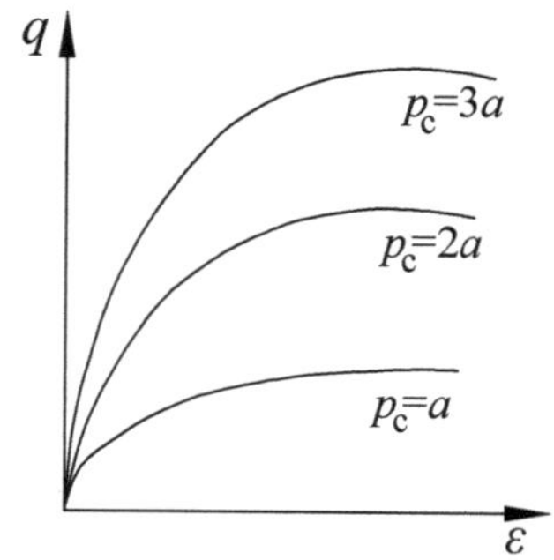

Fig. 6.23 Triaxial compression results of normally consolidated soils under undrained condition

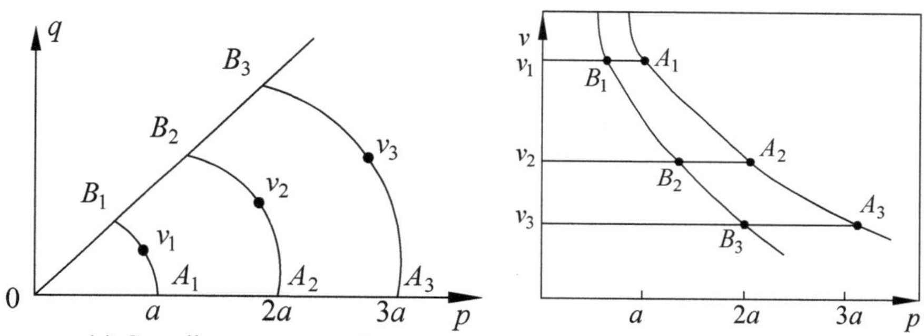

(a) Coordinate system of q-p (b) Coordinate system of v-p

Fig. 6.24 Triaxial undrained test results of normally consolidated soils

is similar, and all the stress paths can be normalized as one curve in the coordinate system of $\frac{q}{p_c} - \frac{p}{p_c}$. The starting point of undrained shear is at the normal consolidation line in the coordinate system of v-p. Because it is undrained, specific volume (volume) remains the same until the soil samples destruct at points B_1, B_2, and B_3. These points B_1, B_2, and B_3 form a smooth curve in the coordinate system of v-p. Appearance of this curve is similar to the normal consolidation line. These points B_1, B_2, and B_3 constitute a straight line in the plane q-p.

There are many identical phenomena at these points B_1, B_2, and B_3: each point is failure point; Stress state (q, p) and volume are constant, and the shear strain keeps increasing, so these points are the point of shear failure. The line formed by these points in the q-p plane is just the Coulomb shear failure line (Eq. (6.2.18)).

$$q_{cs} = Mp_{cs} \tag{6.2.18}$$

(2) Drainage test

Specimens consolidate under different hydraulic stresses at first. Then under the drained condition, axial pressure keeps increasing until the sample fails

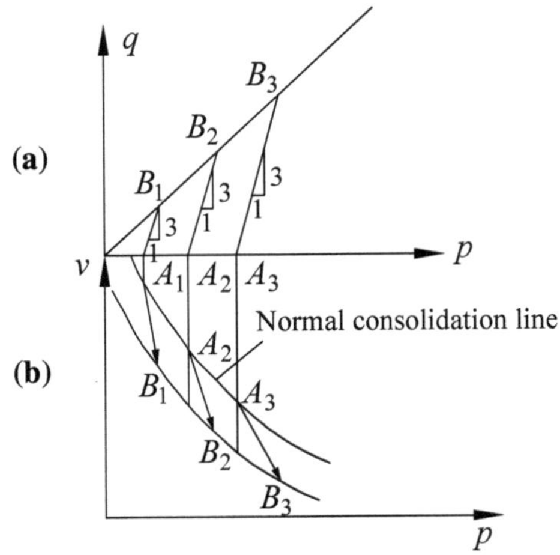

Fig. 6.25 Triaxial drained test results of normally consolidated soils. **a** Coordinate system of *q-p*. **b** Coordinate system of *v-p*

under the condition of constant confining pressure. The results are similar to that in Fig. 6.25.

The stress paths in plane are *q-p* shown in Fig. 6.25 for these drained experiments. As can be seen from the diagram, all of these stress paths are the straight line with the slope of 3 until then reach the shear failure points B_1, B_2, and B_3. The starting point of drained shear is at the normal consolidation line in the coordinate system of *v-p*. Because it is drained, specific volume (volume) remains reduced until the soil samples are damaged at points B_1, B_2, and B_3. These points B_1, B_2, and B_3 form a smooth curve in the coordinate system of *v-p*. Appearance of this curve is similar to the normal consolidation line. These points B_1, B_2, and B_3 constitute a straight line in plane *q-p*.

There are many identical phenomena at these points B_1, B_2, and B_3: each point is failure point; Stress state (q, p) and volume are constant, and the shear strain keeps increasing, so these points are the point of shear failure. The line formed by these points in plane *q-p* is just the Coulomb shear failure line (Eq. (6.2.18)).

In order to highlight the common laws of normally consolidated clay at triaxial shear, the experimental results are put together in Fig. 6.26 under both drained and undrained conditions.

It can be seen that all final damage points fall on the same line under both drained and undrained conditions, and in plane *p-q* and plane *v*-ln*p*. There are many identical phenomena at these points B_1, B_2, and B_3.

Each point is a failure point.

At these points, stress state (q, p) and volume are constant, and the shear strain keeps increasing, so these points are the point of shear failure. The line formed by these points in the plane *q-p* is just the Coulomb shear failure line (Eq. (6.2.18)).

This state is called the critical state, and the line is called the critical state line. Critical state line is Coulomb shear failure line, which is not new. But what needs to

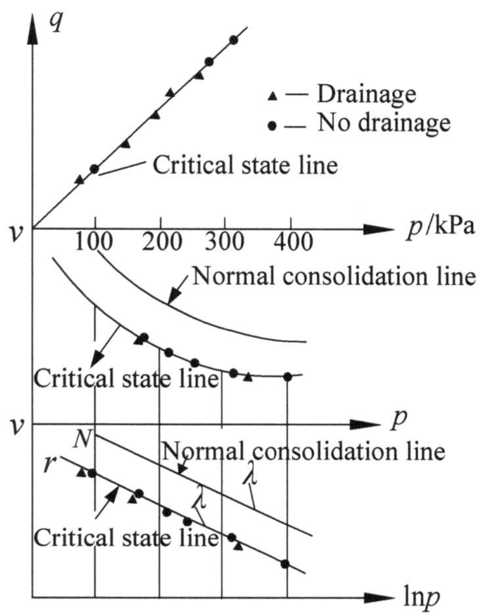

Fig. 6.26 The failure line of normally consolidated soils in triaxial drained and undrained tests

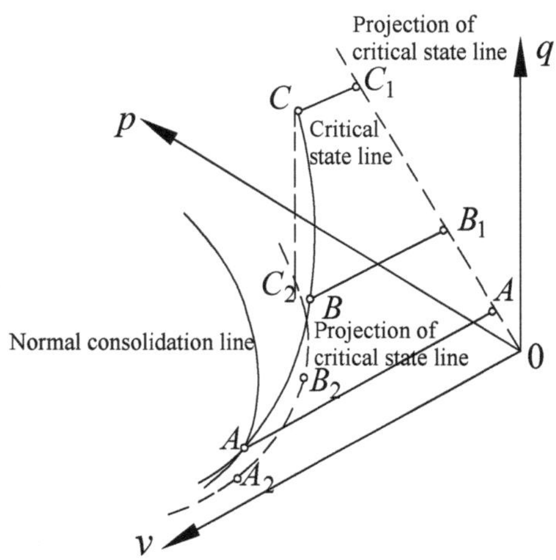

Fig. 6.27 The critical state line in space *v*-*p*-*q*

be emphasized is that critical state is a shear failure state in which stress (q, p) and volume are constant, and the shear strain keeps growing.

The critical state line in space *v*-*p*-*q* is shown in Fig. 6.27.

Due to different consolidation pressures, the volumetric strain and the shear strength are different at the critical state. The curve moves inward and upward in the space with the increase of consolidation pressure. Its projection in plane *v*-*p* is a curve similar to the normal consolidation curve, and the projection in plane *p*-*q* is Coulomb shear failure line.

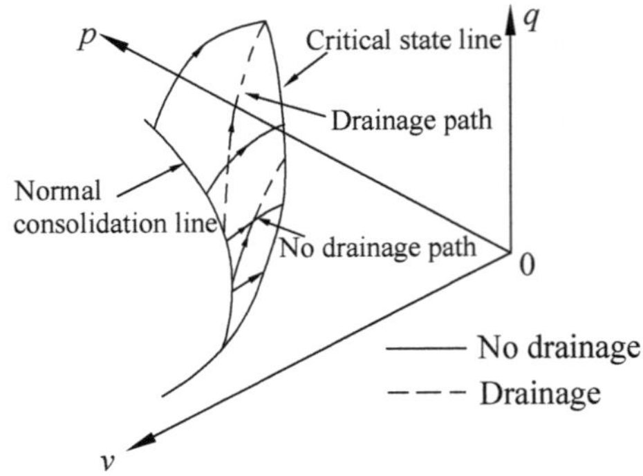

Fig. 6.28 Roscoe surface and the experimental stress path

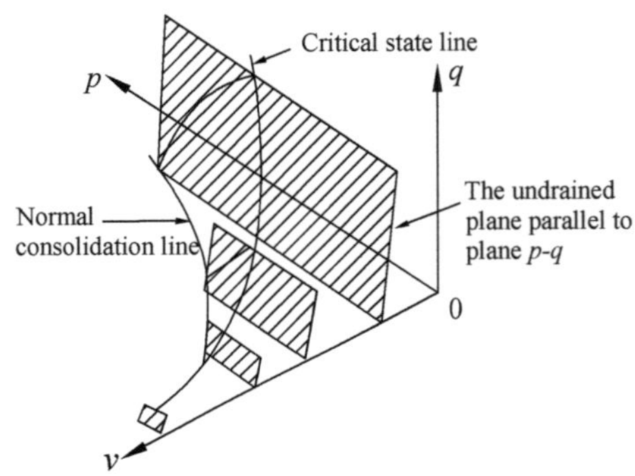

Fig. 6.29 Undrained plane and the experimental stress path in space *v*-*p*-*q*

2. **Roscoe surface**

In space *v*-*p*-*q*, all the stress paths are on the same surface from the normal line to the critical state for normal consolidation soil under drained or undrained condition. This surface is called Roscoe surface (in Fig. 6.28).

Undrained stress path is revealed in Fig. 6.29. With increase of consolidation pressure, the soft clay displays compression and shear shrinkage. The produced positive pore water pressure induces convex of stress path.

The drained stress path is demonstrated in Fig. 6.30. It is in a plane with a slope of 3. With increase of consolidation pressure, the stress path rises upward and inward.

So for triaxial compression, Roscoe surface is unique.

Any tested stress path of normally consolidated soils is not insurmountable to Roscoe surface. Roscoe surface separates the possible and impossible stress state. Therefore, Roscoe surface is also called the state boundary surface.

Characteristics of Roscoe surface:

(1) Roscoe face is constituted by the stress path from normal consolidation condition to the critical state. The stress paths of samples can be determined by the intersection of tested surface and Roscoe surface.
(2) Roscoe surface is a state boundary surface. At the plastic stage, stress path inevitably moves along Roscoe surface to the critical state. The stress path is

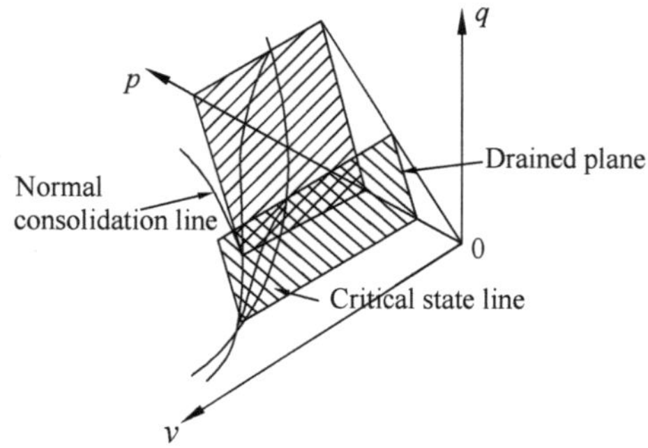

Fig. 6.30 Drained plane and the experimental stress path in space *v*-*p*-*q*

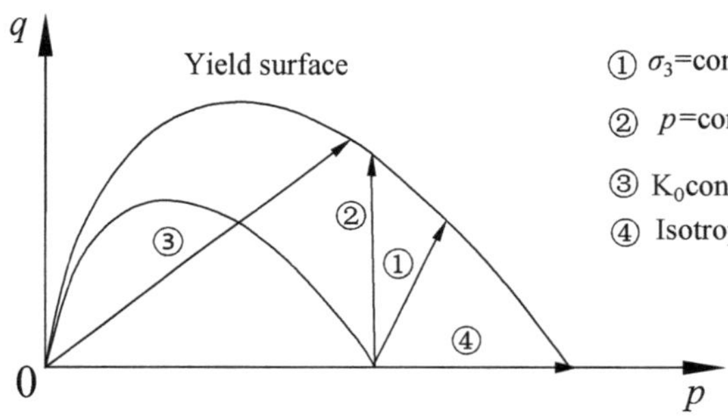

Fig. 6.31 Cam-Clay yield surface

possible on and within the Roscoe surface, while impossible outside Roscoe surface.

(3) The intersection is iso-volume surface for Roscoe surface and is parallel to plane *p*-*q*. When soil samples move alone the stress path in this intersection, the plastic volumetric deformation can be ignored, while the plastic shear strain will be produced. This stress path can be seen as volumetric yield surface.

3. **Volumetric yield surface of normally consolidated soils**

The undrained stress path of Cam-Clay can be regarded as a volumetric yield surface, but Roscoe does not fit the surface directly. On the basis of a large number of triaxial test results, Cambridge yield surface is built on the principle of energy dissipation, and the specific derivation process is provided in Cam-Clay model of the subsequent chapter. Cam-Clay yield surface is a kind of volumetric yield surface [25, 26], expressed as Eq. (6.2.19).

$$p\exp\left(\frac{q}{Mp}\right) = p_c \tag{6.2.19}$$

Cam-Clay yield surface is demonstrated in Fig. 6.31.

Cam-Clay yield surface is of a shape of a bullet. Is there any problem? Let us have a look at the fourth stress path (hydraulic yield). Large shear deformation will

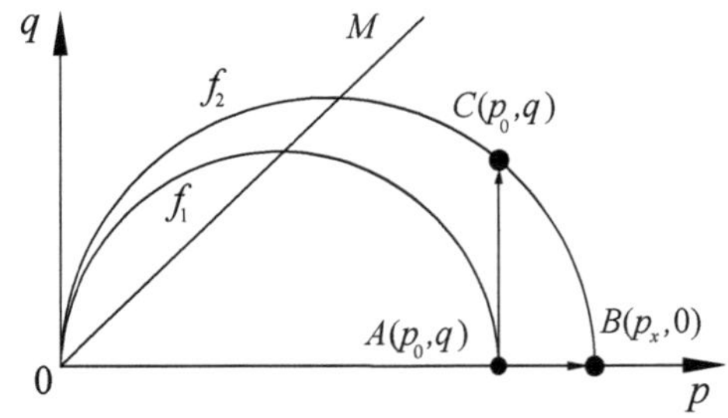

Fig. 6.32 Modified Cam-Clay yield surface

be produced if the associated flow rule is applied. It is not tally with the experimental results that the deformation is mainly the volumetric yield and shear deformation can be ignored.

Cam-Clay yield surface was corrected by Burland [27], and it is called the modified Cam-Clay yield surface (Eq. (6.2.20)).

$$\frac{(p-\frac{p_c}{2})^2}{p_c^2}+\frac{q^2}{M^2p_c^2}=1,\quad p_c=\exp(\frac{e}{\lambda-\kappa}\varepsilon_v^p) \tag{6.2.20}$$

Modified Cam-Clay yield surface is elliptical and revealed in Fig. 6.32. It can be seen that the normal of yield surface is consistent with axis p, with volumetric yield and no shear deformation as the associated flow rule is used under hydraulic loading.

6.2.5 *Yield Surface of Overconsolidated Soil*

1. **Shear failure surface**
   Shear failure surface of overconsolidated soils is called Hvorslev surface. The results of drained triaxial test are displayed in Fig. 6.33 for overconsolidated soil. For the overconsolidated soils exhibit the phenomenon of strain softening, it is mainly different with normal consolidation soil in the two ways:

   (1) Strength decreases after peak (strain soften);
   (2) At the beginning the volume contracts, and then it inflates (dilatancy).

The drained triaxial compression stress path will exceed the critical state line in space v-p-q, and then it will go back from the same path and arrive at the critical state line.

The peaks are located in the same straight line for drained and undrained stress paths for overconsolidated soil (Fig. 6.34).

The failure line can be formulated as

$$\frac{q}{p_c}=g+\frac{hp}{p_c} \tag{6.2.21}$$

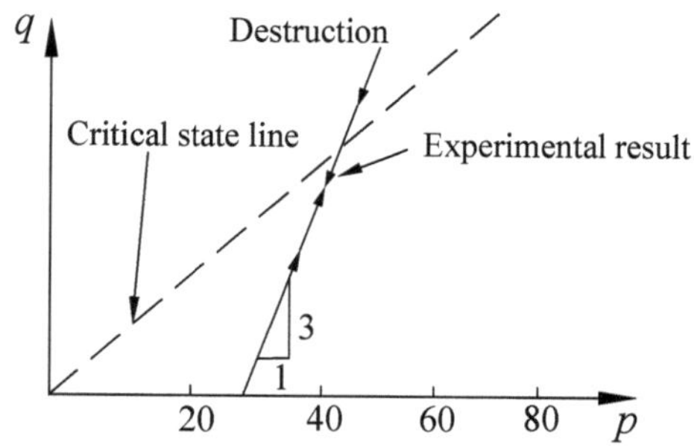

Fig. 6.33 Drained triaxial test for overconsolidated soil

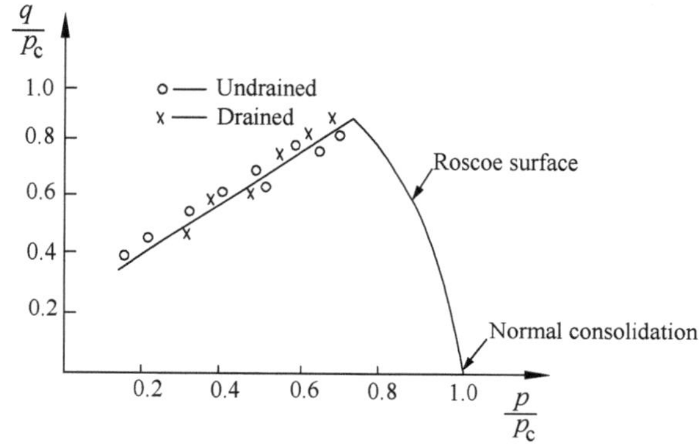

Fig. 6.34 The failure point of drained and undrained triaxial stress path for overconsolidated soil

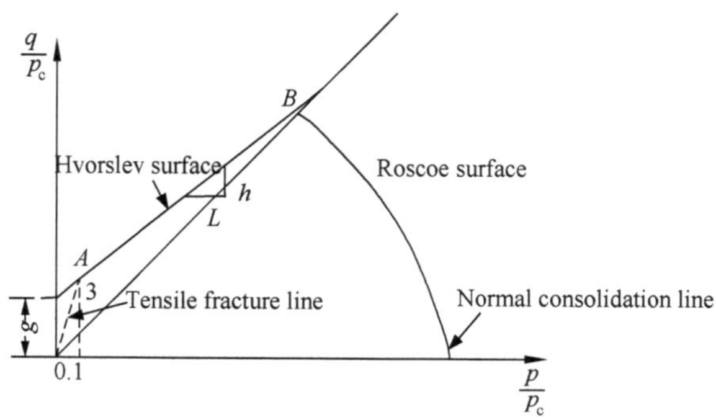

Fig. 6.35 Normalized state boundary surface

where p_c is preconsolidation pressure.

The line is the projection of failure points of Hvorslev surface in plane p-q. The left is cut off by the uniaxial compression line *OA*, and the right is limited by the critical state point *B* (Fig. 6.35). The line is the Coulomb failure line of overconsolidated soils with cohesion, and it is associated with preconsolidation pressure.

Two Hvorslev surface and Roscoe surface are demonstrated in Fig. 6.36 for different specific volumes. The specific volume of the Roscoe surface corresponds to the normal consolidation stress, and the specific volume of the Hvorslev faces corresponds to the preconsolidation stress. It is worth noting that not only shear strain

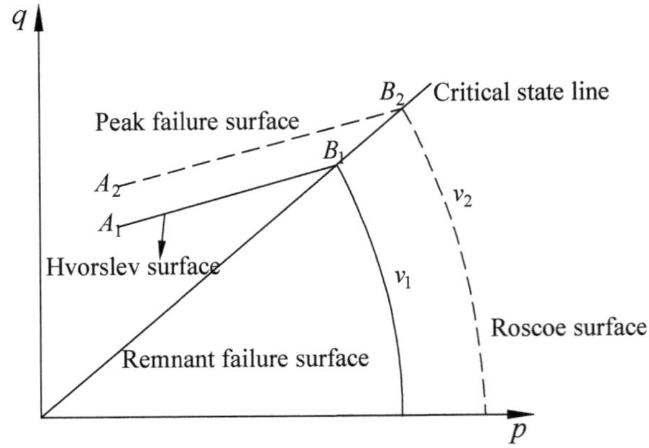

Fig. 6.36 Hvorslev surface and Roscoe surface under different specific volumes

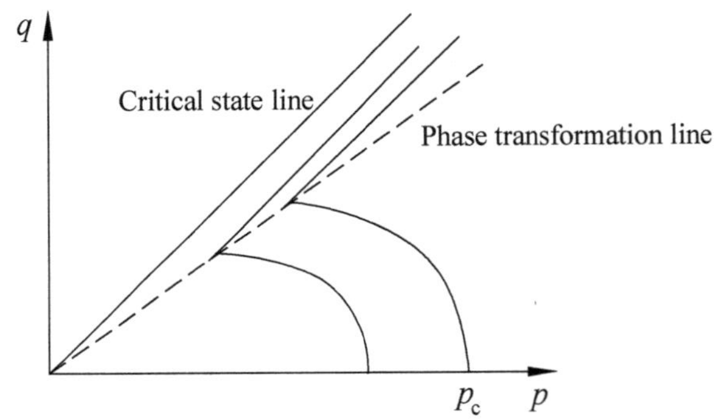

Fig. 6.37 The volumetric yield surface for overconsolidated soils

but also the volumetric strain will be produced on the Hvorslev surface. Hvorslev surface is boundary surface of stress state for overconsolidated soils.

2. **Volumetric yield surface for the overconsolidated soil**

Under triaxial compression of overconsolidated soils, the volume first contracts, and then it inflates. The undrained stress path is revealed in Fig. 6.37, which can be regarded as the volumetric yield surface for overconsolidated soils.

6.2.6 *Part Yield*

Two kinds of yield surface have been introduced in the above section, i.e., shear yield surface and volumetric yield surface. As shown in Fig. 6.38, two yield surfaces divide the stress space into four parts.

In Fig. 6.38, part I is elastic zone, part IV is a completely yield zone for both of yield surface satisfied yield criterion, and parts II, III are only part yield zones

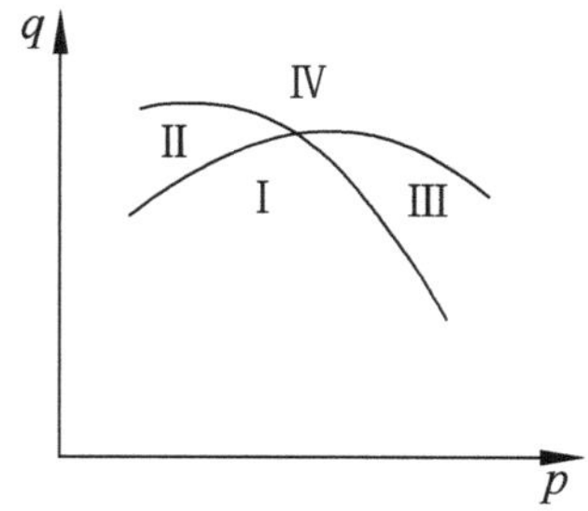

Fig. 6.38 Two yield surfaces in stress space

for only one yield surface satisfied yield criterion. In some zone, only partial yield surface satisfied yield criterion is called part yield.

Multiple yield surface follows the same rule.

6.3 Hardening Laws

According to the Classical Plastic Theory, once the yield surface f is set, we can get the plastic strain increment.

$$\mathrm{d}\boldsymbol{\varepsilon}^{\mathrm{p}} = \mathrm{d}\lambda \frac{\partial f}{\partial \boldsymbol{\sigma}}$$

where $\mathrm{d}\lambda$ is the plastic factor, which means the magnitude of the plastic strain increment. It cannot be calculated yet. The calculation of plastic factor needs introducing the hardening law.

6.3.1 Hardening Theory

Hardening law is a criterion to determine how a yield surface changes into the subsequent yield surface. It can be used to calculate plastic strain under the given stress increment, namely, the magnitude of the plastic strain increment.

Considering the loading–unloading criterion,

under $\frac{\partial f}{\partial \boldsymbol{\sigma}} \mathrm{d}\boldsymbol{\sigma} > 0$, $\mathrm{d}\lambda > 0$

Obviously, the two are related, so the hypothesis is proposed.

$$\mathrm{d}\lambda = h \frac{\partial f}{\partial \boldsymbol{\sigma}} \mathrm{d}\boldsymbol{\sigma} = \frac{1}{A} \frac{\partial f}{\partial \boldsymbol{\sigma}} \mathrm{d}\boldsymbol{\sigma} \tag{6.3.1}$$

where h, A are referred to as hardening functions, and they are calculated through the hardening model.

6.3.2 Hardening Model

Hardening model is the model that describes the change rule for the size, shape, and location of yield surface along with the development of plastic deformation. The unified form can be expressed as

$$f(\sigma, H_{\alpha}) = 0 \tag{6.3.2}$$

where $H_{\alpha} = H_{\alpha}(\varepsilon^{\mathrm{p}})$ is the hardening parameter which is a function of plastic strain.

$$\mathrm{d}f(\sigma, H_{\alpha}) = 0$$

$$\frac{\partial f}{\partial \sigma}\mathrm{d}\sigma + \frac{\partial f}{\partial H_{\alpha}}\mathrm{d}H_{\alpha} = 0$$

$$\mathrm{d}H_{\alpha} = \frac{\partial H_{\alpha}}{\partial \varepsilon^{\mathrm{p}}}\mathrm{d}\boldsymbol{\varepsilon}^{\mathrm{p}} = \frac{\partial H_{\alpha}}{\partial \boldsymbol{\varepsilon}^{\mathrm{p}}}\frac{\partial Q}{\partial \sigma}\mathrm{d}\lambda = \frac{\partial H_{\alpha}}{\partial \boldsymbol{\varepsilon}^{\mathrm{p}}}\frac{\partial Q}{\partial \sigma}h\frac{\partial f}{\partial \sigma}\mathrm{d}\sigma \tag{6.3.3}$$

where Q is the plastic potential function which defines the direction of plastic strain increment.

$$\frac{\partial f}{\partial \sigma}\mathrm{d}\sigma + \frac{\partial f}{\partial H_{\alpha}}\frac{\partial H_{\alpha}}{\partial \boldsymbol{\varepsilon}^{\mathrm{p}}}\frac{\partial Q}{\partial \sigma}h\frac{\partial f}{\partial \sigma}\mathrm{d}\sigma = 0 \tag{6.3.4}$$

$$A = -\frac{\partial f}{\partial H_{\alpha}}\frac{\partial H_{\alpha}}{\partial \boldsymbol{\varepsilon}^{\mathrm{p}}}\frac{\partial Q}{\partial \sigma}, h = -\frac{1}{\frac{\partial f}{\partial H_{\alpha}}\frac{\partial H_{\alpha}}{\partial \boldsymbol{\varepsilon}^{\mathrm{p}}}\frac{\partial Q}{\partial \sigma}} \tag{6.3.5}$$

After the hardening functions (h, A) are confirmed, the plastic strain increment can be calculated.

Hardening in the theory of plasticity means that the yield surface changes in size or location or even in shape, with the loading history (often measured by some form of plastic deformation). When the initial yield condition exists and is identified, the rule of hardening defines its modification during the process of plastic flow.

Most plastic models currently in use assume that the shape of the yield surface remains unchanged, although it may change in size or location. This restriction is largely based on mathematical convenience, rather than upon any physical principle or experimental evidence. The two most widely used rules of hardening are known as isotropic hardening and kinematic (or anisotropic) hardening.

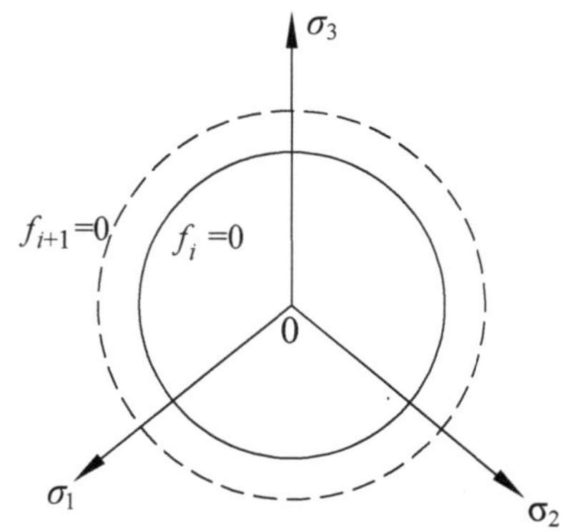

Fig. 6.39 Isotropic hardening with uniform expansion of the yield surface

6.3.3 *Isotropic Hardening*

The rule of isotropic hardening assumes that the yield surface maintains its shape, center, and orientation, but it expands or contracts uniformly about the center of the yield surface.

A yield surface with its center at the origin may be generally described by the following function:

$$f = f(\sigma_{ij}) - R(H_\alpha) = 0 \tag{6.3.6}$$

where R represents the size of the yield surface, depending on plastic strains through the hardening parameter H_α. The two earliest and most widely used hardening parameters are the accumulated equivalent plastic strain

$$H_\alpha = \int \sqrt{\frac{2}{3}} (\mathrm{d}\varepsilon_{ij}^{\mathrm{p}} \mathrm{d}\varepsilon_{ij}^{\mathrm{p}})^{\frac{1}{2}} \tag{6.3.7}$$

and the plastic work

$$H_\alpha = \int \sigma_{ij} \mathrm{d}\varepsilon_{ij}^{\mathrm{p}} \tag{6.3.8}$$

Figure 6.39 shows an example of isotropic hardening where the yield surface is uniformly expanding during the process of plastic flow when a stress increment is applied from step i to $i+1$. The size of the yield surface at any stage of loading is determined as long as an evolution rule defining the relationship between R and H_α is defined.

For isotropic hardening material, the yield function can be described by

$$f(\boldsymbol{\sigma}, H_\alpha) = 0 \tag{6.3.9}$$

Then Prager's consistency condition requires

$$\frac{\partial f}{\partial \boldsymbol{\sigma}}\mathrm{d}\boldsymbol{\sigma} + \frac{\partial f}{\partial H_\alpha}\mathrm{d}H_\alpha = 0 \tag{6.3.10}$$

Since the hardening parameter is a function of plastic strains, the consistency condition Eq. (6.3.10) can be further written as follows:

$$\frac{\partial f}{\partial \boldsymbol{\sigma}}\mathrm{d}\boldsymbol{\sigma} + \frac{\partial f}{\partial H_\alpha}\frac{\partial H_\alpha}{\partial \boldsymbol{\varepsilon}^{\mathrm{p}}}\mathrm{d}\boldsymbol{\varepsilon}^{\mathrm{p}} = 0 \tag{6.3.11}$$

For the special case of perfectly plastic material, the second term of Eq. (6.3.11) will be zero.

A general procedure for deriving a complete stress–strain relation for perfectly plastic and hardening materials is given below:

1. To divide the total strain rate (or increment) into elastic and plastic strain rates, namely,

$$\mathrm{d}\boldsymbol{\varepsilon} = \mathrm{d}\boldsymbol{\varepsilon}^{\mathrm{e}} + \mathrm{d}\boldsymbol{\varepsilon}^{\mathrm{p}} \tag{6.3.12}$$

2. Hooke's law is used to link the stress rate with elastic strain rate by elastic stiffness matrix D_{ijkl}.

$$\mathrm{d}\sigma_{ij} = D_{ijkl}\mathrm{d}\varepsilon_{kl}^{\mathrm{e}} = D_{ijkl}(\mathrm{d}\varepsilon_{kl} - \mathrm{d}\varepsilon_{kl}^{\mathrm{p}}) \tag{6.3.13}$$

3. The general nonassociated plastic flow rule is used to express Eq. (6.3.13) in the following form:

$$\mathrm{d}\sigma_{ij} = D_{ijkl}(\mathrm{d}\varepsilon_{kl} - \mathrm{d}\lambda\frac{\partial g}{\partial \sigma_{kl}}) \tag{6.3.14}$$

4. By substituting Eq. (6.3.1) into the consistency condition Eq. (6.3.11), we obtain

$$\mathrm{d}\lambda = \frac{1}{H}\frac{\partial f}{\partial \sigma_{ij}}D_{ijkl}\mathrm{d}\varepsilon_{kl} \tag{6.3.15}$$

where H is given by

$$H = \frac{\partial f}{\partial \sigma_{ij}}D_{ijkl}\frac{\partial g}{\partial \sigma_{kl}} - \frac{\partial f}{\partial H_\alpha}\frac{\partial H_\alpha}{\partial \varepsilon_{ij}^{\mathrm{p}}}\frac{\partial g}{\partial \sigma_{ij}} \tag{6.3.16}$$

5. Substituting Eq. (6.3.15) into Eq. (6.3.14), we obtain a complete relation between a stress rate and a strain rate.

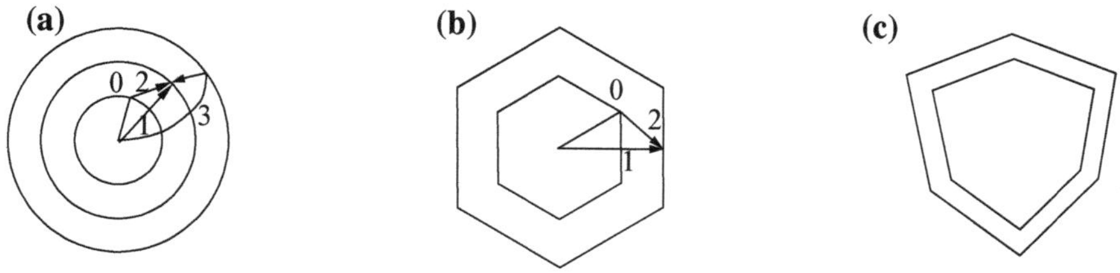

Fig. 6.40 Different types of isotropic hardening model in plane π

$$d\sigma_{ij} = D^{\text{ep}}_{ijkl} d\varepsilon_{kl} \tag{6.3.17}$$

where the elastic–plastic stiffness matrix D^{ep}_{ijkl} is defined by

$$D^{\text{ep}}_{ij} = D_{ijkl} - \frac{1}{H} D_{ijmn} \frac{\partial g}{\partial \sigma_{mn}} \frac{\partial f}{\partial \sigma_{pq}} D_{pqkl} \tag{6.3.18}$$

The above procedure is valid for both strain hardening and perfectly plastic solids. It is noted that for the case of perfectly plastic solids, the yield surface remains unchanged so that Eq. (6.3.16) takes the following simpler form:

$$H = \frac{\partial f}{\partial \sigma_{ij}} D_{ijkl} \frac{\partial g}{\partial \sigma_{kl}} \tag{6.3.19}$$

Different types of isotropic hardening model are displayed in Fig. 6.40 for their performance in plane π.

6.3.4 *Kinematic Hardening*

The term kinematic hardening was introduced by Prager to construct the first kinematic hardening model. In this first model, it was assumed that during plastic flow, the yield surface translated in the stress space and its shape and size remained unchanged. This is consistent with the Bauschinger effect observed in the uniaxial tension–compression.

Assume that the initial yield surface can be described by

$$f = f(\boldsymbol{\sigma} - \boldsymbol{\alpha}) - R_0 = 0 \tag{6.3.20}$$

where $\boldsymbol{\alpha}$ represents the coordinates of the center of the yield surface, which is also known as the back stress. R_0 is a material constant representing the size of the original yield surface. It can be seen that as the back stress $\boldsymbol{\alpha}$ changes due to plastic flow,

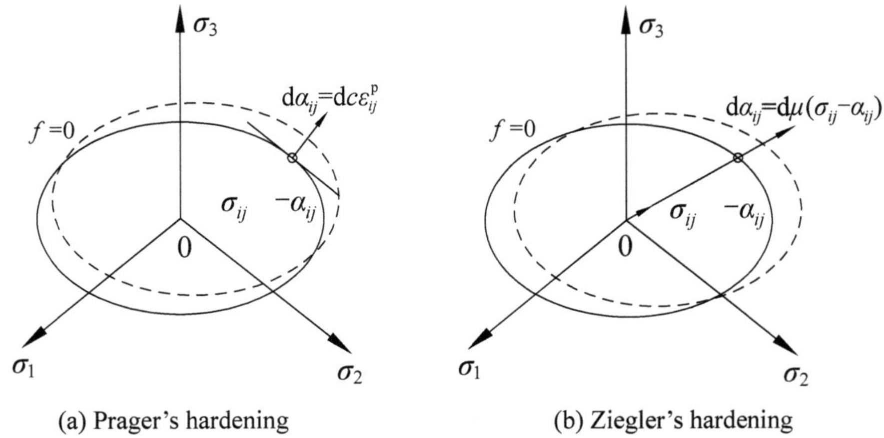

Fig. 6.41 Prager's and Ziegler's kinematic hardening

the yield surface translates in the stress space while maintaining its initial shape and size.

It is clear now that the formulation of a kinematic hardening model involves assuming an evolution rule of the back stress $\boldsymbol{\alpha}$ in terms of $\boldsymbol{\varepsilon}^{\mathrm{P}}$, $\boldsymbol{\sigma}$, or $\boldsymbol{\alpha}$.

The first simple kinematic hardening model was proposed by Prager. This classical model assumes that the yield surface keeps its original shape and size and moves in the direction of plastic strain rate tensor (see Fig. 6.41). Mathematically, it can be expressed by the following linear evolution rule:

$$\mathrm{d}\boldsymbol{\alpha} = c\mathrm{d}\boldsymbol{\varepsilon}^{\mathrm{P}} \tag{6.3.21}$$

where c is a material constant.

While Prager's model is reasonable for one-dimensional problems, it does not seem to give consistent predictions for two- and three-dimensional cases. The reason is that the yield function takes different forms for one-, two-, and three-dimensional cases. To overcome this limitation, Ziegler suggested that the yield surface should move in the direction as determined by the vector $\boldsymbol{\sigma}$-$\boldsymbol{\alpha}$ (see Fig. 6.41). Mathematically, Ziegler's model can be expressed as follows:

$$\mathrm{d}\boldsymbol{\alpha} = \mathrm{d}\mu(\boldsymbol{\sigma} - \boldsymbol{\alpha}) \tag{6.3.22}$$

where $\mathrm{d}\mu$ is a material constant.

For kinematic hardening material, the yield function may be expressed as

$$f = f(\boldsymbol{\sigma} - \boldsymbol{\alpha}) - R_0 = 0 \tag{6.3.23}$$

where α_{ij} denotes the coordinates of the center of the yield surface, often known as the back stress tensor.

First, let us consider the kinematic hardening law proposed by Prager.

$$\mathrm{d}\boldsymbol{\alpha} = c\mathrm{d}\boldsymbol{\varepsilon}^{\mathrm{p}} = c\mathrm{d}\lambda\frac{\partial g}{\partial \boldsymbol{\sigma}} \tag{6.3.24}$$

where g denotes the plastic potential.

With Prager's consistency condition applied to the yield function Eq. (6.3.23), we have

$$\frac{\partial f}{\partial \boldsymbol{\sigma}}\mathrm{d}\boldsymbol{\sigma} + \frac{\partial f}{\partial \boldsymbol{\alpha}}\mathrm{d}\boldsymbol{\alpha} = 0 \tag{6.3.25}$$

By the assumed form of the yield function (6.3.23),

$$\frac{\partial f}{\partial \boldsymbol{\sigma}} = -\frac{\partial f}{\partial \boldsymbol{\alpha}} \tag{6.3.26}$$

With Eqs. (6.3.24) and (6.3.25), Eq. (6.3.26) can be rewritten as

$$\frac{\partial f}{\partial \boldsymbol{\sigma}}\mathrm{d}\boldsymbol{\sigma} = \frac{\partial f}{\partial \boldsymbol{\sigma}}c\mathrm{d}\lambda\frac{\partial g}{\partial \boldsymbol{\sigma}} \tag{6.3.27}$$

The plastic multiplier

$$\mathrm{d}\lambda = \frac{1}{c}\frac{\frac{\partial f}{\partial \boldsymbol{\sigma}}\mathrm{d}\boldsymbol{\sigma}}{\frac{\partial f}{\partial \boldsymbol{\sigma}}\frac{\partial g}{\partial \boldsymbol{\sigma}}} = \frac{1}{c}\frac{\mathrm{d}f}{\frac{\partial f}{\partial \boldsymbol{\sigma}}\frac{\partial g}{\partial \boldsymbol{\sigma}}} \tag{6.3.28}$$

Then, the increments of the back stress tensor and the plastic strains are determined by

$$\mathrm{d}\boldsymbol{\alpha} = c\mathrm{d}\boldsymbol{\varepsilon}^{\mathrm{p}} = c\mathrm{d}\lambda\frac{\partial g}{\partial \boldsymbol{\sigma}}\frac{\mathrm{d}f}{\frac{\partial f}{\partial \boldsymbol{\sigma}}\frac{\partial g}{\partial \boldsymbol{\sigma}}}\frac{\partial g}{\partial \boldsymbol{\sigma}} \tag{6.3.29}$$

$$\mathrm{d}\boldsymbol{\varepsilon}^{\mathrm{p}} = \frac{1}{c}\frac{\frac{\partial g}{\partial \boldsymbol{\sigma}}}{\frac{\partial f}{\partial \boldsymbol{\sigma}}\frac{\partial g}{\partial \boldsymbol{\sigma}}}\mathrm{d}f \tag{6.3.30}$$

By using an elastic stress–strain relation, we can determine the elastic strain rate

$$\mathrm{d}\boldsymbol{\varepsilon}_{ij}^{\mathrm{e}} = C_{jkl}\mathrm{d}\boldsymbol{\sigma}_{kl} \tag{6.3.31}$$

where C_{jkl} is the elastic compliance matrix. The total strain rate is the sum of the elastic and plastic parts

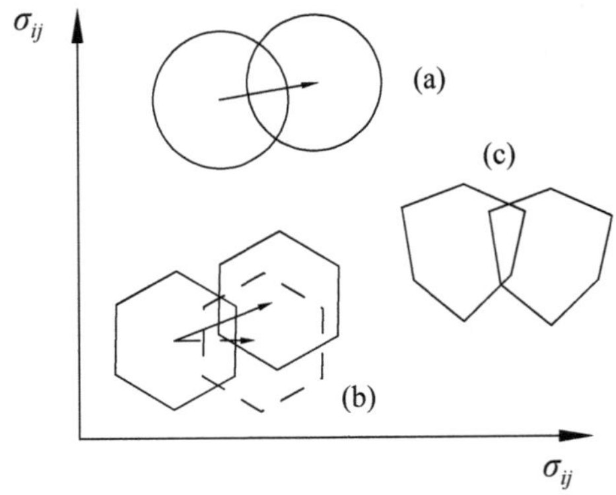

Fig. 6.42 Different types of kinematic hardening in plane π

$$d\boldsymbol{\varepsilon}_{ij} = C_{jkl}d\boldsymbol{\sigma}_{kl} + \frac{1}{c}\frac{\frac{\partial g}{\partial \boldsymbol{\sigma}_{ij}}}{\frac{\partial f}{\partial \boldsymbol{\sigma}_{ij}}\frac{\partial g}{\partial \boldsymbol{\sigma}_{ij}}}df \tag{6.3.32}$$

which can be further written as

$$d\boldsymbol{\varepsilon}_{ij} = C_{ijkl}^{ep}d\boldsymbol{\sigma}_{kl} \tag{6.3.33}$$

where the elastic–plastic compliance matrix is

$$C_{ijkl}^{ep} = C_{jkl} + \frac{1}{c}\frac{\frac{\partial g}{\partial \boldsymbol{\sigma}_{ij}}\frac{\partial f}{\partial \boldsymbol{\sigma}_{kl}}}{\frac{\partial f}{\partial \boldsymbol{\sigma}_{ij}}\frac{\partial g}{\partial \boldsymbol{\sigma}_{ij}}} \tag{6.3.34}$$

It is worth noting that Eq. (6.3.33) can be inverted into

$$d\boldsymbol{\sigma}_{ij} = [C_{ijkl}^{ep}]^{-1}d\boldsymbol{\varepsilon}_{kl} = D_{ijkl}^{ep}d\boldsymbol{\varepsilon}_{kl} \tag{6.3.35}$$

Different types of kinematic hardening model are displayed in Fig. 6.42 for their performance in plane π.

6.3.5 Mixed Hardening

The term mixed hardening is used to indicate cases when the yield surface expands, contracts, and translates in the stress space upon plastic loading (see Fig. 6.43). This means that both the center and size of the yield surface will depend on plastic strain. In this case, the yield function can be expressed by

$$f = f(\boldsymbol{\sigma} - \boldsymbol{\alpha}) - R(H_{\alpha}) = 0 \tag{6.3.36}$$

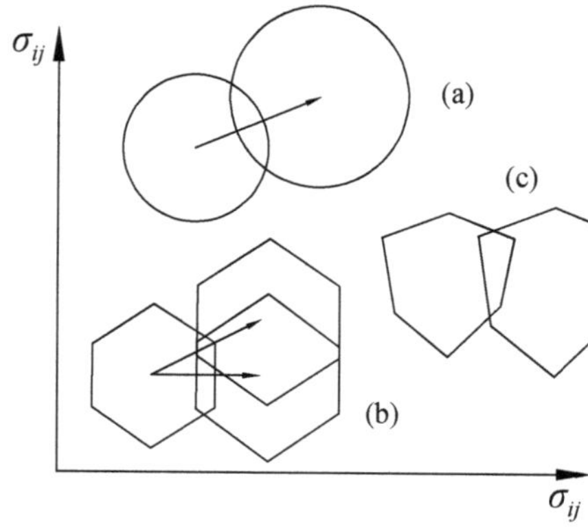

Fig. 6.43 Different types of mixed hardening in plane π

where the size of the yield surface can be assumed to be a function of either plastic strain or plastic work, while either Prager's rule (Eq. (6.3.36)) or Ziegler's rule (Eq. (6.3.23)) may be used to control the translation of the yield surface upon loading.

Of course, this mixed hardening model is relatively simple, because it does not consider rotation movement of yield surface and the change of the shape of the yield surface.

6.3.6 The General Form of Hardening Model

The general form of hardening model should be mixed hardening, and it can be expressed as

$$f(\boldsymbol{\sigma} - \boldsymbol{\alpha}, H_\alpha) = 0 \tag{6.3.37}$$

The differential is

$$\mathrm{d}f(\boldsymbol{\sigma} - \boldsymbol{\alpha}, H_\alpha) = 0$$

$$\frac{\partial f}{\partial \boldsymbol{\sigma}}\mathrm{d}\boldsymbol{\sigma} - \frac{\partial f}{\partial \boldsymbol{\sigma}}\mathrm{d}\boldsymbol{\alpha} + \frac{\partial f}{\partial H_\alpha}\mathrm{d}H_\alpha = 0 \tag{6.3.38}$$

where

$$\mathrm{d}\boldsymbol{\alpha} = \frac{\partial \boldsymbol{\alpha}}{\partial \boldsymbol{\varepsilon}^{\mathrm{p}}}\mathrm{d}\boldsymbol{\varepsilon}^{\mathrm{p}} = \frac{\partial \boldsymbol{\alpha}}{\partial \boldsymbol{\varepsilon}^{\mathrm{p}}}\frac{\partial Q}{\partial \boldsymbol{\sigma}}h\frac{\partial f}{\partial \boldsymbol{\sigma}}\mathrm{d}\boldsymbol{\sigma}$$

$$\mathrm{d}H_\alpha = \frac{\partial H_\alpha}{\partial \boldsymbol{\varepsilon}^{\mathrm{p}}}\mathrm{d}\boldsymbol{\varepsilon}^{\mathrm{p}} = \frac{\partial H_\alpha}{\partial \boldsymbol{\varepsilon}^{\mathrm{p}}}\frac{\partial Q}{\partial \boldsymbol{\sigma}}h\frac{\partial f}{\partial \boldsymbol{\sigma}}\mathrm{d}\boldsymbol{\sigma}$$

Equation (6.3.38) can be rewritten as

$$\frac{\partial f}{\partial \boldsymbol{\sigma}}\mathrm{d}\boldsymbol{\sigma} - \frac{\partial f}{\partial \boldsymbol{\sigma}}\frac{\partial \boldsymbol{\alpha}}{\partial \boldsymbol{\varepsilon}^{\mathrm{p}}}\frac{\partial Q}{\partial \boldsymbol{\sigma}}h\frac{\partial f}{\partial \boldsymbol{\sigma}}\mathrm{d}\boldsymbol{\sigma} + \frac{\partial f}{\partial H_{\alpha}}\frac{\partial H_{\alpha}}{\partial \boldsymbol{\varepsilon}^{\mathrm{p}}}\frac{\partial Q}{\partial \boldsymbol{\sigma}}h\frac{\partial f}{\partial \boldsymbol{\sigma}}\mathrm{d}\boldsymbol{\sigma} = 0$$

$$1 - \frac{\partial f}{\partial \boldsymbol{\sigma}}\frac{\partial \boldsymbol{\alpha}}{\partial \boldsymbol{\varepsilon}^{\mathrm{p}}}\frac{\partial Q}{\partial \boldsymbol{\sigma}}h + \frac{\partial f}{\partial H_{\alpha}}\frac{\partial H_{\alpha}}{\partial \boldsymbol{\varepsilon}^{\mathrm{p}}}\frac{\partial Q}{\partial \boldsymbol{\sigma}}h = 0$$

$$h = \frac{1}{\frac{\partial f}{\partial \boldsymbol{\sigma}}\frac{\partial \boldsymbol{\alpha}}{\partial \boldsymbol{\varepsilon}^{\mathrm{p}}}\frac{\partial Q}{\partial \boldsymbol{\sigma}} - \frac{\partial f}{\partial H_{\alpha}}\frac{\partial H_{\alpha}}{\partial \boldsymbol{\varepsilon}^{\mathrm{p}}}\frac{\partial Q}{\partial \boldsymbol{\sigma}}} \tag{6.3.39}$$

$$A = \frac{1}{h} = \frac{\partial f}{\partial \boldsymbol{\sigma}}\frac{\partial \boldsymbol{\alpha}}{\partial \boldsymbol{\varepsilon}^{\mathrm{p}}}\frac{\partial Q}{\partial \boldsymbol{\sigma}} - \frac{\partial f}{\partial H_{\alpha}}\frac{\partial H_{\alpha}}{\partial \boldsymbol{\varepsilon}^{\mathrm{p}}}\frac{\partial Q}{\partial \boldsymbol{\sigma}} = A_1 + A_2 \tag{6.3.40}$$

Equation (6.3.40) reflects the influence of kinematic hardening and isotropic hardening, the former being called the modulus of kinematic hardening, and the latter being called the modulus of isotropic hardening. Q is the plastic potential function, characterizing the direction of plastic strain increment.

1. **The commonly used model of isotropic hardening**

(1) Plastic work hardening

$$H_{\alpha} = H_{\alpha}(W^{\mathrm{p}}) = \int \boldsymbol{\sigma}\,\mathrm{d}\boldsymbol{\varepsilon}^{\mathrm{p}}$$

$$A = A_2 = -\frac{\partial f}{\partial H_{\alpha}}\frac{\partial H_{\alpha}}{\partial \boldsymbol{\varepsilon}^{\mathrm{p}}}\frac{\partial Q}{\partial \boldsymbol{\sigma}} = -\frac{\partial f}{\partial H_{\alpha}}\boldsymbol{\sigma}\frac{\partial Q}{\partial \boldsymbol{\sigma}} \tag{6.3.41}$$

(2) Plastic strain hardening

$$H_{\alpha} = H_{\alpha}(\boldsymbol{\varepsilon}^{\mathrm{p}}) = \boldsymbol{\varepsilon}^{\mathrm{p}}$$

$$A = A_2 = -\frac{\partial f}{\partial H_{\alpha}}\frac{\partial H_{\alpha}}{\partial \boldsymbol{\varepsilon}^{\mathrm{p}}}\frac{\partial Q}{\partial \boldsymbol{\sigma}} = -\frac{\partial f}{\partial \boldsymbol{\varepsilon}^{\mathrm{p}}}\frac{\partial Q}{\partial \boldsymbol{\sigma}} \tag{6.3.42}$$

(3) Plastic volumetric strain hardening

$$H_{\alpha} = H_{\alpha}(\boldsymbol{\varepsilon}^{\mathrm{p}}) = \boldsymbol{\varepsilon}_{\mathrm{v}}^{\mathrm{p}}$$

$$A = A_2 = -\frac{\partial f}{\partial H_{\alpha}}\frac{\partial H_{\alpha}}{\partial \boldsymbol{\varepsilon}^{\mathrm{p}}}\frac{\partial Q}{\partial \boldsymbol{\sigma}} = -\frac{\partial f}{\partial \boldsymbol{\varepsilon}_{\mathrm{v}}^{\mathrm{p}}}\delta\frac{\partial Q}{\partial \boldsymbol{\sigma}} = -\frac{\partial f}{\partial \boldsymbol{\varepsilon}_{\mathrm{v}}^{\mathrm{p}}}\frac{\partial Q}{\partial p} \tag{6.3.43}$$

(4) Plastic shear strain hardening

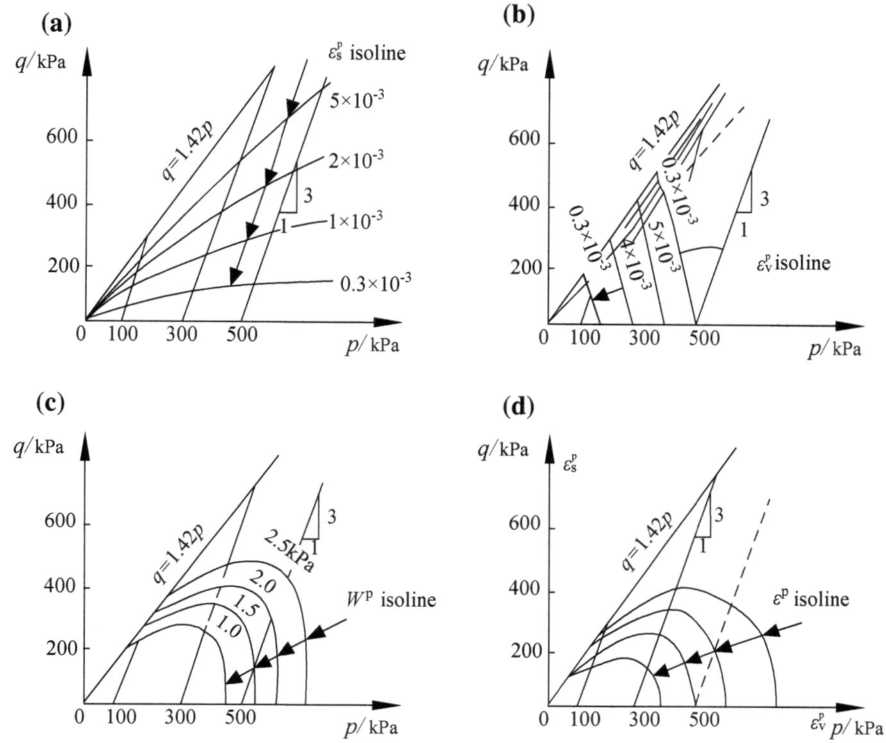

Fig. 6.44 The yield surface for the iso-surface of different hardening parameters

$$H_\alpha = H_\alpha(\boldsymbol{\varepsilon}^p) = \boldsymbol{\varepsilon}_s^p$$

$$A = A_2 = -\frac{\partial \varphi}{\partial H_\alpha}\frac{\partial H_\alpha}{\partial \boldsymbol{\varepsilon}^p}\frac{\partial Q}{\partial \boldsymbol{\sigma}} = -\frac{\partial \varphi}{\partial \boldsymbol{\varepsilon}_s^p}\frac{\partial \boldsymbol{\varepsilon}_s^p}{\partial \boldsymbol{\varepsilon}^p}\frac{\partial Q}{\partial \boldsymbol{\sigma}} = -\frac{\partial \varphi}{\partial \boldsymbol{\varepsilon}_s^p}\frac{\partial Q}{\partial q} \tag{6.3.44}$$

The yield surface selected from the iso-surface of different hardening parameters is demonstrated in Fig. 6.44.

2. **Kinematic hardening model**

Linear kinematic hardening model of Prager is

$$\mathrm{d}\boldsymbol{\alpha} = c\mathrm{d}\boldsymbol{\varepsilon}^p$$

$$A = A_1 = \frac{\partial \varphi}{\partial \boldsymbol{\sigma}}\frac{\partial \boldsymbol{\alpha}}{\partial \boldsymbol{\varepsilon}^p}\frac{\partial Q}{\partial \boldsymbol{\sigma}} = c\frac{\partial \varphi}{\partial \boldsymbol{\alpha}}\frac{\partial Q}{\partial \boldsymbol{\sigma}} \tag{6.3.45}$$

3. **Mixed hardening model**

Yield surface of the modified Cam-Clay model is a typical example of mixed hardening.

Yield surface of the modified Cam-Clay model is

$$\frac{(p-\frac{p_c}{2})^2}{p_c^2}+\frac{q^2}{M^2p_c^2}=1,\quad p_c=\exp(\frac{e}{\lambda-\kappa}\varepsilon_v^p)$$

The yield surface is an ellipse. The center of the ellipse ($\frac{p_c}{2}$, 0) is changing, which means it is kinematic hardening.

Both the horizontal axis $a=p_c$ and vertical axis $b=Mp_c$ of the ellipse are changing. That is to say, the size of yield surface is also changing.

Therefore, yield surface of the modified Cam-Clay model is mixed hardening.

6.4 Plastic Flow Rule

After material reaching yield, the rule used to determine the direction of plastic flow is called the flow rule.

A key question that the theory of plasticity sets out to answer is how to determine the plastic deformation (or plastic strains) once the stress state is on the yield surface. The most widely used theory is to assume that the plastic strain rate (or increment) can be determined by the following formula:

$$\mathrm{d}\boldsymbol{\varepsilon}^{\mathrm{p}}=\mathrm{d}\lambda\frac{\partial g}{\partial \boldsymbol{\sigma}} \tag{6.4.1}$$

where $\mathrm{d}\lambda$ is a plastic scalar, and

$$g=g(\boldsymbol{\sigma}) \tag{6.4.2}$$

where g is known as the plastic potential, which may or may not be the same as the yield surface.

Equation (6.4.1) is referred to as a plastic flow rule that basically defines the ratios of the components of the plastic strain rate. This plastic flow rule is based on the observation by de Saint-Venant that for metals the principal axes of the plastic strain rate coincide with those of the stress. This is the so-called coaxial assumption, which has been the foundation of almost all the plastic models used in engineering. It must be noted that recent experimental data suggests that the coaxial assumption is generally not valid for geomaterial.

6.4.1 Associated Flow Rule

If the plastic potential is the same as the yield surface, then the plastic flow rule (Eq. (6.4.1)) is called the associated flow (or normality) rule. The associated flow rule follows considerations of the plastic deformation of polycrystalline aggregates in which individual crystals deform by slipping over preferred planes.

After the yield surface is determined, and the associated flow rule is used, the plastic strain increment can be calculated as follows:

$$\mathrm{d}\boldsymbol{\varepsilon}^{\mathrm{p}} = \mathrm{d}\lambda \frac{\partial f}{\partial \boldsymbol{\sigma}} \tag{6.4.3}$$

$$\mathrm{d}\lambda = \frac{1}{A} \frac{\partial f}{\partial \boldsymbol{\sigma}} \mathrm{d}\boldsymbol{\sigma} \tag{6.4.4}$$

When the associated flow rule is applied on the Coulomb yield surface, there will be a very outstanding phenomenon of dilatancy, and the significant dilatancy will appear at the beginning of yield. Obviously, it does not accord with the basic mechanical characteristics of geomaterial.

When the associated flow rule is applied to the volumetric yield surface like Cam-Clay model, what will appear? Prominent volumetric contraction will be calculated. It is suitable for soft clay, not for sand.

Thus, the nonassociated flow rule is put forward.

6.4.2 Nonassociated Flow Rule

For nonassociated flow rule, yield surface f determining the material yield is different with the plastic potential surface Q which decides the direction of the plastic deformation. How to calculate the plastic deformation in this situation?

$$\mathrm{d}\boldsymbol{\varepsilon}^{\mathrm{p}} = \mathrm{d}\lambda \frac{\partial Q}{\partial \boldsymbol{\sigma}} \tag{6.4.5}$$

$$\mathrm{d}\lambda = \frac{1}{h} \frac{\partial f}{\partial \boldsymbol{\sigma}} \mathrm{d}\boldsymbol{\sigma} \tag{6.4.6}$$

6.4.3 Mixed Flow Rule

For the multi-yield surface theory, the associated flow rule is used on part of the yield surface and the nonassociated flow rule is applied on other part of the yield surface. It is called mixed flow rule.

Let us take double-yield surface model as an example: for yield surface f_1 associated flow rule is used, while for yield surface f_2 nonassociated flow rule is used, and the corresponding plastic potential surface is Q_2. How can we calculate the plastic deformation?

According to the concept of part yield, the space can be divided into four zones (Fig. 6.38).

1. Part I is elastic zone, no plastic deformation.
2. Part II is a part yield zone where only yield surface f_1 satisfies yield criterion. When only the yield surface f_1 yields, the associated flow rule should be adopted.

$$d\boldsymbol{\varepsilon}^{\mathrm{p}} = d\lambda_1 \frac{\partial f_1}{\partial \boldsymbol{\sigma}} \tag{6.4.7}$$

$$d\lambda_1 = \frac{1}{A_1} \frac{\partial f_1}{\partial \boldsymbol{\sigma}} d\boldsymbol{\sigma} \tag{6.4.8}$$

3. Part III is a part yield zone where only yield surface f_2 satisfies yield criterion. When only the yield surface f_2 yields, the nonassociated flow rule should be adopted.

$$d\boldsymbol{\varepsilon}^{\mathrm{p}} = d\lambda_2 \frac{\partial Q_2}{\partial \boldsymbol{\sigma}} \tag{6.4.9}$$

$$d\lambda_2 = \frac{1}{A_2} \frac{\partial f_2}{\partial \boldsymbol{\sigma}} d\boldsymbol{\sigma} \tag{6.4.10}$$

4. Part IV is a completely yield zone where both yield surfaces satisfy yield criterion, and the mixed flow rule should be used.

$$d\boldsymbol{\varepsilon}^{\mathrm{p}} = d\lambda_1 \frac{\partial f_1}{\partial \boldsymbol{\sigma}} + d\lambda_2 \frac{\partial Q_2}{\partial \boldsymbol{\sigma}} \tag{6.4.11}$$

$$d\lambda_1 = \frac{1}{A_1} \frac{\partial f_1}{\partial \boldsymbol{\sigma}} d\boldsymbol{\sigma} \tag{6.4.12}$$

$$d\lambda_2 = \frac{1}{A_2} \frac{\partial f_2}{\partial \boldsymbol{\sigma}} d\boldsymbol{\sigma} \tag{6.4.13}$$

This rule can be applied to other cases, too.

Generally, the nonassociated flow rule is applied on shear yield surface, while the associated flow rule is used on volumetric yield surface.

6.5 Loading and Unloading Rule

After material enters the plastic deformation stage, stress–strain response is completely different for loading and unloading. Only elastic deformation occurs under nonplastic loading stage. Plastic strain increment can be produced only if the stress increment meets the plastic loading rule. When unloading, only elastic deformation recovers, but plastic deformation remains the same. It can be seen the importance of loading–unloading criterion in elastoplastic analysis.

The commonly used rules for loading–unloading of geomaterial are described and analyzed below [28].

6.5.1 Loading–Unloading Rule Based on Yield Surface

According to the Classical Plastic Theory, loading–unloading could be decided based on the yield surface. Assume the yield surface is known as

$$f(\boldsymbol{\sigma}) = 0$$

In Eq. (6.5.1), the loading and unloading are analyzed.

$$\begin{cases} \dfrac{\partial F}{\partial \boldsymbol{\sigma}} \mathrm{d}\boldsymbol{\sigma} > 0 & \text{loading} \\ \dfrac{\partial F}{\partial \boldsymbol{\sigma}} \mathrm{d}\boldsymbol{\sigma} = 0 & \text{neutral loading} \\ \dfrac{\partial F}{\partial \boldsymbol{\sigma}} \mathrm{d}\boldsymbol{\sigma} < 0 & \text{unloading} \end{cases} \tag{6.5.1}$$

If multiple yield surface is used, the loading–unloading judgment is also performed.

Seemingly, the rule is very perfect, but what are the problems?

1. Determining yield function reasonably is a difficult and complicated work. The stress–strain relationship of geomaterial relies on the stress path. Construction of the yield function will be very difficult if stress path is a bit complicated. Especially, many yield functions are not only very complicated, but some parameters lack clear physical meaning. This is why some scholars avoid yield surface when they explore constitutive relationship of geomaterial.

2. Based on the yield function in the stress space, the stress path of hydraulic stress decreases while the deviatoric stress remains constant, which will be deemed to be unloading for it is a movement within the yield surface. But along the path, the plastic shear strain increases unceasingly, while the dilatancy phenomena occur obviously until the destruction of geomaterial.

For this reason, nowadays, the loading–unloading judgment is developed for many constitutive models without considering yield surface.

6.5.2 Loading–Unloading Rule with Stress Type

Some constitutive models use the loading–unloading rule with stress parameters (such as hydraulic stress *p*, generalized shear stress *q*). A loading–unloading rule is illustrated below in stress style of plane *p*-*q*.

$$\begin{cases} p = p_{\max} \ \mathrm{d}p > 0 \ \text{ loading } \ K = K_{\mathrm{LD}} \\ p \le p_{\max} \ \mathrm{d}p < 0 \ \text{unlaoding } K = K_{\mathrm{UN}} \\ p < p_{\max} \ \mathrm{d}p > 0 \ \text{reloading } K = K_{\mathrm{RL}} \\ q = q_{\max} \ \mathrm{d}q > 0 \ \text{ loading } \ G = G_{\mathrm{LD}} \\ q \le q_{\max} \ \mathrm{d}q < 0 \ \text{unloading } G = G_{\mathrm{UN}} \\ q < q_{\max} \ \mathrm{d}q > 0 \ \text{reloading } G = G_{\mathrm{RL}} \end{cases} \tag{6.5.2}$$

where $p_{\max}$, $q_{\max}$ are the largest values of *p*, *q* in the stress history; *K*, *G* are bulk modulus and shear modulus, respectively; the subscript labels *LD*, *UN*, *RL* mean loading, unloading, and reloading, respectively.

What iss the problem of loading–unloading rule with stress style?

No corresponding relationship exists between plastic deformation and stress, and thus there is no theoretical basis for the stress style rule.

It does not take into account simultaneous change of *p* and *q*, and the influence of the stress Lode's Angle. Thus, it is an incomplete criterion. For example, it should be loading under $q=q_{\max}$, $\mathrm{d}p<0$, but it is not included.

If all situations are considered, such criteria will be lengthy and complex, and even contradictory.

6.5.3 Loading–Unloading Rule with Strain Style

Loading–unloading rule with strain style is a criterion based on strain parameters. A loading–unloading rule with the parameters of elastic strain increment and total strain is presented below:

$$\begin{cases} \boldsymbol{\varepsilon} + \mathrm{d}\boldsymbol{\varepsilon}^{\mathrm{e}} < \boldsymbol{\varepsilon}_{\mathrm{M}} \quad \mathrm{d}\boldsymbol{\varepsilon}^{\mathrm{e}} < 0 \ \text{ elastical unloading } \ \mathrm{d}\boldsymbol{\varepsilon} = \mathrm{d}\boldsymbol{\varepsilon}^{\mathrm{e}} = [\mathbf{C}_{\mathrm{e}}]\mathrm{d}\boldsymbol{\sigma} \\ \boldsymbol{\varepsilon} + \mathrm{d}\boldsymbol{\varepsilon}^{\mathrm{e}} < \boldsymbol{\varepsilon}_{\mathrm{M}} \quad \mathrm{d}\boldsymbol{\varepsilon}^{\mathrm{e}} > 0 \ \text{ elastic loading } \ \mathrm{d}\boldsymbol{\varepsilon} = \mathrm{d}\boldsymbol{\varepsilon}^{\mathrm{e}} = [\mathbf{C}_{\mathrm{e}}]\mathrm{d}\boldsymbol{\sigma} \\ \boldsymbol{\varepsilon} = \boldsymbol{\varepsilon}_{\mathrm{M}} \quad \mathrm{d}\boldsymbol{\varepsilon}^{\mathrm{e}} = 0 \ \text{ neural loading } \ \mathrm{d}\boldsymbol{\varepsilon} = \mathrm{d}\boldsymbol{\varepsilon}^{\mathrm{e}} = 0 \\ \boldsymbol{\varepsilon} + \mathrm{d}\boldsymbol{\varepsilon}^{\mathrm{e}} > \boldsymbol{\varepsilon}_{\mathrm{M}} \quad \boldsymbol{\varepsilon} = \boldsymbol{\varepsilon}_{\mathrm{M}} \ \text{ plastic loading } \ \mathrm{d}\boldsymbol{\varepsilon} = \left[\mathbf{C}_{\mathrm{ep}}\right]\mathrm{d}\boldsymbol{\sigma} > \mathrm{d}\boldsymbol{\varepsilon}^{\mathrm{e}} \\ \boldsymbol{\varepsilon} + \mathrm{d}\boldsymbol{\varepsilon}^{\mathrm{e}} > \boldsymbol{\varepsilon}_{\mathrm{M}} \quad \boldsymbol{\varepsilon} < \boldsymbol{\varepsilon}_{\mathrm{M}} \ \text{ platic reloading } \\ \mathrm{d}\boldsymbol{\varepsilon} = (\boldsymbol{\varepsilon}_{\mathrm{M}} - \boldsymbol{\varepsilon}) + \left[\mathbf{C}_{\mathrm{ep}}\right]\left[[\mathbf{C}_{\mathrm{e}}]^{-1}(\mathrm{d}\boldsymbol{\varepsilon}^{\mathrm{e}} - \boldsymbol{\varepsilon}_{\mathrm{M}} + \boldsymbol{\varepsilon})\right] > \mathrm{d}\boldsymbol{\varepsilon}^{\mathrm{e}} \end{cases} \tag{6.5.3}$$

Because the elastic constants of geomaterial must be given for no matter which kind of geomaterial model, thus the elastic flexibility matrix and the elastic strain increment can be obtained quickly. And the criterion can be used for the analysis of arbitrary stress increment under any stress state. So it is a widely used criterion for geomaterial at the hardening stage.

Questions

1. What is the concept of yield?
2. What are the similarities and differences of shear yield surface and failure surface?
3. What is the reason for the reasonable calculation results of introducing the equal-area circle in plane π?
4. How many types of shear yield surface are there which consider the number of the principal shear stress?
5. What is the yield condition of Zienkweiz-Pande and how can you prove its rationality?
6. What are the advantages and disadvantages of the yield condition of double shear?
7. How can you describe critical state and volume yield surface?
8. Will stress path overstep the critical state line, and fall down to the critical state line after reaching the Hvorslev surface in the undrained test of overconsolidated soils?
9. The following is the hyperbolic yield surface with plastic shear strain hardening for Chongqing red clay.

   $$q = \frac{p}{a + bp}$$

   where $a = -0.11\,\varepsilon_s^p + 2.2$, $b = 8 \times 10^{-5}\varepsilon_s^p - 0.005$.
   What is the hardening model of this yield surface? How can you calculate the hardening modulus?
10. How can you calculate plastic deformation of double-yield surface using mixed flow rule for geomaterial?
11. What are the classification of loading–unloading rule and their advantages and disadvantages?

References

1. Shen ZJ, Sheng SX (1982) The uniqueness assumption of the constitutive theory of soil. Chin Sci Study Irrig Water Carriage 2:11–19
2. Anandarajah A, Sobhan K, Kuganenthira N (1995) Incremental stress-strain behavior of granular soil. J Geotech Eng 121(1):57–68
3. Tatsuoka F, Sonada S (1986) Failure and deformation of sand in torsional shear. Soils Found 26(4):79–97
4. Liu YX (1997) The general stress strain relation of soils involving principal stress axes rotation [Doctoral Dissertation]. Logistical Engineering University, Chongqing, China
5. Liu YX, Zheng YR, Chen ZH (1998) The general stress-strain relation of soils involving principal stress axes rotation. Appl Math Mech 19(5):407–413
6. Liu YX, Zheng YR (1998) A new method for considering the influences of principal stress axes rotation on soils stress strain relation. Chin J Geotech Eng 20(2):45–47
7. Liu YX (2001) Study of several basic problems of geomaterial constitutive theory. Chin J Geotech Eng 23(1):45–48
8. Shen ZJ (1985) Elastoplastic analysis of consolidation deformation of soft clay foundation. Chin Sci (A) (11):1050–1060
9. Yin ZZ (1998) A two yield surface stress strain model of soils. Chin J Geotech Eng 10(4):64–71
10. Kiyama S, Hasegawa T (1998) A two-surface model with anisotropic hardening and nonassociated flow rule for geomaterials. Soils Found 38(1):45–59
11. Zheng YR (1991) Multi-yielding surface theory for soils. Comput Methods Adv Geomech 715–720
12. Lade PV, Kim MK (1988) Single hardening constitutive model for frictional materials. Comput Geotech 6:1–47
13. Huang SJ (1988) The thermodynamics principle of stability postulate in plastic mechanics. Chin J Solid Mech 9(2):95–101
14. Drucker DC, Gibson RE, Henkel DH (1957) Soil mechanics and work-hardening theories of plasticity. Trans ASCE 122:94–112
15. Shen ZJ (1998) The basic problems in modem soil mechanics. Mech Pract 6:1–6
16. Liu YX, Zheng YR (2000) The generalized plastic theory involving principal stress axes rotation. Chin Q J Mech 21(1):119–123
17. Lu HH, Yin ZZ (1994) Analysis and Improvement of the flexible matrix of two yield surface model. In: Zheng YR (ed) Proceeding of 5th Chinese conference of numerical analysis and analysis methods in geomechanics. Press of Wuhan Survey Science and Technology University, Wuhan, pp 139–144
18. Yin ZZ, Zhu JG, Lu HH (1994) The elastoplastic flexible matrix and experimental study by true three triaxial test. In: Ye SL (ed) Proceeding of 7th Chinese conference of soil mechanics and foundation engineering. Chinese Press of Architecture Engineering, Beijing, pp 21–25
19. Yoshimine M, Ishihara K, Vargas W (1998) Effects of principal stress direction and intermediate principal stress on undrained shear behaviour of sands. Soils Found 38(3):179–188
20. Drucker DC, Prager W (1952) Soil mechanics and plastic analysis on limit design. J Appl Math 10(2):157–165
21. Yu MH (1992) New system of strength theory. Xi'an Jiaotong University Press, Xi'an
22. Hoek E, Brown ET (1980) Empirical strength criterion for rock masses. J Geotech Eng Div ASCE 1013–1025
23. Hoek E, Brown ET (1988) The Hoek-Brown failure criterion-a 1988 update. In: Proceeding of 15th Canadian rock mechanics symposium, pp 31–38
24. Hoek E, Carranza-Torres C T, Corkum B (2002) Hoek-Brown failure criterion-2002 edition. In: Proceedings of the 5th North American rock mechanics symposium, vol 1, pp 267–273
25. Roscoe KH, Schofield AN, Wroth CP (1958) On the yielding of soils. Geotechnique 8(1):22–53
26. Roscoe KH, Schofield AN, Thurairajah A (1963) Yielding of clays in states wetter than critical. Geotechnique 13(3):211–240

27. Roscoe KH, Burland JB (1968) On the generalized stress strain behavior of "wet" clay. In: Hoyman J, Leekie FA (eds) Engineering plasticity. Cambridge University Press, Cambridge, pp 535–609
28. Liu YX, Zheng YR (2001) The loading-unloading rule for elastoplastic theory of geomaterial. Chin J Rock Mech Eng 20(6):768–771

Chapter 7
The Static Elastoplastic Model for Geomaterial

Elastoplastic theory assumes that elastic deformation coexists with plastic deformation. The total strain can be divided into two parts: elastic and plastic.

$$d\boldsymbol{\varepsilon} = d\boldsymbol{\varepsilon}^{\mathrm{e}} + d\boldsymbol{\varepsilon}^{\mathrm{p}} \tag{7.0.1}$$

The elastic part is researched rarely in general constitutive models, while the Generalized Hooke's Law is adopted. The emphasis is on the description of the plastic deformation.

With the development of numerical methods, such as finite element methods, it has become feasible to analyze and predict the behavior of complex geotechnical structure problems. Such analyses depend considerably on the representation of the relations between stresses and strains for the various materials involved in the geotechnical structure. In numerical computations, the relations between stresses and strains in a given material are represented by a so-called constitutive model, consisting of mathematical expressions that model the behavior of the geomaterial in a single element. Because geomaterial is a relatively weak material involved in common geotechnical problems, they determine the deformations and the possibility of failure of the structure, and it is therefore important to characterize these materials accurately over the entire range of stresses and strains to which they will become exposed. Other construction material, such as concrete and steel, may remain stiff in comparison with geomaterial, and it may be sufficient to characterize these materials as elastic or elastic-perfectly plastic. Thus, the purpose of a constitutive model is to simulate the geomaterial behavior with sufficient accuracy under all loading conditions in the numerical computations.

Significant developments of constitutive models have occurred over the past five decades. Naturally, the initial models are relatively simple, and a progression in complexity and capabilities of the models have led to much improved abilities to capture the behavior of geotechnical structure under complex loading conditions. Simple as well as advanced models have been formulated on the basis of principles of

Y. Liu and Y. Zheng, *Plastic Mechanics of Geomaterial*, Springer Geophysics,
https://doi.org/10.1007/978-981-13-3753-6_7

mechanics, some being more rigorous than others, some being based on experimental evidence, and others have been based on theoretical principles. Characterization of geomaterial behavior can become quite involved, because the stress–strain relations are nonlinear in nature, the geomaterial is fundamentally frictional materials, and volume changes occur during drained shearing.

It is clear that constitutive models have already gone through substantial improvements, and there will be more to come. Thus, development of new and improved constitutive models is a continuing endeavor over a foreseeable future. However, the principles, main characteristic features, and components in the constitutive models for geomaterial are currently and widely available to practitioners and researchers in packaged numerical software and in the technical literature are presented and reviewed.

7.1 Cam-Clay and Modified Cam-Clay Model

7.1.1 The Concept of Critical States

In the book of Critical State Soil Mechanics, Schofield and Wroth [1] describe the concept of critical states in the following way.

The kernel of their idea is the concept that soil and other granular materials, if continuously distorted until they flow as a frictional fluid, will come into a well-defined critical state determined by two equations.

$$q = Mp \tag{7.1.1}$$

$$\Gamma = v + \lambda \ln p \tag{7.1.2}$$

where the constants M, Γ, and λ represent basic soil material properties; the parameters q, v, and p are defined in due course.

As a result, the critical states depend on the mean effective stress p, shear stress q, and soil specific volume v. As shown graphically in Fig. 7.1, in two straight lines (now known as the critical state lines *CSL*), e denotes the void ratio.

Schofield and Wroth further explain that at the critical state, soils behave as a frictional fluid so that yielding occurs at constant volume and constant stresses. In other words, the plastic volumetric strain increment is zero at the critical state, since elastic strain increments will be zero due to the constant stress condition at the critical state. Also, it is assumed that the critical state lines are unique for a given soil regardless of stress paths used to bring them about from any initial conditions.

In many ways, the critical states defined or assumed above may be regarded as the ultimate states. It is also noted that the concept of steady states proposed later is also similar to the concept of critical states. In addition, a concept of disturbed states is proposed for use in constitutive modeling. In effect, the critical states correspond to fully disturbed states. Given the uniqueness of the critical state lines, they are used

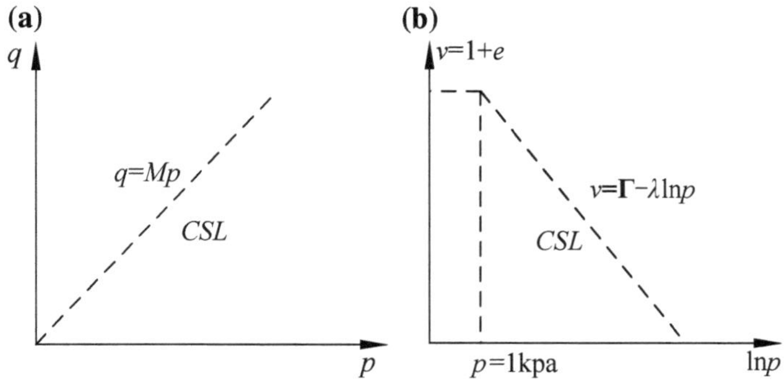

Fig. 7.1 The concept of critical states

as a convenient base of reference in formulating a strain hardening/softening plastic model to describe the measured behavior of soil and other granular materials.

The concept of critical states was initially developed based on limited triaxial test data obtained on reconstituted Cam-Clay [2]. However, over the last 40 years, a lot of additional experimental data for many other types of soil and granular material (e.g., sand, rock, natural clay, unsaturated soil, and sugar) have been obtained which seems to support, at least to a very large extent, the general concept of critical states.

7.1.2 Cam-Clay Model

As the earliest elastoplastic critical state model, Cam-Clay model was developed by Roscoe et al. [2] at Cambridge University. The Cam-Clay model was modified by Roscoe and Burland [3]. It marks the beginning of the new development stage for the constitutive theory of geomaterial. It is the only elastoplastic model recognized by the world until now.

The shear yield is connected with volumetric yield for the first time in the constitutive model of geomaterial. Their mechanisms used to be regarded as being different and they were studied separately by different methods.

1. **Elastic deformation**

It is assumed that there is only elastic volumetric strain of soil and no elastic shear strain.

$$d\varepsilon_s^e = 0 \tag{7.1.3}$$

In hydraulic consolidation test, the relations between total volumetric strain and elastic part can be expressed as follows:

$$\varepsilon_{\mathrm{v}} = \lambda \ln \frac{p}{p_i} \tag{7.1.4}$$

$$\varepsilon_{\mathrm{v}}^{\mathrm{e}} = k \ln \frac{p}{p_i} \tag{7.1.5}$$

where λ, k are compression and swell slope, respectively, for hydraulic consolidation and unloading of soil; p_i is the initial consolidation pressure.

Then, the plastic volumetric deformation can be obtained.

$$\varepsilon_{\mathrm{v}}^{\mathrm{p}} = (\lambda - k) \ln \frac{p}{p_i} \tag{7.1.6}$$

The increment of elastic volumetric strain is

$$\mathrm{d}\varepsilon_{\mathrm{v}}^{\mathrm{e}} = k \frac{\mathrm{d}p}{p} \tag{7.1.7}$$

The increment of elastic strain is

$$\mathrm{d}\boldsymbol{\varepsilon}^{\mathrm{e}} = k \frac{\mathrm{d}p}{p} \frac{\delta}{3} \tag{7.1.8}$$

2. **Plastic deformation**

Plastic deformation complies with the Classical Plastic Theory in Cam-Clay model. Its core is the yield surface.

(1) Yield surface

We know that the critical state concept of soil is one important contribution of Roscoe. The undrained stress path can approximate to volumetric yield surface for soil from the normal consolidation state to the critical state, as demonstrated in Fig. 7.2.

Not fitting the undrained curve directly, Roscoe deduced the yield surface from the viewpoint of energy dissipation.

Work of external force can be divided into elastic work and the plastic work.

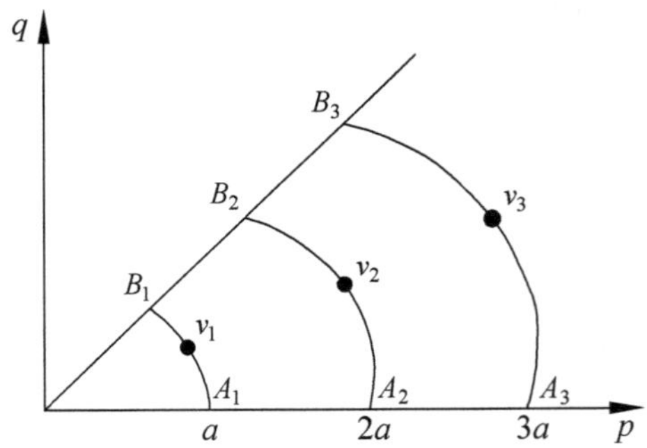

Fig. 7.2 Undrained stress path of normally consolidated soils

$$dW = dW^{e} + dW^{p} \tag{7.1.9}$$

According to the hypothesis of Cam-Clay model, elastic work is

$$dW^{e} = pd\varepsilon_{v}^{e} + qd\varepsilon_{s}^{e} = pd\varepsilon_{v}^{e} = pk\frac{dp}{p} \tag{7.1.10}$$

The plastic work per unit volume of a triaxial sample with the externally applied mean and shear stresses p and q is

$$dW_{p} = pd\varepsilon_{v}^{p} + qd\varepsilon_{s}^{p} \tag{7.1.11}$$

where ε_{v}^{p} and ε_{s}^{p} are volumetric and shear plastic strains, respectively.

To determine how this plastic energy is dissipated, Schofield and Wroth [1] follow the simple analysis of shear box test results. All the plastic works, defined by (7.1.11), are dissipated entirely in friction, namely

$$dW_{\text{dis}} = Mpd\varepsilon_{s}^{p} \tag{7.1.12}$$

where M is the ratio of $\frac{q}{p}$ at the critical state.

$$M = \frac{6\sin\varphi}{3-\sin\varphi}$$

where φ is the angle of internal friction of soil.

The energy conservation then requires

$$dW_{p} = dW_{\text{dis}} \tag{7.1.13}$$

which leads to the following work equation for the Cam-Clay model.

$$pd\varepsilon_{v}^{p} + qd\varepsilon_{s}^{p} = Mpd\varepsilon_{s}^{p} \tag{7.1.14}$$

The above work equations can be further rearranged as follows:

$$\frac{q}{p} + \frac{d\varepsilon_{v}^{p}}{d\varepsilon_{s}^{p}} = M \tag{7.1.15}$$

The associated flow rule is adopted in Cam-Clay model. Now let us assume that there exists a yield surface for the soil that depends on both the mean and shear stresses.

$$f = f(p, q) = 0 \tag{7.1.16}$$

From which we can write the following:

$$d\varepsilon_v^p = d\lambda \frac{\partial f}{\partial p} \text{ and } d\varepsilon_s^p = d\lambda \frac{\partial f}{\partial q} \tag{7.1.17}$$

$$\frac{d\varepsilon_v^p}{d\varepsilon_s^p} = \frac{\partial f}{\partial p} / \frac{\partial f}{\partial q} \tag{7.1.18}$$

The differential of Eq. (7.1.16)

$$df = df(p, q) = \frac{\partial f}{\partial p} dp + \frac{\partial f}{\partial q} dq = 0$$

$$\frac{\partial f}{\partial p} / \frac{\partial f}{\partial q} = -\frac{dq}{dp} \tag{7.1.19}$$

Simultaneous Eqs. (7.1.18) and (7.1.19),

$$\frac{dq}{dp} = -\frac{d\varepsilon_v^p}{d\varepsilon_s^p} \tag{7.1.20}$$

By substituting Eqs. (7.1.20) into (7.1.15), we obtain

$$\frac{q}{p} - \frac{dq}{dp} = M \tag{7.1.21}$$

which may be integrated to give an equation for the yield surface.

$$\frac{q}{Mp} + \ln p = C \tag{7.1.22}$$

What is C in Eq. (7.1.22)? It is the hardening parameter of yield surface. The intersection is $(p_0, 0)$ for the curve and axis p.

$$q = 0, C = \ln p_0$$

The yield surface will be

$$\frac{q}{p} = M \ln \frac{p_0}{p} \tag{7.1.23}$$

How can we express p_0 which is also a hardening parameter? It can be determined from the hydraulic consolidation stress path.

$$\varepsilon_v^p = (\lambda - k) \ln \frac{p}{p_i} \bigg|_{p=p_0} = (\lambda - k) \ln \frac{p_0}{p_i}$$

$$p_0 = p_i \exp^{\frac{\varepsilon_v^p}{\lambda - k}} \tag{7.1.24}$$

p_0 is the preconsolidation pressure served as the hardening parameter that changes with the plastic strain.

(2) Plastic factor

$$\begin{aligned}
\mathrm{d}p_0 &= \mathrm{d}(p_i \exp^{\frac{\varepsilon_v^p}{\lambda-k}}) \\
&= p_i \exp^{\frac{\varepsilon_v^p}{\lambda-k}} \frac{\mathrm{d}\varepsilon_v^p}{\lambda-k} \\
&= p_0 \frac{\mathrm{d}\varepsilon_v^p}{\lambda-k} \\
&= p_0 \frac{\mathrm{d}\lambda \frac{\partial F}{\partial p}}{\lambda-k} \\
&= \frac{\mathrm{d}\lambda p_0}{\lambda-k} \frac{\partial F}{\partial p}
\end{aligned}$$

The plastic factor will be

$$\mathrm{d}\lambda = \frac{\lambda-k}{p_0 \frac{\partial F}{\partial p}} \mathrm{d}p_0 \tag{7.1.25}$$

where

$$\begin{aligned}
\mathrm{d}(\frac{q}{p}) &= \mathrm{d}(M \ln \frac{p_0}{p}) \\
\frac{p\mathrm{d}q - q\mathrm{d}p}{p^2} &= M \frac{p}{p_0} \frac{p_0}{p} (\frac{p_0\mathrm{d}p - p\mathrm{d}p_0}{p^2}) \\
&= \frac{M(p_0\mathrm{d}p - p\mathrm{d}p_0)}{p^2} \\
p\mathrm{d}q - q\mathrm{d}p &= M(p_0\mathrm{d}p - p\mathrm{d}p_0)
\end{aligned}$$

$$\mathrm{d}p_0 = \frac{Mp_0\mathrm{d}p - p\mathrm{d}q + q\mathrm{d}p}{Mp} \tag{7.1.26}$$

$$\begin{aligned}
\frac{\partial F}{\partial p} &= \frac{\partial(\frac{q}{p} - M \ln \frac{p_0}{p})}{\partial p} \\
&= \frac{-q}{p^2} - M \frac{p}{p_0} p_0 (-\frac{1}{p^2}) \\
&= \frac{Mp - q}{p^2}
\end{aligned} \tag{7.1.27}$$

Combining Eqs. (7.1.25), (7.1.26), and (7.1.27), the plastic factor is gotten.

$$\mathrm{d}\lambda = \frac{(\lambda - k)p}{p_0(Mp - q)} \frac{Mp_0\mathrm{d}p - p\mathrm{d}q + q\mathrm{d}p}{M} \tag{7.1.28}$$

(3) Plastic deformation

Plastic deformation for general stress space is computed as follows:

$$\mathrm{d}\boldsymbol{\varepsilon}^{\mathrm{p}} = \mathrm{d}\lambda \frac{\partial F}{\partial \boldsymbol{\sigma}} \tag{7.1.29}$$

$$\begin{aligned}\frac{\partial F}{\partial \boldsymbol{\sigma}} &= \frac{\partial(\frac{q}{p} - M\ln\frac{p_0}{p})}{\partial \boldsymbol{\sigma}} \\ &= \frac{p\frac{\partial q}{\partial \boldsymbol{\sigma}} - q\frac{\partial p}{\partial \boldsymbol{\sigma}}}{p^2} - M\frac{p}{p_0}p_0(-\frac{\frac{\partial p}{\partial \boldsymbol{\sigma}}}{p^2}) \\ &= \frac{p\frac{\partial q}{\partial \boldsymbol{\sigma}} - q\frac{\partial p}{\partial \boldsymbol{\sigma}} + Mp\frac{\partial p}{\partial \boldsymbol{\sigma}}}{p^2}\end{aligned} \tag{7.1.30}$$

$$\frac{\partial p}{\partial \boldsymbol{\sigma}} = \frac{\partial(\frac{\boldsymbol{\sigma}\boldsymbol{\delta}}{3})}{\partial \boldsymbol{\sigma}} = \frac{\boldsymbol{\delta}}{3} \tag{7.1.31}$$

$$\begin{aligned}\frac{\partial q}{\partial \boldsymbol{\sigma}} &= \frac{\partial\sqrt{\frac{3}{2}S_{ij}S_{ij}}}{\partial \boldsymbol{\sigma}} = \frac{\partial\sqrt{\frac{3}{2}(\boldsymbol{\sigma} - p\boldsymbol{\delta})(\boldsymbol{\sigma} - p\boldsymbol{\delta})}}{\partial \boldsymbol{\sigma}} \\ &= \frac{\frac{3}{2}\times 2(\boldsymbol{\sigma} - p\boldsymbol{\delta})(1 - \frac{\boldsymbol{\delta}}{3}\boldsymbol{\delta})}{\sqrt{\frac{3}{2}(\boldsymbol{\sigma} - p\boldsymbol{\delta})(\boldsymbol{\sigma} - p\boldsymbol{\delta})}} \\ &= \frac{(\boldsymbol{\sigma} - p\boldsymbol{\delta})(3 - \boldsymbol{\delta}\boldsymbol{\delta})}{\sqrt{\frac{3}{2}(\boldsymbol{\sigma} - p\boldsymbol{\delta})(\boldsymbol{\sigma} - p\boldsymbol{\delta})}}\end{aligned} \tag{7.1.32}$$

7.1.3 Modified Cam-Clay Model

Modified Cam-Clay model was developed by Roscoe and Burland [3], and it represents a slight extension of Cam-Clay model by adopting a revised work equation to derive the yield function and plastic potential.

1. **Elastic deformation**

It is assumed that there is only elastic volumetric strain of soil and no elastic shear strain.

2. **Plastic deformation**

The Classical Plastic Theory is also adopted, and the core is in the yield surface. It is revised for the energy dissipation formula. Energy dissipation is studied for the two special points.

(1) Isotropic consolidation

$$dW^p = pd\varepsilon_v^p$$

Critical state point

$$dW^p = qd\varepsilon_s^p = Mpd\varepsilon_s^p$$

Combining these two equations, the following can be presented.

$$\begin{aligned} dW^p &= pd\varepsilon_v^p + qd\varepsilon_s^p = \sqrt{(pd\varepsilon_v^p)^2 + (Mpd\varepsilon_s^p)^2} \\ &= p\sqrt{(d\varepsilon_v^p)^2 + (M\,d\varepsilon_s^p)^2} \end{aligned} \tag{7.1.33}$$

The obtained yield surface is an ellipse.

$$\frac{(p-\frac{p_0}{2})^2}{(\frac{p_0}{2})^2} + \frac{q^2}{(\frac{Mp_0}{2})^2} = 1$$

where $p_0 = p_i \exp^{\frac{\varepsilon_v^p}{\lambda - k}}$.

In the same way, it is put forward for the stress–strain relationship.

7.1.4 Comment on Cam-Clay Model

1. **Advantages**

(1) Its hypothesis has experimental basis, and the basic concept is clear, such as critical state, etc.
(2) It takes into consideration the basic mechanical properties of geomaterial, i.e., isotropic yield, dilatancy, and pressure-hardening.
(3) The model is relatively simple with only three parameters. The calculation result is accurate for normal consolidation soil. The model parameters can be decided by conventional triaxial isotropic consolidation and unloading, and triaxial shear test. With more experience in it, it has an important impact on the world.

2. **Disadvantages**

(1) Based on the Classical Plastic Theory, the results obtained using the associated flow rule and the single yield surface do not accord with a lot of actual situations.
(2) Only the hardening of plastic volumetric strain is considered in the model, and the hardening of shear plastic strain is taken into account inadequately, such as the dilatancy of dense sand.
(3) The structural of natural soil cannot be simulated properly.

(4) Anisotropy is not considered.
(5) The effect of principal stress is not involved.

7.2 Lade Model

An elastoplastic constitutive model with a single yield surface has been developed by Lade [4–11] for the behavior of frictional materials such as sand, clay, concrete, and rock. Hooke's law is employed to model the elastic strains, and the framework for the plastic behavior consists of a failure criterion, a nonassociated flow rule, a yield criterion that describes contours of equal plastic work, and a work-hardening/softening law. For soils, the model incorporates eleven parameters that can all be determined from simple experiments. The components of the model are reviewed, determination of all material parameters is demonstrated, typical values of material parameters are given for selected sands, and the capabilities of the model are described.

7.2.1 Components of Constitutive Model

The total strain increments observed in a material when loaded are divided into elastic and plastic components.

$$d\boldsymbol{\varepsilon} = d\boldsymbol{\varepsilon}^{e} + d\boldsymbol{\varepsilon}^{p} \tag{7.2.1}$$

These strains are then calculated separately, the elastic strains being by Hooke's law, and the plastic strains by a plastic stress–strain law. Both are expressed in terms of effective stresses.

7.2.2 Elastic Behavior

The elastic strain increments, which are recoverable upon unloading, are calculated from Hooke's law, using a model for the nonlinear variation of Young's modulus with stress state. The value of Poisson's ratio, ν, being limited between zero and one half for most materials, is assumed to be constant. The expression for Young's modulus is derived from theoretical considerations based on the principle of conservation of energy. According to this derivation, Young's modulus E can be expressed in terms of a power law involving nondimensional material constants and stress functions as follows:

$$E = Mp_a\left[\left(\frac{I_1}{p_a}\right)^2 + 6\frac{1+v}{1-2v}\left(\frac{J_2}{p_a^2}\right)\right]^{\lambda} \quad (7.2.2)$$

where I_1 is the first invariant of the stress tensor; J_2 is the second invariant of the deviatoric stress tensor. They are given as follows:

$$I_1 = \sigma_x + \sigma_y + \sigma_z \quad (7.2.3)$$

$$J_2 = \frac{1}{6}\left[\left(\sigma_x - \sigma_y\right)^2 + \left(\sigma_y - \sigma_z\right)^2 + \left(\sigma_z - \sigma_x\right)^2\right] + \tau_{xy}^2 + \tau_{yz}^2 + \tau_{zx}^2 \quad (7.2.4)$$

The parameter p_a is atmospheric pressure expressed in the same units as E, I_1 and $\sqrt{J_2}$, and the modulus number M and the exponent λ are constant, dimensionless numbers. The three material parameters v, M, and λ may be obtained from the unloading–reloading cycles of simple tests such as triaxial compression tests. The model can be used for materials with effective cohesion, as explained below.

7.2.3 Failure Criterion

A general, three-dimensional failure criterion has been developed for soils, concrete, and rock. The criterion is expressed in terms of the first and third stress invariants of the stress tensor.

$$f_n = \left(\frac{I_1^3}{I_3} - 27\right)\left(\frac{I_1}{p_a}\right)^m \quad (7.2.5)$$

$$f_n = \eta_1 \text{ at failure} \quad (7.2.6)$$

where

$$I_3 = \sigma_x\sigma_y\sigma_z + \tau_{xy}\tau_{yz}\tau_{zx} + \tau_{yx}\tau_{zy}\tau_{xz} - \left(\sigma_x\tau_{yz}\tau_{zy} + \sigma_y\tau_{zx}\tau_{xz} + \sigma_z\tau_{xy}\tau_{yx}\right) \quad (7.2.7)$$

The parameters η_1 and m are constant dimensionless numbers.

Figure 7.3 shows that in principal stress space, the failure criterion is shaped like an asymmetric bullet with the pointed apex at the origin of the stress axes, and the cross-sectional shape in the octahedral plane is triangular with smoothly rounded edges in a fashion that conforms to experimental evidence. The apex angle increases with the value of η_1. The failure surface is always concave toward the hydrostatic axis, and its curvature increases with the value of m. If $m=0$, the failure surface is straight, and the shape of the cross sections does not change with the value of I_1. If $m>0$, the cross-sectional shape of the failure surface changes from triangular to become more circular with increasing value of I_1. Similar changes in cross-sectional shape are observed from experimental studies on soil, concrete, and rock. If $m>1.979$, the

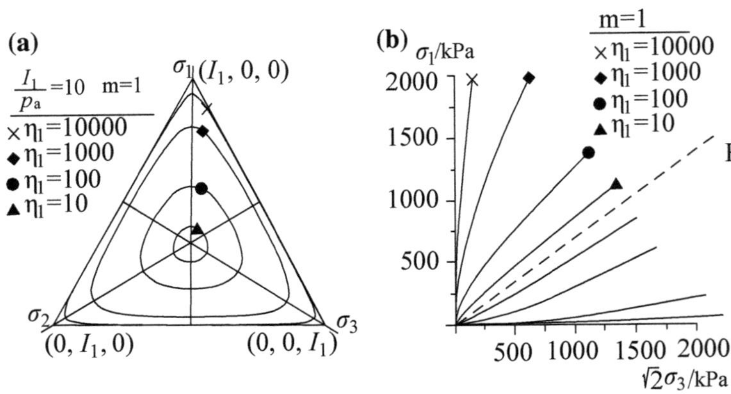

Fig. 7.3 Characteristics of failure criterion in principal stress space: Traces shown in octahedral plane (**a**), and triaxial plane (**b**)

failure surface becomes convex toward the hydrostatic axis. Analysis of numerous sets of data for concrete and rock indicates that m-values rarely exceed 1.5.

In order to include the effective cohesion and the tension which can be sustained by cemented soils, concrete, and rock, a translation of the principal stress space along the hydrostatic axis is performed. Thus, a constant stress p_a is added to the normal stresses before substitution in Eq. (7.2.5).

$$\bar{\sigma}_i = \sigma_i + a \cdot p_a \quad (i = 1, 2, 3) \tag{7.2.8}$$

where a is a dimensionless parameter. The value of ap_a reflects the effect of the tensile strength of the material. The three material parameters η_1, m, and a may be determined from results of simple tests such as triaxial compression tests.

7.2.4 Plastic Potential and Flow Rule

The plastic strain increments are calculated from the flow rule.

$$d\boldsymbol{\varepsilon}^p = d\lambda \frac{\partial Q}{\partial \boldsymbol{\sigma}} \tag{7.2.9}$$

where Q is a plastic potential function and dλ is a scalar factor of proportionality.

A suitable plastic potential function for frictional materials is developed and presented. This function is different from the yield function and nonassociated flow is consequently obtained. The plastic potential function is written in terms of the three invariants of the stress tensor.

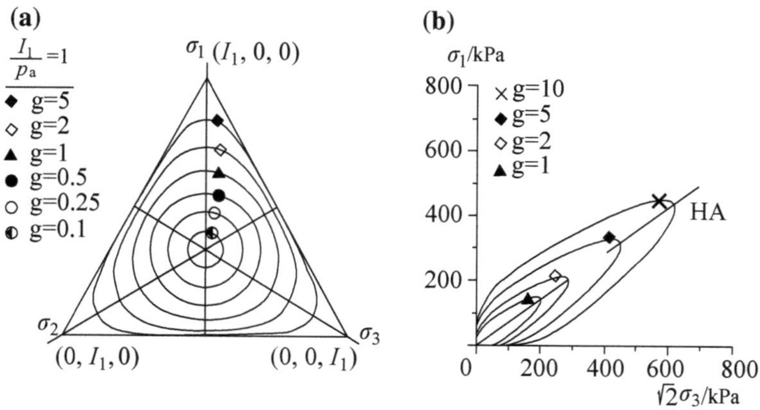

Fig. 7.4 Characteristics of the plastic potential function in principal stress space: Traces shown in octahedral plane (**a**), and in triaxial plane (**b**)

$$Q = \left(\psi_1 \frac{I_1^3}{I_3} - \frac{I_1^2}{I_2} + \psi_2\right)\left(\frac{I_1}{p_a}\right)^{\mu} \tag{7.2.10}$$

where I_1 and I_3 are given in Eqs. (7.2.3) and (7.2.7), and the second stress invariant is defined as

$$I_2 = \tau_{xy}\tau_{yx} + \tau_{yz}\tau_{zy} + \tau_{zx}\tau_{xz} - \left(\sigma_x\sigma_y + \sigma_y\sigma_z + \sigma_z\sigma_x\right) \tag{7.2.11}$$

The material parameters ψ_2 and μ are dimensionless constants that may be determined from triaxial compression tests. The parameter ψ_1 is related to the curvature parameter m of the failure criterion as follows:

$$\psi_1 = 0.00155m^{-1.27} \tag{7.2.12}$$

The parameter ψ_1 acts as a weighting factor between the triangular shape (from the I_3 term) and the circular shape (from the I_2 term). The parameter ψ_2 controls the intersection with the hydrostatic axis, and the exponent μ determines the curvature of meridians. The corresponding plastic potential surfaces are shown in Fig. 7.4. They are shaped as asymmetric cigars with smoothly rounded triangular cross sections similar but not identical to those for the failure surfaces.

The derivatives of Q with regard to the stresses are

$$\frac{\partial Q}{\partial \sigma} = \left(\frac{I_1}{p_a}\right)^{\mu} \begin{Bmatrix} G - (\sigma_y + \sigma_z)\frac{I_1^2}{I_2^2} - \psi_1\left(\sigma_y\sigma_z - \tau_{yz}^2\right)\frac{I_1^3}{I_3^2} \\ G - (\sigma_z + \sigma_x)\frac{I_1^2}{I_2^2} - \psi_1\left(\sigma_z\sigma_y - \tau_{zx}^2\right)\frac{I_1^3}{I_3^2} \\ G - (\sigma_x + \sigma_y)\frac{I_1^2}{I_2^2} - \psi_1\left(\sigma_x\sigma_y - \tau_{xy}^2\right)\frac{I_1^3}{I_3^2} \\ 2\frac{I_1^2}{I_2^2}\tau_{yz} - 2\psi_1\left(\tau_{xy}\tau_{zx} - \sigma_x\tau_{yz}\right)\frac{I_1^3}{I_3^2} \\ 2\frac{I_1^2}{I_2^2}\tau_{zx} - 2\psi_1\left(\tau_{xy}\tau_{yz} - \sigma_y\tau_{zx}\right)\frac{I_1^3}{I_3^2} \\ 2\frac{I_1^2}{I_2^2}\tau_{xy} - 2\psi_1\left(\tau_{yz}\tau_{zx} - \sigma_z\tau_{xy}\right)\frac{I_1^3}{I_3^2} \end{Bmatrix} \tag{7.2.13}$$

where

$$G = \psi_1(\mu + 3)\frac{I_1^2}{I_3} - (\mu + 2)\frac{I_1}{I_2} + \frac{\mu}{I_1}\psi_2 \tag{7.2.14}$$

7.2.5 *Yield Criterion and Work-Hardening/Softening Relation*

The yield surfaces are intimately associated with and derived from surfaces of constant plastic work. The isotropic yield function is expressed as follows:

$$f_p = f_p'(\sigma_{ij}) - f_p''(W_p) = 0 \tag{7.2.15}$$

where

$$f_p' = \left(\psi_1\frac{I_1^3}{I_3} - \frac{I_1^2}{I_2}\right)\left(\frac{I_1}{p_a}\right)^h \cdot e^z \tag{7.2.16}$$

where h is constant and z varies from zero at the hydrostatic axis to unity at the failure surface; the parameter ψ_1 acts as a weighting factor between the triangular shape (from the I_3 term) and the circular shape (from the I_2 term); the constant parameter h is determined on the basis that the plastic work is constant along a yield surface.

The value of z varies with stress level S which is defined as

$$S = \frac{f_n}{\eta_1} = \frac{1}{\eta_1}\left(\frac{I_1^3}{I_3} - 27\right)\left(\frac{I_1}{p_a}\right)^m \tag{7.2.17}$$

where the stress level S varies from zero at the hydrostatic axis to unity at the failure surface, and the variation of z with S is expressed as

$$z = \frac{\alpha S}{1 - (1 - \alpha)S} \tag{7.2.18}$$

where α is constant.

For hardening, the yield surface inflates isotropically with plastic work according to

$$f_p'' = \left(\frac{1}{D}\right)^{\frac{1}{y}} \left(\frac{W_p}{p_a}\right)^{\frac{1}{y}} y \tag{7.2.19}$$

where the values of y and D are constants for a given material. Thus, f_p'' varies with the plastic work only. The values of D and y are given by

$$D = \frac{C}{(27\psi_1 + 3)^y} \tag{7.2.20}$$

and

$$y = \frac{x}{h} C \tag{7.2.21}$$

The parameters C and y in Eq. (7.2.20) are used to model the plastic work during isotropic compression.

$$W_p = Cp_a \left(\frac{I_1}{p_a}\right)^x ABS \tag{7.2.22}$$

The yield surfaces are shaped as asymmetric tear drops with smoothly rounded triangular cross sections and traces in the triaxial plane as shown in Fig. 7.5. As the plastic work increases, the isotropic yield surface inflates until the current stress point reaches the failure surface. The relation between f_p'' and W_p is described by a monotonically increasing function whose slope decreases with increasing plastic work, as shown in Fig. 7.6.

For softening, the yield surface deflates isotropically according to an exponential decay function.

$$f_p'' = Ae^{-B\frac{W_p}{p_a}} \tag{7.2.23}$$

where A and B are positive constants to be determined on the basis of the slope of the hardening curve at the point of peak failure, $S = 1$, as indicated in Fig. 7.5. Thus,

$$A = \left[f_p'' e^{B\frac{W_p}{p_a}}\right]_{S=1} \tag{7.2.24}$$

and

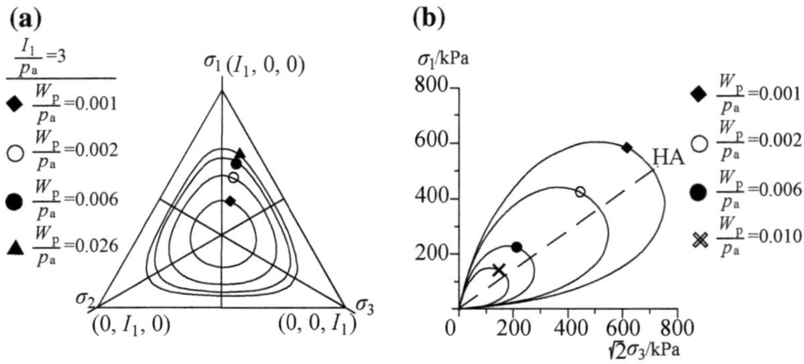

Fig. 7.5 Characteristics of the yield function in principal stress space: Traces shown in **a** octahedral plane, and in **b** triaxial plane

Fig. 7.6 Modeling of work-hardening and softening

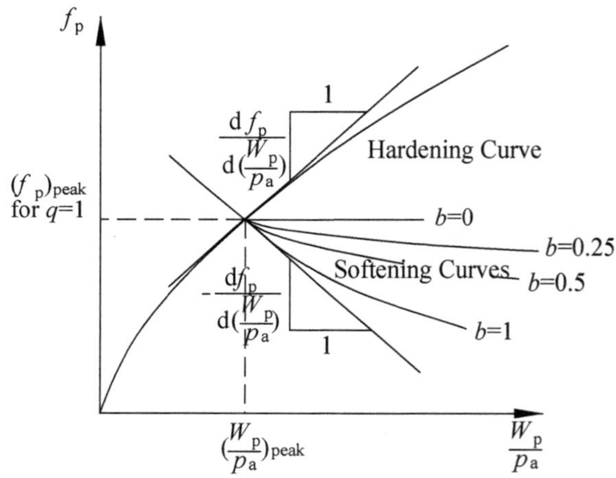

$$B = \left[b \frac{\mathrm{d}f_{\mathrm{p}}''}{\mathrm{d}\left(\frac{W_{\mathrm{p}}}{p_{\mathrm{a}}}\right)} \frac{1}{f_{\mathrm{p}}'} \right]_{S=1} \tag{7.2.25}$$

where both the size of the yield surface f_{p}'' and the derivative $\frac{\mathrm{d}f_{\mathrm{p}}''}{\mathrm{d}\left(\frac{W_{\mathrm{p}}}{p_{\mathrm{a}}}\right)}$ are obtained from the hardening curve at peak failure, indicated by $S=1$. The value of $\mathrm{d}f_{\mathrm{p}}''$ is negative during softening. The parameter b is greater than or equal to zero, where the lower limit corresponds to that of a perfect plastic material. A default value of $b=1$ is built into the model, so b occurs as an optional material parameter.

Using the expression for the plastic potential in Eq. (7.2.10), the relation between plastic work increment and the scalar factor of proportionality $\mathrm{d}\lambda$ in Eq. (7.2.9) may be expressed as

$$d\lambda = \frac{dW_p}{\mu Q} \tag{7.2.26}$$

where the increment of plastic work can be determined by differentiation of the hardening and softening equations.

Combining Eqs. (7.2.24) and (7.2.25) with Eq. (7.2.26), and substituting it into Eq. (7.2.9), the expression for the incremental plastic strain increments is produced.

7.2.6 Determination of Material Parameters

The governing functions of the single hardening model have been presented and the material parameters identified. The material parameters depend on the specific material and may be calibrated to results of isotropic and triaxial compression tests. The calibration procedure is also demonstrated below for Sand No. D, a fine sand, dredged from the ocean bottom and tested in isotropic compression and drained triaxial compression tests.

1. **Elastic behavior**

The constant value of Poisson's ratio v may be determined easiest from the initial slope of the reloading branch of the volume change curve $\frac{\Delta\varepsilon_v}{\Delta\varepsilon_1}$, as shown in Fig. 7.7. The strains in this portion of the curve are considered to be entirely elastic and Poisson's ratio is determined as

$$v = -\frac{\Delta\varepsilon_3}{\Delta\varepsilon_1} = \frac{1}{2}\left(1 - \frac{\Delta\varepsilon_v}{\Delta\varepsilon_1}\right) \tag{7.2.27}$$

The average value obtained from the unloading–reloading curves from tests with different confining pressures is representative of the elastic Poisson's ratio for the material.

The dimensionless, constant values of the modulus number M, and the exponent λ may be determined from the initial slopes of the unloading–reloading cycles in the triaxial compression tests, as also illustrated in Fig. 7.7. These initial slopes are considered to represent the elastic Young's moduli of the material. The corresponding values of the stress invariants in Eq. (7.2.2) are calculated from the stresses at the points of reversal. In order to determine the values of M and λ, Eq. (7.2.2) is rearranged and logs are taken on both sides of the equation.

$$\log\left(\frac{E}{p_a}\right) = \log M + \lambda \log\left[\left(\frac{I_1}{p_a}\right)^2 + 6\frac{1+v}{1-2v}\left(\frac{J_2'}{p_a^2}\right)\right] \tag{7.2.28}$$

Thus, by plotting $\frac{E}{p_a}$ versus the stress function on the right-hand side of Eq. (7.2.28) on log–log scales, as shown in Fig. 7.8, the value of M is determined as the intercept

Fig. 7.7 Determination of Young's moduli and Poisson's ratio from stress–strain and volumetric strain relations from unloading–reloading cycles in triaxial compression tests on Sand No. D (1 psi = 6.895 kPa)

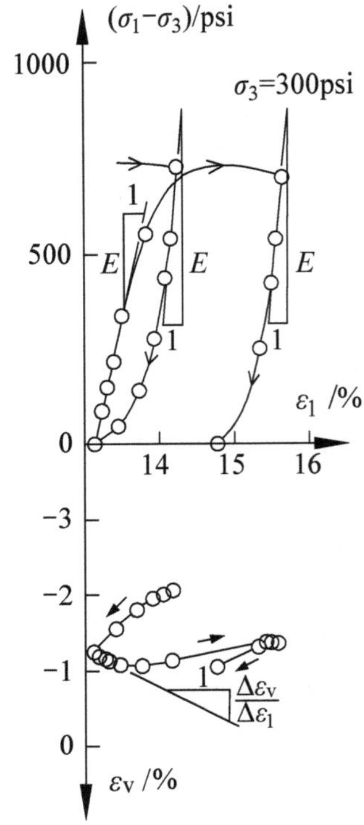

Fig. 7.8 Determination of *M* and λ for Young's modulus variation on Sand No. D

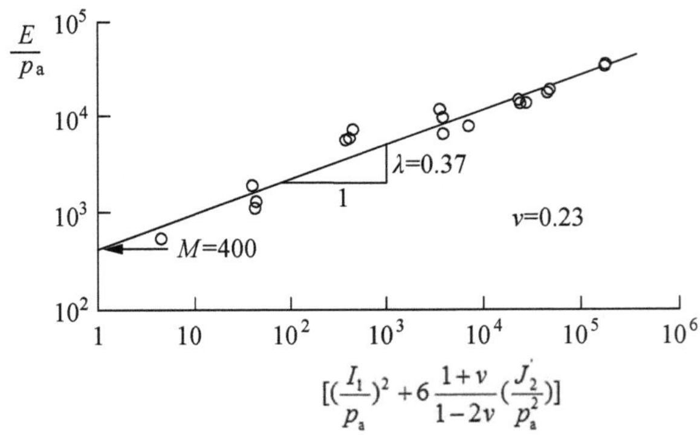

between the best fitting straight line and the vertical line corresponding to unity of the stress function. The slope of the straight line corresponds to the exponent λ.

2. **Failure criterion**

The expression for the failure criterion in Eq. (7.2.27) is rearranged and logs are taken on both sides of the equation.

$$\log\left(\frac{I_1^3}{I_3} - 27\right) = \log \eta_1 + m \log\left(\frac{p_a}{I_1}\right) \tag{7.2.29}$$

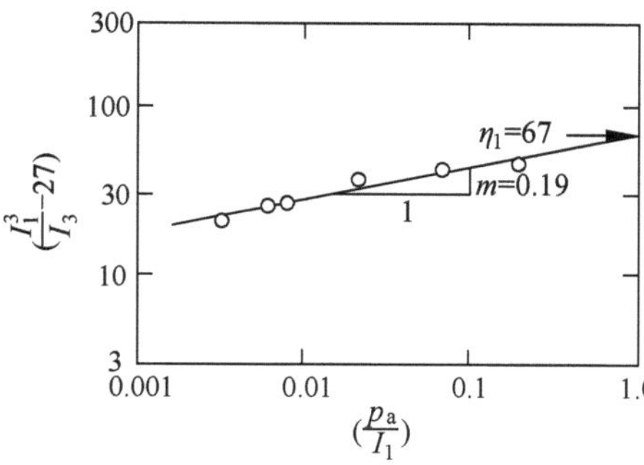

Fig. 7.9 Determination of η_1 and m for failure criterion for Sand No. D

By plotting $\left(\frac{I_1^3}{I_3} - 27\right)$ versus $\left(\frac{p_a}{I_1}\right)$ on log–log scales, as shown in Fig. 7.9, the value of η_1 is determined as the intercept between the best fitting straight line and the vertical line corresponding to $\left(\frac{p_a}{I_1}\right) = 1$. The slope of the straight line is the exponent m.

3. **Plastic potential parameters**

In the plastic potential, the parameter ψ_1 is determined from Eq. (7.2.12), and the other parameters, ψ_2 and μ, can be determined using triaxial compression test data. This is done by expressing the incremental plastic strain ratio defined as

$$v_p = -\frac{d\varepsilon_3^p}{d\varepsilon_1^p} \tag{7.2.30}$$

The plastic strain increments in Eq. (7.2.30) are calculated from the results of the triaxial compression tests by subtracting the elastic strain increments from the measured strain increments. Substitution of Eq. (7.2.13) for the plastic strain increments under triaxial compression conditions ($\sigma_2 = \sigma_3$) produces the following equation:

$$\xi_y = \frac{1}{\mu}\xi_x - \psi_2 \tag{7.2.31}$$

where

$$\xi_x = \frac{1}{1+v_p}\left[\frac{I_1^3}{I_2^2}(\sigma_1 + \sigma_3 + 2v_p\sigma_3) + \psi_1 \cdot \frac{I_1^4}{I_3^2}(\sigma_1\sigma_3 + v_p\sigma_3^2)\right] - 3\psi_1 \cdot \frac{I_1^3}{I_3} + 2\frac{I_1^2}{I_2} \tag{7.2.32}$$

and

$$\xi_y = \psi_1 \cdot \frac{I_1^3}{I_3} - \frac{I_1^2}{I_2} \tag{7.2.33}$$

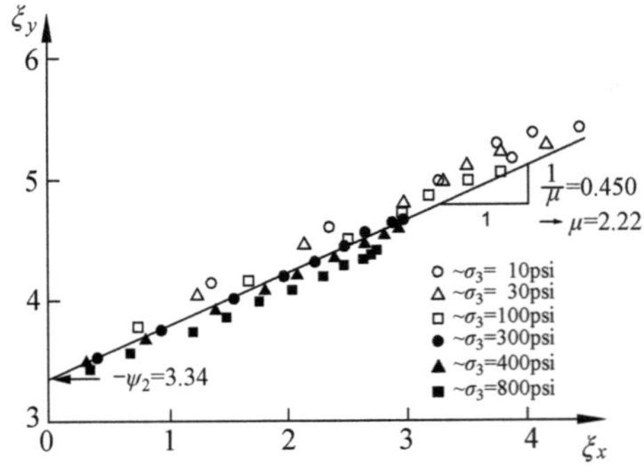

Fig. 7.10 Determination of ψ_2 and μ for the plastic potential function for Sand No. D (1 psi = 6.895 kPa)

Thus, $\frac{1}{\mu}$ and $-\psi_2$ can be determined by linear regression between ζ_x and ζ_y determined from several data points. Figure 7.10 is a plot of ζ_x and ζ_y for the triaxial compression tests on Sand No. D. All data points are consistent with Eq. (7.2.32). The value of $-\psi_2$ is the intercept value of ζ_y at $\zeta_x = 0$, and the value of $\frac{1}{\mu}$ is the slope of the best fitting straight line.

4. **Yield criterion and work-hardening/softening relation**

The work-hardening relation along the hydrostatic axis, expressed in Eq. (7.2.22), is determined first because the parameter values of C and x are required in determination of z in the yield criterion.

The plastic work along the hydrostatic axis is calculated from

$$W_{\mathrm{p}} = \int [\sigma_{ij}]^{\mathrm{T}}[\mathrm{d}\varepsilon_{ij}^{\mathrm{p}}] \tag{7.2.34}$$

in which an isotropic compression test reduces to

$$W_{\mathrm{p}} = \int \sigma_3 \cdot \mathrm{d}\varepsilon_{\mathrm{v}}^{\mathrm{p}} \tag{7.2.35}$$

Compression tests are expressed by subtracting the elastic strains from the measured strains. $\frac{W_{\mathrm{p}}}{p_{\mathrm{a}}}$ is then plotted as a function of $\frac{I_1}{p_{\mathrm{a}}}$, in which isotropic compression is equal to $\frac{3\sigma_3}{p_{\mathrm{a}}}$.

The diagram in Fig. 7.11 shows the relationship between $\frac{W_{\mathrm{p}}}{p_{\mathrm{a}}}$ and $\frac{I_1}{p_{\mathrm{a}}}$, plotted on log–log scales for Sand No. D. This relationship is modeled by Eq. (7.2.22) in which C and x are determined as shown in Fig. 7.11. On this diagram, C is the intercept with $\frac{I_1}{p_{\mathrm{a}}}=1$ and x is the slope of the straight line.

The yield criterion in Eq. (7.2.16) requires two parameter values. The value of h is determined on the basis that the plastic work is constant along a yield surface. Thus, for two stress points, A on the hydrostatic axis and B on the failure surface, the following expression is obtained for h:

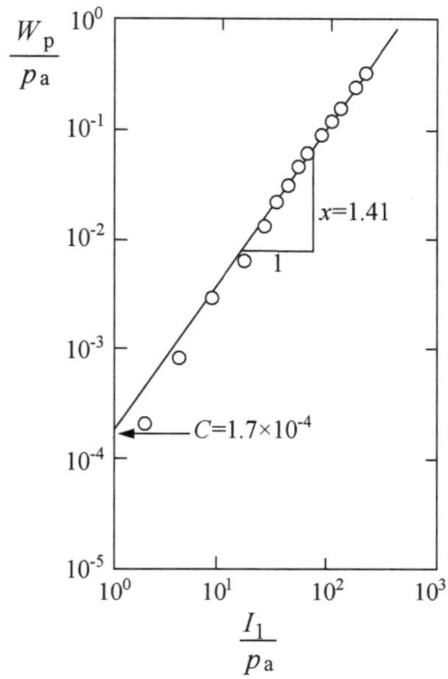

Fig. 7.11 Determination of C and x for work-hardening relation for Sand No. D

$$h = \frac{\dfrac{\ln\left(\psi_1 \cdot \frac{I_{1B}^3}{I_{3B}} - \frac{I_{1B}}{I_{2B}}\right) \cdot e}{27\psi_1 + 3}}{\ln \frac{I_{1A}}{I_{1B}}} \tag{7.2.36}$$

where e is the base of natural logarithms.

Substituting Eqs. (7.2.16) and (7.2.19) into Eq. (7.2.15) and solving for z produces

$$z = \ln \frac{\left(\frac{W_p}{D \cdot p_a}\right)^{\frac{1}{y}}}{\left(\psi_1 \cdot \frac{I_1^3}{I_3} - \frac{I_1^2}{I_2}\right)\left(\frac{I_1}{p_a}\right)^h} \tag{7.2.37}$$

The variation of z from Eq. (7.2.37) with S from Eq. (7.2.17) is shown in Fig. (7.12) for Sand No. D. This variation may be expressed by Eq. (7.2.18) in which α is constant. The best fitting value of α is determined from Eq. (7.2.18) using simultaneous values of z and S at S=0.80.

5. **Typical parameter values**

The components of the constitutive model and the corresponding 11 parameter values have been determined by isotropic compression tests and drained triaxial compression tests for many different sands without effective cohesion. Thus, a=0 and this parameter is not listed. The characteristics of these sands (void ratios, relative densities) and their model parameter values are listed in Table 7.1. None of the parameters have dimensions. All dimensions are controlled, where it is appropriate, by the dimension of the atmospheric pressure p_a, as in Eq. (7.2.2).

Table 7.1 Parameter values for select sands for Lade model [10]

Sand type	Vold ratio, e	Rel. density D_r (%)	v	M	λ	η_1	m	$C \cdot 10^4$	x	ψ_2	μ	h	α
Sand No. D	0.534	89	0.23	400	0.37	67	0.19	1.7	1.41	−3.34	2.22	0.60	0.41
Fine Silica sand	0.76	30	0.27	440	0.22	24.7	0.10	3.24	1.25	−3.69	2.26	0.355	0.515
Sacramento River sand	0.61	100	0.20	900	0.28	80	0.23	0.396	1.82	−3.09	2.01	0.765	0.229
Sacramento River sand	0.87	38	0.20	510	0.28	28	0.093	1.27	1.65	−3.72	2.36	0.534	0.794
Painted Rock material	0.40	100	0.20	920	0.24	101	0.21	3.51	1.25	−3.26	2.82	0.501	0.313
Painted Rock material	0.48	70	0.20	350	0.33	67	0.16	0.46	1.78	−3.39	2.72	0.698	0.386
Crushed Napa basalt	0.53	100	0.20	1050	0.17	280	0.423	0.814	1.61	−2.97	2.80	0.546	0.727
Crushed Napa basalt	0.66	70	0.20	590	0.19	130	0.30	4.57	1.39	−2.90	2.55	0.542	0.851
Monterey No.0 sand	0.57	98	0.17	1120	0.33	104	0.16	0.269	1.44	−3.38	2.30	0.49	0.896
Monterey No.0 sand	0.78	27	0.17	800	0.26	36	0.12	2.14	1.26	−3.60	2.50	0.43	0.577
Santa Monica Beach sand	0.613	88.5	0.15	1270	0.23	107	0.25	1.44	1.39	−3.16	2.07	0.56	0.49

(continued)

Table 7.1 (continued)

Sand type	Vold ratio, e	Rel. density D_r (%)	ν	M	λ	η_1	m	$C \cdot 10^4$	x	ψ_2	μ	h	α
Santa Monica Beach sand	0.681	64.9	0.19	1050	0.24	59.1	0.165	2.12	1.37	−3.34	2.20	0.57	0.58
Santa Monica Beach sand	0.755	39.2	0.22	820	0.26	37.7	0.105	2.26	1.42	−3.62	2.27	0.58	0.68
Santa Monica Beach sand	0.815	18.4	0.26	600	0.27	31.2	0.095	2.36	1.55	−3.74	2.36	0.67	0.46
Eastern Scheldt sand	0.67	73	0.20	460	0.41	70.2	0.288	1.27	1.61	−3.15	2.06	0.553	0.617

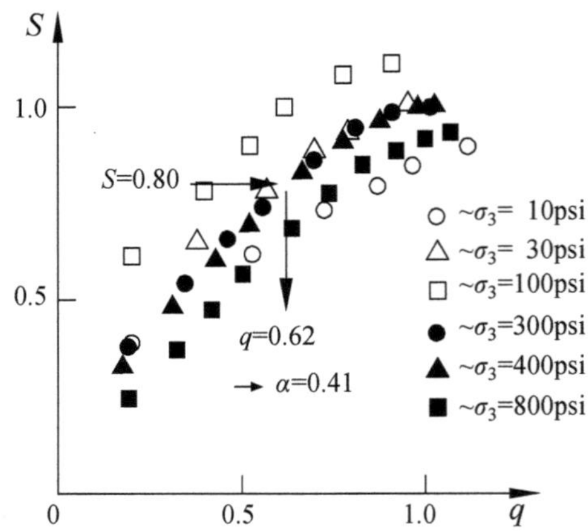

Fig. 7.12 Determination of α for yield criterion for Sand No. D (1 psi = 6.895 kPa)

From Table 7.1, it may be seen that several parameter values vary in a systematic manner with relative density. Thus, the M-values increase with increasing relative density, while the values of Poisson's ratio for sands are typically around 0.2. The values of η_1 and m increase with relative density, but it is more difficult to discern a consistent pattern in the remaining parameters. It should be noted that the parameters appear in the mathematical expressions as pairs (λ and M, m and η_1, C and x, ψ_2 and μ, h and α), and they are combined to produce moduli or strengths through the magnitudes of the stresses. Scatters in the experimental data therefore play a role in the derived parameter values. Although there are ranges of values in which each parameter tends to fall, it does not appear to be a general pattern of variation for sands that may be helpful in indicating approximate values of parameters in the absence of actual test data.

7.2.7 Model Comments

Lade model is an elastoplastic model with a single yield surface of plastic work-hardening adopting the nonassociated flow rule. Its advantage is to consider the influence of shear yield and stress Lode angle properly. But the disadvantage is too more calculation parameters, excessive dilatancy phenomena are produced even with the nonassociated flow rule. It is more suitable for cohesionless soil.

7.3 A Unified Hardening Constitutive Model for Soils

7.3.1 Introduction

The various complex features of a thing are the results of the joint action of its internal cause and external cause. Soil is not exceptional, too. Soil is a three-phase mixture consisting of soil particle, pore water, and pore gas, and this is the internal cause. Due to the fact that the composition of soil is different, it has many features, such as a lot of differences between unsaturated soil and saturated soil. The external cause is the different loading conditions of the soil, such as normal consolidated soil and overconsolidated soil differing as a result of stress history which leads to its different nature. Because of the composition of soil and the different load conditions, soil shows many unique physical and mechanical properties. The principal characteristics for the soil mainly include yield of hydraulic pressure, dilatancy, strain hardening and softening and overconsolidated, dependent on stress path, anisotropy, structural, and creep. In order to describe soil stress–strain relationship reasonably, numerous different constitutive models are put forward for geotechnical material by the domestic and foreign scholars, such as the Cam-Clay model, modified Cam-Clay model, and Lade model. But these constitutive models can simulate the behavior of only one type of soil, and they are difficult to reflect comprehensively the various features of soil. For example, the modified Cam-Clay model is mainly suitable for normal consolidated clay, and not suitable to the overconsolidated soil because it cannot reflect the dilatancy. Lade model is mainly suitable for sand, mainly reflecting the shear yield, but it does not reflect the volumetric yield reasonably. Over the years, many scholars have been looking for a soil constitutive model which could be able to fully describe the mechanical characteristics of geomaterial. In recent years, using modified Cam-Clay model as the basic framework, a unified hardening model of soil has been established mainly by Yao [12–15], which can reflect a variety of basic mechanical properties of soil. Yang-ping Yao is a Professor at Beijing University of Aeronautics and Astronautics, the former doctor candidate of Ding-yi Xi in xi'an, and the junior fellow apprentice of Professor Zheng-han Chen in Logistical Engineering University (LEU). He later cooperates with professor Matsuoka in Japan.

So let us take a look at the description of the unified hardening model. Based on the theoretical basis of the modified Cam-Clay model, the unified hardening model is built by introducing the unified hardening parameter that is independent of the stress path. Through the appropriate adjustments of unified hardening parameter, a series of constitutive models are put forward for soil to be compatible with the characteristics of different types of soil. These models can describe the deformation characteristics of dilatancy, hardening, strain-softening, overconsolidated, stress-induced anisotropy, material anisotropy, structural, creep, cyclic loading, and the particle breakage under high stress. These models include the unified hardening model for clay, sandy soil, overconsolidated soils, K_0 consolidation, strain-softening, and the material anisotropy. It is more than ten kinds, and it shows its strong function.

The name of the constitutive model of soil is very interesting, such as Cam-Clay model, Tsinghua model, LEU model. According to the general thought, the unified hardening model should be named Beihang model. But it is easy to be misunderstood as a model of plane, rocket, or missile. The name of the unified hardening model sounds like the feeling of very domineering, but it will take a long time to complete covering all the characteristics of soil. Until now, we have talked about the unified hardening model of soil is on the basis of the modified Cam-Clay model, by introducing the unified hardening parameter that is independent of the stress path. So now let us look at what the unified hardening parameter is which has nothing to do with the stress path.

7.3.2 *The Unified Hardening Parameter Which Has Nothing to Do with the Stress Path*

1. **Hardening and hardening parameter**

In the previous chapter, we have learned that the core of the plastic mechanics of geomaterial is to build the stress–strain relations and calculate the plastic strain. Three problems need to be solved, namely, when the plastic strain appears, direction of plastic strain, and the magnitude of the plastic strain. In order to answer these problems, three corresponding theories are created, namely, the yield criterion, flow rule, and hardening law. Yield criterion is to determine the start point of stress condition where plastic deformation appears. It is described by the yield surface F. Stress state point is the elastic state when it is inside the yield surface, and the plastic strain will be produced when it is on the yield surface. Flow rule is to determine the direction of plastic strain increment. When the associated flow rule is adopted, the plastic potential surface Q will be the same as the yield surface, namely, $Q=F$. The direction of plastic strain increment is identical with outer normal of the yield surface. For nonassociated flow rule, plastic potential surface is different with the yield surface, namely, $F\neq Q$, direction of plastic strain increment is determined by the plastic potential surface. Hardening rule is to study the evolution of the yield surface, which can be used to determine the magnitude of the plastic strain increment. After loading stress arrives at the yield surface, plastic strain appears, and the shape, size, and center of yield surface will change, and the new yield surface occurs. This feature is called hardening. The new yield surface is also called the loading surface, which is related with the stress state and strain history. So, in order to reflect change of the internal microstructure caused by plastic deformation, a parameter is introduced which is known as the hardening parameter H_α. It is a parameter to reflect the strain history. The loading surface is represented as

$$F(\sigma, H_\alpha) = 0$$

where $H_\alpha = H_\alpha(\varepsilon^p)$ is the hardening parameter.

After the loading surface is determined, the magnitude of the plastic strain can be calculated. In other words, the future plastic deformation is determined by what occurred before. One will feel very speculative, as one of the past will determine the future of a person.

Hardening parameter H_α is a parameter which reflects plastic strain history. This concept is very abstract, including many contents, but there is no real expression. It is unlike other parameters which can be measured directly, and there is a clear definition formula. So we have to do some limitations of hardening parameter H_α, to meet our request. We know that the hardening parameter H_α is a variable which describes the change rule of yielding surface. Two basic conditions should be satisfied:

(1) The hardening history of the material should be reflected fully.
(2) Its value has nothing to do with the stress path. That is, from the same starting point, it loads along different stress paths to the same point on another yield surface, and the value of H_α should be the same; or it starts from the same stress point loading along different stress paths to the different points on another yield surface, and the value of H_α should be the same. That is to say, the value of H_α should be equivalent in the same yield surface.

These two conditions are the basis, on which we look for hardening parameters, as well as the standard for us to verify the reasonability of hardening parameter.

In the current state of plastic mechanics of geomaterial, the choice is not unified for the hardening parameter. For example, for the Cam-Clay model and modified Cam-Clay model, the plastic volumetric strain ε_v^p is chosen as the hardening parameter, and the plastic work W_p is chosen as the hardening parameter for Lade model, and the combination of the plastic volumetric strain and the plastic shear strain is selected as the hardening parameter for other models. Because the hardening parameter is related to the plastic strain history, it is easier for people to choose plastic volumetric strain ε_v^p, plastic shear strain ε_s^p, or their combination as the hardening parameter of soil, but these parameters cannot fully simulate the hardening process of different types of geomaterial. So how to select or construct a suitable hardening parameter has been a focus in the study of international scholars of constitutive relation.

2. **The problems existing in construction of hardening parameter**

As shown in Fig. 7.13, f_1 and f_2 are two volumetric yield surfaces in the *p*-*q* coordinates for the modified Cam-Clay model. While the stress point is located within the yield surface, the stress–strain relationship is linear elastic. When stress is located on the yield surface, the plastic strain will come into being, the yield surface will expand into a new yield surface, and elastic domain will expand accordingly. We choose two typical stress paths, such as path of constant *p* and that of constant *q*. When loading along a path of constant *p*, i.e., for the path of constant hydrostatic pressure, only the generalized shear stress changes, namely, loading along the stress path of *abc*. Point *b* is on the initial yield surface f_1, and the new plastic strain will be generated when the generalized shear stress beyond the point *b* along line *bc*. The new yield surface f_2 will be formed when unloading at point *c*. The stress state backs away along line *abc* to point *a*′(*a*), namely, $q=0$. The plastic shear strain increment $d\varepsilon_s^p$

Fig. 7.13 Loading along different paths and unloading

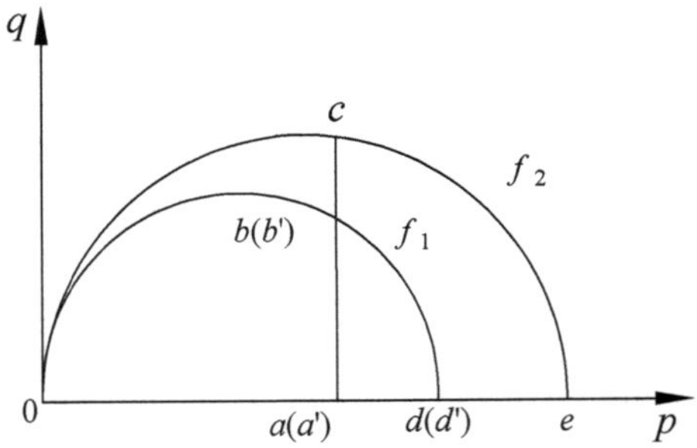

Fig. 7.14 Shear stress–shear strain

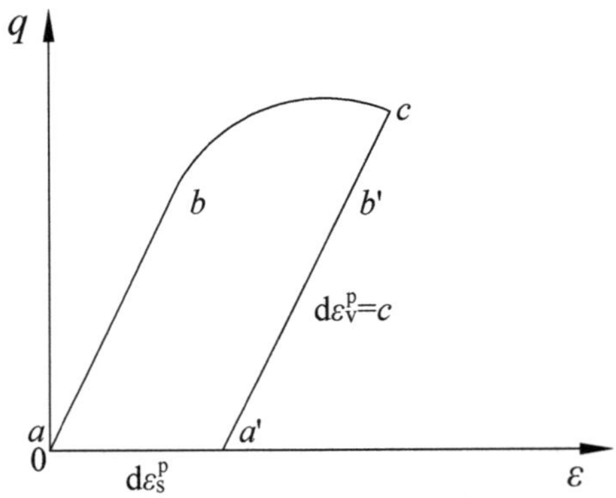

and the plastic volumetric strain $d\varepsilon_v^p$ produced is shown in Fig. 7.14 for the whole stress path *abcb′a′*. When loading along the constant q stress path *aded′a′*, a similar stress–strain curve will be got too, as shown in Fig. 7.15. Point *d* is on the original yield f_1, and point *e* on the new yield surface f_2. Unloading to point *a′*, the plastic strain increment $d\varepsilon_v^p$ is also produced. Here, it should be noted that point *b*(*b′*) and point *d*(*d′*) are located on the same initial yield surface f_1, and points *c*(*c′*) and *e*(*e′*) are on the same new yield surface f_2. For normally consolidated clay, the plastic volumetric strain increment of the same magnitude will be brought about to the two stress path *abcb′a′* and *aded′a′*. Therefore, the plastic volumetric strain can be used as the hardening parameter for normal consolidated clay, because it has nothing to do with stress path. But the plastic shear strain cannot be chosen as the hardening parameter, because the plastic shear strain is not equal for these two stress paths of *abc* and *ade*, and plastic shear strain increment is not equal along the stress path *abcb′a′* and *aded′a′*. It is zero for stress path *aded′a′*, and it is larger than zero for stress path *abcb′a′*, i.e., the plastic shear strain is related to the stress path.

Let us see constitutive response dependent on stress path for sands which always displays the dilatancy. The different stress paths are illustrated in Fig. 7.16 for triaxial compression test of sand, i.e., stress path *ADEF*, *ABCF*, *AF*, and *ABEF*. The volumetric strain ε_v^p relative to the mean stress p is given in Fig. 7.17 for the above four kinds of stress path, and that for the shear strain is provided in Fig. 7.18. As can be

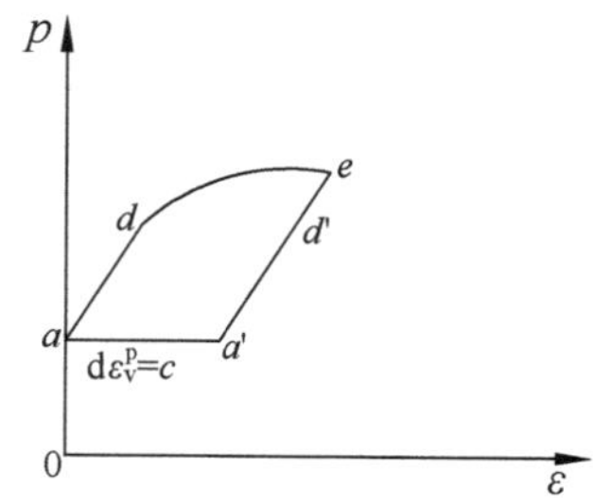

Fig. 7.15 Hydraulic stress–volumetric strain

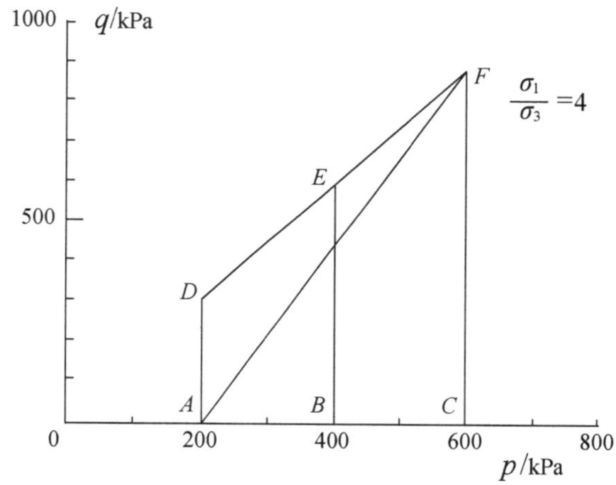

Fig. 7.16 Loading along different stress paths

seen from Figs. 7.17 and 7.18, the plastic strain is very different in magnitude when loading to point F along different paths, respectively. For sand, the plastic strain ε_v^p and ε_s^p are dependent on not only the initial stress state and final stress state but also the stress path. Thus, the plastic strain ε_v^p and ε_s^p have strong correlation with stress path, and cannot be used as the hardening parameter directly.

3. **The composition method of the hardening parameter has nothing to do with stress path for geomaterial**

Further analysis of the above test results can be found in the stress path of AD and CF. Although p and q change differently, the plastic volumetric strain increment and the plastic shear strain increment are almost the same, this is because stress ratio η ($\eta = \frac{q}{p}$) is the same at the end point of stress path D and F. At the stress path of AC and DF, although the change of p values is the same, the end points correspond to different stress ratios, and different plastic strain increments bring about on the stress path of AC and DF. From the previous analysis of stress path, the induced different plastic volumetric strains and plastic shear strains lie in the different stress paths corresponding to the different stress ratios η of stress path. The longer the experienced stress path of great stress ratio η, the more the plastic strain produced. If a correction factor R can be used to get rid of influence of stress path,

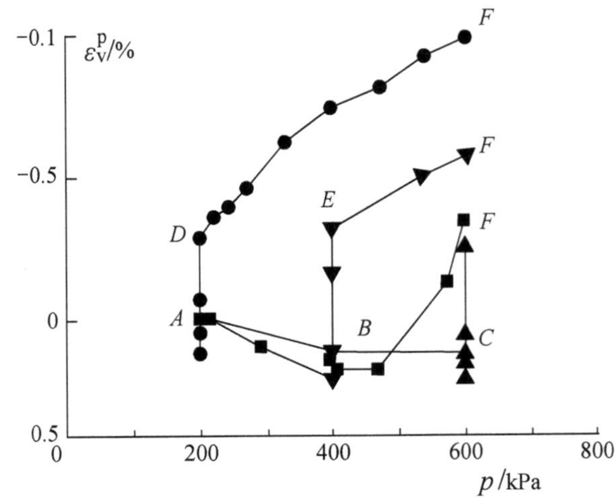

Fig. 7.17 The plastic volumetric strain and mean stress

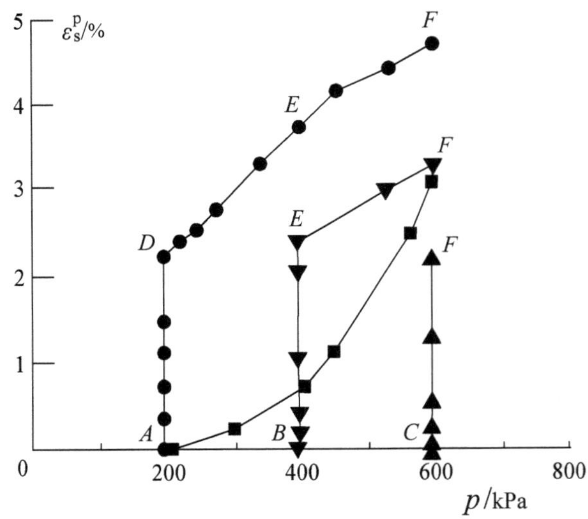

Fig. 7.18 The plastic shear strain and mean stress

which is equivalent to the plastic strain increment of these stress paths processed by this correction coefficient, the stress path independent of hardening parameter can be constructed by integral of the corrected plastic strain. Of course, the correction coefficient R is closely related to the stress ratio η.

This is regarded as a key idea, an innovation, and also the core of the research. The plastic strain increment cannot be used as the hardening parameter because the plastic strain increment is relative to stress path. So it should be put aside, and other new ones for the hardening parameter should be researched. But the idea here is to look for a correction coefficient R to get rid of components of plastic strain increment that related to the stress path. Behind, this there is a problem which we should think about and discuss: how is the plastic strain increment partially related to the stress path? For instance, one part relates to the stress path, and the other part has nothing to do with stress path. How can we remove the plastic strain relative to the stress path. It is a bit like the deformation under researching. The deformation can be divided into two parts, the elastic deformation can recover, and the other part (plastic deformation) is unrecoverable. The plastic strain increment partly relates to the stress path: one part has nothing to do with stress path, while the other part relates to the stress path. What is the mechanism behind them? How can we remove the portion of the plastic strain that related to the stress path?

The following content is aiming at the simplification of the problems, and expression of unified hardening parameter is simplified based on the modified Cam-Clay model. We know that soil is a material with the characteristics of dilatancy. Through the dilatancy equation, the relation can be established for plastic volumetric strain increment and the plastic shear strain increment.

$$\frac{\mathrm{d}\varepsilon_{\mathrm{v}}^{\mathrm{p}}}{\mathrm{d}\varepsilon_{\mathrm{s}}^{\mathrm{p}}} = f(\eta)$$

For example, the dilatancy equation of the modified Cam-Clay model

$$\frac{\mathrm{d}\varepsilon_{\mathrm{v}}^{\mathrm{p}}}{\mathrm{d}\varepsilon_{\mathrm{s}}^{\mathrm{p}}} = \frac{M^2 - \eta^2}{2\eta}$$

where M is the stress ratio at critical state of normally consolidated clay, and the stress ratio of the phase transformation $(\frac{q}{p})_{\mathrm{pt}}$ for sand.

Obviously, $\mathrm{d}\varepsilon_{\mathrm{v}}^{\mathrm{p}}$ and $\mathrm{d}\varepsilon_{\mathrm{s}}^{\mathrm{p}}$ are not two independent variables, so the constitution of hardening parameters need only one plastic strain increment, which contains only the plastic volumetric strain increment $\mathrm{d}\varepsilon_{\mathrm{v}}^{\mathrm{p}}$, making the problem simple further. The constitution idea of hardening parameters is simplified as follows: To find a factor $R(\eta)$ associated with stress path to remove component related of stress path for the plastic volumetric strain increment $\mathrm{d}\varepsilon_{\mathrm{v}}^{\mathrm{p}}$. $\frac{\mathrm{d}\varepsilon_{\mathrm{v}}^{\mathrm{p}}}{R(\eta)}$ has nothing to do with stress path, and it is integral along the stress path. Then, a hardening parameter will be obtained which has nothing to do with stress path, namely, the integral results are only related to the stress level of the starting point and end point of stress path. The following is the general expressions of hardening parameter that has nothing to do with the stress path:

$$H_{\alpha} = \int H_{\alpha} = \int \frac{\mathrm{d}\varepsilon_{\mathrm{v}}^{\mathrm{p}}}{R(\eta)} \quad \text{or} \quad \mathrm{d}H_{\alpha} = \frac{\mathrm{d}\varepsilon_{\mathrm{v}}^{\mathrm{p}}}{R(\eta)} \tag{7.3.1}$$

Following the above ideas, the key problem becomes how to construct the correction factor $R(\eta)$. For normal consolidated clay, $\mathrm{d}\varepsilon_{\mathrm{v}}^{\mathrm{p}}$ is independent of stress path, and the correction factor $R(\eta)$ of the stress path should be degraded to 1. On the basis of experiment and theoretical research, a unified hardening parameter H_α is deduced by Yao which is suitable for different types of soil and has nothing to do with the stress path.

7.3.3 *Unified Hardening Model for Natural Consolidation Soil*

1. **Analysis of the Modified Cam-Clay Model**

The three basic elements of elastoplastic model include yield surface, flow rule, and hardening model. The unified hardening model adopts oval form of the yield surface like that of the modified Cam-Clay model, the associated flow rule, and the unified hardening parameter for the hardening model. So, let us have a new look at the modified Cam-Clay model.

The modified Cam-Clay model is a widely used model on the soil mechanics field, and it is mainly suitable for the normal consolidated soil and lightly overconsolidated soils. It is a simple model with clear basic concept, less parameters which can be obtained through conventional triaxial test. The important and basic characteristics of geomaterial can be simulated by the Cam-Clay model including yield of hydrostatic pressure, dilatancy, and hardening with confining stress. Its yield surface is expressed as

$$f = g = \ln \frac{p}{p_0} + \ln\left(1 + \frac{q^2}{M^2 p^2}\right) - \frac{1}{c_\mathrm{p}} \varepsilon_\mathrm{v}^\mathrm{p} = 0 \tag{7.3.2}$$

where $c_\mathrm{p} = \frac{\lambda - k}{1+e_0}$; f is the yield surface; g is the plastic potential surface; M is the stress ratio at critical state or phase transformation; p is the hydrostatic stress (mean stress); q is the generalized shear stress; p_0 is the initial mean stress; λ is slope of isotropic compression line; k is slope of unloading line of isotropic compression; and e_0 is the initial void ratio.

Critical state line (CSL) is the trajectory of damage point in p-q-v coordinate. When the stress state of geomaterial arrives at this line, the mean stress p, generalized shear stress q, and volume are not changed, and the shear deformation is increasing until failure, only shear failure occurs.

Stress ratio of critical state or phase transformation M: the slope of critical state line or phase transformation line, namely, the ratio of generalized shear stress q and the mean stress p, also known as the normalization strength.

Roscoe surface is the surface formed by the trajectory from the normal consolidation state to the critical state, and the undrained Roscoe surface will be the volumetric yield surface.

Thus it can be seen in Fig. 7.19, the elastic region of the modified Cam-Clay model is enclosed by the yield surface equation $f(p, q, H_\alpha)=0$, the critical line $\frac{q}{p} = M$,

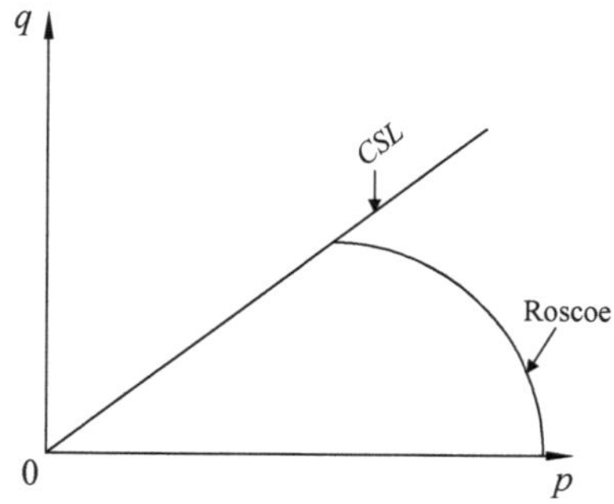

Fig. 7.19 Yield surface of the modified Cam-Clay model (Undrain Roscoe surface)

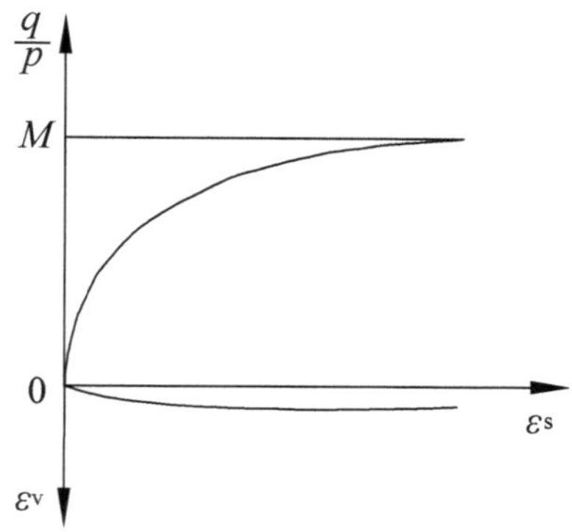

Fig. 7.20 The relationship of stress–strain (clay)

and the hydraulic consolidation line $q=0$. It should be mentioned that the description of modified Cam-Clay model is relatively accurate for the stress–strain relationship of normally consolidated soils (in Fig. 7.20). But its following limitations are well known:

(1) It is subject to the Classical Plastic Theory, using Drucker's Postulate and associated flow rule, and it does not tally with the actual status of some soil.
(2) Adopting the plastic bulk strain as the hardening parameter, it does not fully consider the shear deformation, and it can only reflect the volumetric hardening of soil, failing to reflect the positive dilatancy and softening of the soil. The use range is limited.
(3) Concept of Cam-Clay model is from reconstituted clays. It does not consider the cement of natural clay, and it cannot describe the inherent anisotropy of soil too.
(4) There is no meter of the second principal stress effect on strength and deformation.
(5) There is no consideration of viscosity of clayey soil and the time-dependent stress–strain relationship.

Let us have a look at the dilatancy characteristics of sands (in Fig. 7.21). It shows a changing process from the shear shrinkage to dilatancy, so its critical state is not

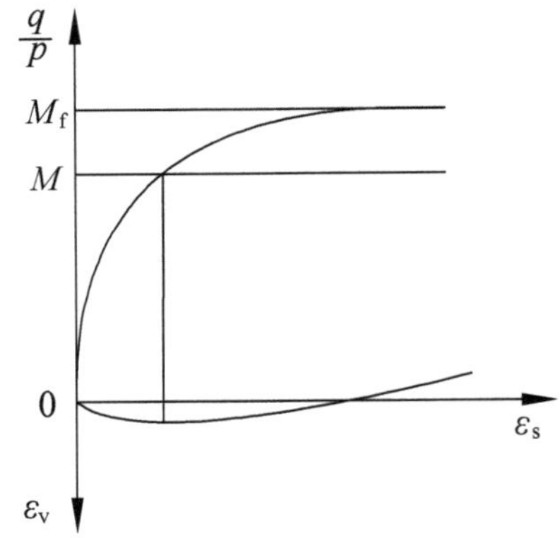

Fig. 7.21 Stress–strain relationship (sand)

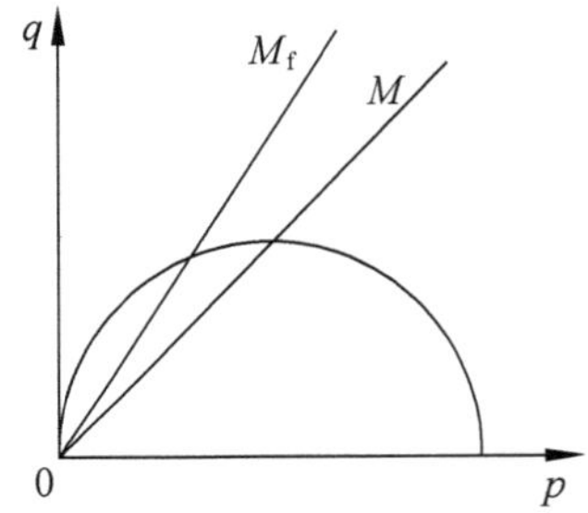

Fig. 7.22 Yield surface (or plastic potential surface)

the same as the peak state. Critical state refers to the point where the shear shrinkage transits to dilatancy, and the corresponding stress ratio is M. Peak state refers to the point where damage occurs, and the corresponding stress ratio is M_f, also known as the peak strength. The elastic zone (in Fig. 7.22) is surrounded by the yield equation $f(p, q, H_\alpha)=0$, the failure line $\frac{q}{p} = M_f$, and the isotropic consolidation line $q=0$.

7.3.4 *The Unified Hardening Model of the Normal Consolidated Soil*

The yield surface is the same as the plastic potential surface in the unified hardening model, and expressed as

$$f = g = \ln \frac{p}{p_0} + \ln\left(1 + \frac{q^2}{M^2 p^2}\right) - \frac{1}{c_p} H = 0 \tag{7.3.3}$$

It is different with the modified Cam-Clay model in the hardening parameter H_α. H_α is no longer ε_v^p, being represented as

$$H_\alpha = \int \frac{\mathrm{d}\varepsilon_\mathrm{v}^\mathrm{p}}{R(\eta)}$$

1. **The derivation process of the $R(\eta)$**

The principal idea is that the hardening parameter H_α is equivalent in the same yield surface.

By Eq. (7.3.3), it is obtained

$$\mathrm{d}f = \frac{\partial f}{\partial p}\mathrm{d}p + \frac{\partial f}{\partial q}\mathrm{d}q - \frac{1}{c_\mathrm{p}}\mathrm{d}H_\alpha = 0$$

namely,

$$\mathrm{d}f = \frac{\partial f}{\partial p}\mathrm{d}p + \frac{\partial f}{\partial q}\mathrm{d}q - \frac{1}{c_\mathrm{p}}\frac{\mathrm{d}\varepsilon_\mathrm{v}^\mathrm{p}}{R(\eta)} = 0 \tag{7.3.4}$$

According to the associated flow rule,

$$\mathrm{d}\varepsilon_\mathrm{v}^\mathrm{p} = \mathrm{d}\lambda\frac{\partial g}{\partial p} = \mathrm{d}\lambda\frac{\partial f}{\partial p}$$

Substituting it into Eq. (7.3.3), we can get

$$\mathrm{d}f = \frac{\partial f}{\partial p}\mathrm{d}p + \frac{\partial f}{\partial q}\mathrm{d}q - \frac{1}{c_\mathrm{p}}\frac{1}{R(\eta)}\mathrm{d}\lambda\frac{\partial f}{\partial p} = 0 \tag{7.3.5}$$

Then, the following is obtained:

$$\mathrm{d}\lambda = \frac{\frac{\partial f}{\partial p}\mathrm{d}p + \frac{\partial f}{\partial q}\mathrm{d}q}{\frac{1}{c_\mathrm{p}}\frac{1}{R(\eta)}\frac{\partial f}{\partial p}}$$

$$\mathrm{d}\varepsilon_\mathrm{v}^\mathrm{p} = \mathrm{d}\lambda\frac{\partial f}{\partial p} = c_\mathrm{p}R(\eta)(\frac{\partial f}{\partial p}\mathrm{d}p + \frac{\partial f}{\partial q}\mathrm{d}q) \tag{7.3.6}$$

$$\mathrm{d}\varepsilon_\mathrm{s}^\mathrm{p} = \mathrm{d}\lambda\frac{\partial f}{\partial q} = c_\mathrm{p}R(\eta)[\frac{\partial f}{\partial p}\mathrm{d}p + \frac{(\frac{\partial f}{\partial q})^2}{(\frac{\partial f}{\partial p})}\mathrm{d}q] \tag{7.3.7}$$

By Eq. (7.3.3), we could obtain

$$\frac{\partial f}{\partial p} = \frac{1}{p}\frac{M^2 - \eta^2}{M^2 + \eta^2}, \quad \frac{\partial f}{\partial q} = \frac{1}{p}\frac{2\eta}{M^2 + \eta^2}$$

Two typical stress paths (Fig. 7.23) are chosen for calculation of $R(\eta)$, the factor related to stress path. One is equivalent q stress path AB ($\mathrm{d}p=0$), and another is equivalent p stress path AC ($\mathrm{d}q=0$).

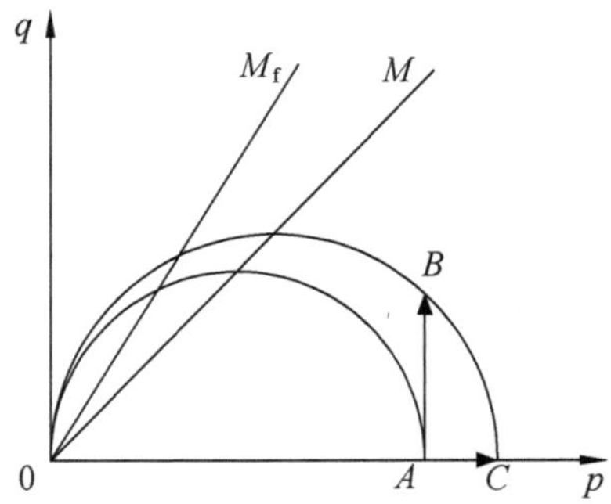

Fig. 7.23 Different loading paths for determining of hardening parameter

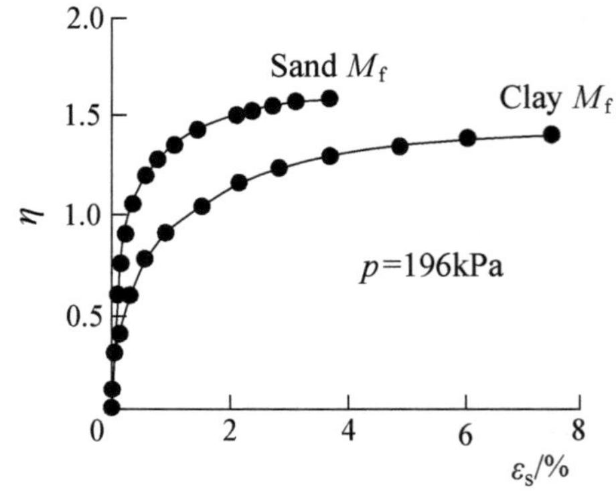

Fig. 7.24 Curve of ε_s^p-η

(1) Analysis of stress path of equivalent *p*.

By Eq. (7.3.7),

$$d\varepsilon_s^p = c_p R(\eta) \frac{4\eta^2}{M^4 - \eta^4} \frac{1}{p} dq$$

When *p* keeps constant,

$$d\varepsilon_s^p = c_p R(\eta) \frac{4\eta^2}{M^4 - \eta^4} d\eta \tag{7.3.8}$$

For normal consolidation clay, $R(\eta) = 1$, then

$$d\varepsilon_s^p = c_p \frac{4\eta^2}{M^4 - \eta^4} d\eta$$

Because in the ε_s^p-η coordinate system (Fig. 7.24), the curve shapes are similar for sand and clay. So for the sand,

$$\mathrm{d}\varepsilon_{\mathrm{s}}^{\mathrm{p}} = \rho c_{\mathrm{p}} \frac{4\eta^2}{M_{\mathrm{f}}^4 - \eta^4} \mathrm{d}\eta \tag{7.3.9}$$

where ρ is the proportional relation of plastic shear strain increment for sand and clay; M_{f} is the failure stress ratio of sand.

Simultaneous Eqs. (7.3.8) and (7.3.9),

$$R(\eta) = \rho \frac{M^4 - \eta^4}{M_{\mathrm{f}}^4 - \eta^4}$$

The following can be obtained:

$$H_{\alpha} = \int \frac{1}{\rho} \frac{M_{\mathrm{f}}^4 - \eta^4}{M^4 - \eta^4} \mathrm{d}\varepsilon_{\mathrm{v}}^{\mathrm{p}} \tag{7.3.10}$$

(2) Analysis of stress path of equivalent q ($\mathrm{d}q=0$ and $q=0$).

By Eq. (7.3.10), at the time $\eta=0$,

$$H_{\alpha} = \int \frac{1}{\rho} \frac{M_{\mathrm{f}}^4}{M^4} \mathrm{d}\varepsilon_{\mathrm{v}}^{\mathrm{p}} \tag{7.3.11}$$

By the modified Cam-Clay model, at the time $\eta=0$,

$$H_{\alpha} = \int \mathrm{d}\varepsilon_{\mathrm{v}}^{\mathrm{p}} \tag{7.3.12}$$

Simultaneous Eqs. (7.3.11) and (7.3.12),

$$\rho = \frac{M_{\mathrm{f}}^4}{M^4}$$

So the unified hardening parameter of normal consolidated soil is expressed as

$$H_{\alpha} = \int \mathrm{d}H_{\alpha} = \int \frac{M^4}{M_{\mathrm{f}}^4} \frac{M_{\mathrm{f}}^4 - \eta^4}{M^4 - \eta^4} \mathrm{d}\varepsilon_{\mathrm{v}}^{\mathrm{p}} \tag{7.3.13}$$

where M is the stress ratio of critical state, M_{f} is the failure stress ratio, η is the stress ratio, and $\mathrm{d}\varepsilon_{\mathrm{v}}^{\mathrm{p}}$ is the increment of plastic volumetric strain.

(3) The characteristics of unified hardening parameter of normally consolidated soils.

The first one is that it has nothing to do with the stress path. The relationship is illustrated in Fig. 7.25 between the average stress and the unified hardening parameter H_{α} for the different stress paths presented before. We can see that the hardening parameter H_{α} keeps equivalent value for the different loading stress paths starting from point A to point F.

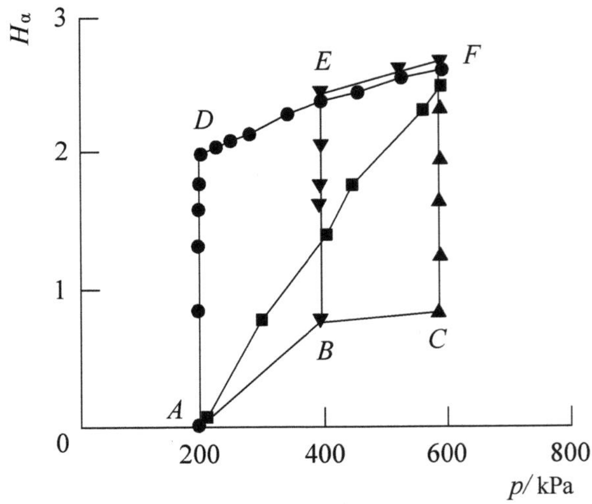

Fig. 7.25 Relationship between unified hardening parameter with average stress

The second is that the different characteristics can be reflected for sand and clay. For the normal consolidation clay, $M_f = M$, so the unified hardening parameter is the same as the hardening parameter of the modified Cam-Clay model, namely

$$H_\alpha = \int \mathrm{d}H_\alpha = \int \mathrm{d}\varepsilon_v^p = \varepsilon_v^p$$

To sand, $M_f \neq M$, so the unified hardening parameter is chosen as what expressed in Eq. (7.3.13).

The third is that the dilatancy can be simulated reasonably for sand.

Equation (7.3.13) can be transformed as

$$\mathrm{d}\varepsilon_v^p = \frac{M_f^4}{M^4}\frac{M^4 - \eta^4}{M_f^4 - \eta^4}\mathrm{d}H_\alpha, \quad \mathrm{d}\varepsilon_s^p = \frac{2\eta}{(M^2 - \eta^2)}\mathrm{d}\varepsilon_v^p$$

(1) $\eta = 0$ (isotropic compression conditions), $\mathrm{d}H_\alpha = \mathrm{d}\varepsilon_v^p > 0$, $\mathrm{d}\varepsilon_v^p > 0$, $\mathrm{d}\varepsilon_s^p = 0$. It is the isotropic consolidation.
(2) $0 < \eta < M$ (shear hardening stage), $\mathrm{d}H_\alpha > 0$, $\mathrm{d}\varepsilon_v^p > 0$, $\mathrm{d}\varepsilon_s^p > 0$, it is the shear shrinkage deformation.
(3) $\eta = M$ (critical state), $\mathrm{d}H_\alpha > 0$, $\mathrm{d}\varepsilon_v^p = 0$, $\mathrm{d}\varepsilon_s^p > 0$, at this time it is a state transforming from the shear shrinkage into dilatancy.
(4) $M < \eta < M_f$ (hardening at dilatancy), $\mathrm{d}H_\alpha > 0$, $\mathrm{d}\varepsilon_v^p < 0$, $\mathrm{d}\varepsilon_s^p > 0$, it is dilatancy deformation.

The fourth is the relation between the correction coefficient of plastic volumetric strain increment and stress ratio.

By Eq. (7.3.13), the following will be obtained:

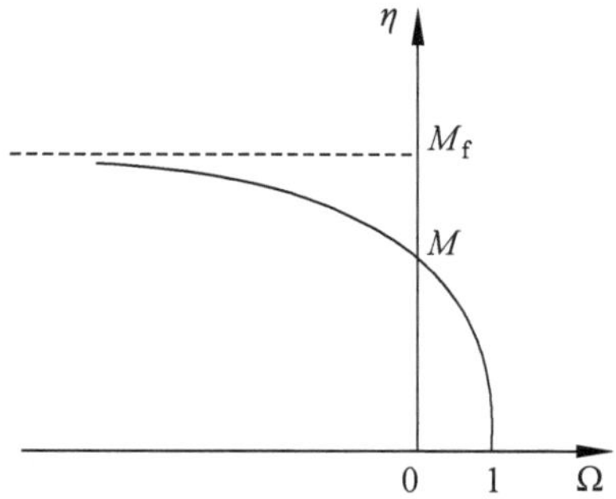

Fig. 7.26 The relation between the correction coefficient of plastic volumetric strain increment and stress ratio

$$dH_\alpha = \frac{M^4}{M_f^4}\frac{M_f^4 - \eta^4}{M^4 - \eta^4}d\varepsilon_v^p$$

Let the correction coefficient of plastic volumetric strain increment Ω be

$$\Omega = \frac{M^4}{M_f^4}\frac{M_f^4 - \eta^4}{M^4 - \eta^4}$$

Let us see the relation between the correction coefficient of plastic volumetric strain increment and stress ratio (Fig. 7.26). It can be seen in Fig. 7.26, at $\eta = 0$, $\Omega = 1$, the plastic volumetric strain increment is independent on stress path, and it does not need to be modified. At $\eta = M$, $\Omega = 0$, it is critical state. At $\eta = M_f$, $\Omega \to -\infty$, M_f is the asymptote of η.

7.3.5 The Stress–Strain Relationship

According to elastic–plastic theory,

$$d\boldsymbol{\varepsilon} = d\boldsymbol{\varepsilon}^e + d\boldsymbol{\varepsilon}^p$$

1. **Elastic stress–strain relationship**

According to the generalized Hooke's law,

$$d\varepsilon_{ij}^e = \frac{1+\nu}{E}d\sigma_{ij} - \frac{\nu}{E}d\sigma_{kk}\delta_{ij}$$

The corresponding components of elastic strain will be

$$d\varepsilon_v^e = d\varepsilon_{11}^e + d\varepsilon_{22}^e + d\varepsilon_{33}^e = \frac{3(1-2\nu)}{E}dp$$

$\mathrm{d}\varepsilon_{\mathrm{s}}^{\mathrm{e}} = \frac{\sqrt{2}}{3}\sqrt{(\mathrm{d}\varepsilon_{11}^{\mathrm{e}} - \mathrm{d}\varepsilon_{22}^{\mathrm{e}})^2 + (\mathrm{d}\varepsilon_{22}^{\mathrm{e}} - \mathrm{d}\varepsilon_{33}^{\mathrm{e}})^2 + (\mathrm{d}\varepsilon_{33}^{\mathrm{e}} - \mathrm{d}\varepsilon_{11}^{\mathrm{e}})^2} = \frac{2(1+\nu)}{3E}\mathrm{d}q$ By $E = \frac{3(1-2\nu)(1+e_0)}{k}p$, we can get

$$\mathrm{d}\varepsilon_{\mathrm{v}}^{\mathrm{e}} = c_{\mathrm{k}}\frac{1}{p}\mathrm{d}p$$

$$\mathrm{d}\varepsilon_{\mathrm{s}}^{\mathrm{e}} = \frac{2}{9}\frac{1+\nu}{1-2\nu}c_{\mathrm{k}}\frac{1}{p}\mathrm{d}q$$

where $c_{\mathrm{k}} = \frac{k}{1+e_0}$, and k is the slope of unloading line of normally consolidation curve of e-ln$p$.

2. **The relations of stress and plastic strain**

The yield function of unified hardening model is expressed as

$$f = c_{\mathrm{p}}\ln\frac{p}{p_0} + c_{\mathrm{p}}\ln(1 + \frac{q^2}{M^2p^2}) - H = 0$$

where $c_{\mathrm{p}} = \frac{\lambda}{1+e_0}$, and λ is the slope of loading line of normally consolidation curve of e-ln$p$.

$$\begin{aligned}
f &= c_{\mathrm{p}}\ln\frac{p}{p_0} + c_{\mathrm{p}}\ln(1 + \frac{q^2}{M^2p^2}) - \int\frac{M^4}{M_{\mathrm{f}}^4}\frac{M_{\mathrm{f}}^4 - \eta^4}{M^4 - \eta^4}\mathrm{d}\varepsilon_{\mathrm{v}}^{\mathrm{p}} = 0\\
\mathrm{d}f &= \frac{\partial f}{\partial p}\mathrm{d}p + \frac{\partial f}{\partial q}\mathrm{d}q + \frac{\partial f}{\partial \varepsilon_{\mathrm{v}}^{\mathrm{p}}}\mathrm{d}\varepsilon_{\mathrm{v}}^{\mathrm{p}} = 0\\
\frac{\partial f}{\partial p} &= c_{\mathrm{p}}\frac{1}{p}(\frac{M^2p^2 - q^2}{M^2p^2 + q^2})\\
\frac{\partial f}{\partial q} &= c_{\mathrm{p}}(\frac{2q}{M^2p^2 + q^2})\\
\frac{\partial f}{\partial \varepsilon_{\mathrm{v}}^{\mathrm{p}}} &= -\frac{M^4}{M_{\mathrm{f}}^4}\frac{M_{\mathrm{f}}^4 - \eta^4}{M^4 - \eta^4}\mathrm{d}\varepsilon_{\mathrm{v}}^{\mathrm{p}}\\
\mathrm{d}\varepsilon_{\mathrm{v}}^{\mathrm{p}} &= c_{\mathrm{p}}\frac{M_{\mathrm{f}}^4}{M^4}\frac{M^4 - \eta^4}{M_{\mathrm{f}}^4 - \eta^4}\frac{1}{p}[(\frac{M^2 - \eta^2}{M^2 + \eta^2})\mathrm{d}p + \frac{2\eta}{M^2 + \eta^2}\mathrm{d}q]
\end{aligned}$$

Substituting the dilatancy equation,

$$\frac{\mathrm{d}\varepsilon_{\mathrm{v}}^{\mathrm{p}}}{\mathrm{d}\varepsilon_{\mathrm{d}}^{\mathrm{p}}} = \frac{M^2 - \eta^2}{2\eta}$$

The following can be obtained:

$$\mathrm{d}\varepsilon_{\mathrm{s}}^{\mathrm{p}} = c_{\mathrm{p}}\frac{M_f^4}{M^4}\frac{M^4 - \eta^4}{M_f^4 - \eta^4}\frac{1}{p}(\frac{2\eta}{M^2 + \eta^2}\mathrm{d}p + \frac{4\eta^2}{M^4 - \eta^4}\mathrm{d}q)$$

3. **The total stress–strain relationship**

$$\begin{bmatrix} d\varepsilon_v \\ d\varepsilon_s \end{bmatrix} = [D^{ep}] \begin{bmatrix} dp \\ dq \end{bmatrix} = \frac{1}{p} \begin{bmatrix} D_{pp} & D_{pq} \\ D_{qp} & D_{qq} \end{bmatrix} \begin{bmatrix} dp \\ dq \end{bmatrix}$$

where

$$D_{pp} = c_k + c_p \frac{M_f^4}{M^4} \frac{(M^2 - \eta^2)^2}{M_f^4 - \eta^4}$$

$$D_{pq} = c_p \frac{M_f^4}{M^4} \frac{(M^2 - \eta^2)2\eta}{M_f^4 - \eta^4}$$

$$D_{qp} = c_p \frac{M_f^4}{M^4} \frac{(M^2 - \eta^2)2\eta}{M_f^4 - \eta^4}$$

$$D_{qq} = \frac{2}{9} c_k \frac{1 + v}{1 - 2v} + c_p \frac{M_f^4}{M^4} \frac{4\eta^2}{M_f^4 - \eta^4}$$

4. **Verification**

As can be seen from the experimental results (Fig. 7.27), the unified hardening model can better reflect the deformation characteristics of normally consolidated soils. For the normal consolidation of clay, $M = M_f$, $R(\eta) = 1$, hardening parameter is the plastic volumetric strain, and the unified hardening model degrades for the modified Cam-Clay model. For sand, its dilatancy and critical state properties can be described better by using the unified hardening model.

7.3.6 Model Comments

The constitutive relation dependent on stress path is not only an important mechanical properties of geomaterial, but also very difficult to be described for constitutive model. The unified hardening mode is very clever in the simulation of influence of stress path on the constitutive relation of geomaterial. The unified hardening parameter eliminates the influence of stress path through plastic deformation, depends on the stress path divided by the stress parameter which also relies on the stress path, and it is obtained for the hardening parameter that has nothing to do with the stress path. In view of the different properties of different geomaterials, the corresponding yield surface is established, and a series of models are formed for geomaterial. If we can further unify the yield surface of geomaterial and establish a unified constitutive model of geomaterial, the theoretical significance will be greater.

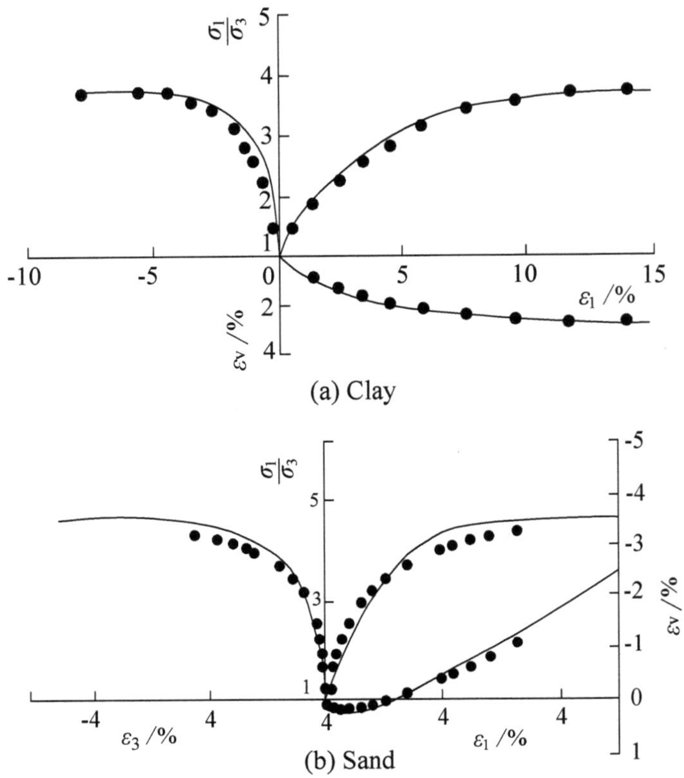

Fig. 7.27 Prediction is compared with the test curve for triaxial compression

Questions

1. What is the theoretic basis of the Cam-Clay model?
2. How can you calculate the elastoplastic deformation for the modified Cam-Clay model in general stress space?
3. What is the theoretic basis of Lade model?
4. How can you calculate the elastoplastic deformation for Lade model in principal stress space?
5. How do you understand the correlation of plastic strain increment and stress path, and how can you remove the component of plastic strain increment which is related to stress path?

References

1. Schofield A, Wroth P (1968) Critical state soil mechanics (European civil engineering series). McGraw-Hill, London
2. Roscoe KH, Schofield AN, Thurairajah A (1963) Yielding of clays in states wetter than critical. Geotechnique 13(3):211–240
3. Roscoe KH, Burland JB (1968) On the generalized stress strain behavior of 'wet' clay. In: Heyman J, Lekie FA (eds) Engineering plasticity. Cambridge University Press, Cambridge, pp 535–609
4. Kim MK, Lade PV (1988) Single hardening constitutive model for frictional materials, I. plastic potential function. Comput Geotech 5(4):307–324
5. Lade PV, Kim MK (1988) Single hardening constitutive model for frictional materials, II. yield criterion and plastic work contours. Comput Geotech 6(1):13–29
6. Lade PV, Kim MK (1988) Single hardening constitutive model for frictional materials, III. comparisons with experimental data. Comput Geotech 6(1):31–47
7. Lade PV (1990) Single-hardening model with application to NC clay. J Geotech Eng 116(3):394–414
8. Lade PV, Kim MK (1995) Single hardening constitutive model for soil, rock and concrete. Int J Solids Struct 32(14):1963–1978
9. Lade PV, Inel S (1997) Rotational kinematic hardening model for sand, part I. concept of rotating yield and plastic potential surfaces. Comput Geotech 21(3):183–216
10. Lade PV, Jakobsen KP (2002) Incremental realization of a single hardening constitutive model for frictional materials. Int J Numer Anal Meth Geomech 26:647–659
11. Lade PV (2005) Calibration of the single hardening model for clays. In: Proceeding of 11th international conference of the international association for computer methods and advances in geomechanics, Turin, Italy, pp 45–68
12. Yao YP, Hou W, Zhou AN (2007) The over-consolidated soils constitutive model based on the Hvorslev. Sci China Ser E 37(11):1417–1429
13. Yao YP, Hou W (2008) A unified hardening model for K_0 overconsolidated clays. Chin J Geotech Eng 30(3):316–322
14. Yao YP (2008) Unified hardening model for soils and its development. Ind Constr 38(8):1–5
15. Yao YP, Hou W, Luo D (2009) Unified hardening model for soils. Chin J Rock Mechan Eng 28(10):2135–2151

Chapter 8
Generalized Plastic Mechanics Considering the Rotation of Principal Axis of Stress

Traditional plastic mechanics ignore the rotation of principal axis of stress, that is, the direction of principal stress axis does not change in the hardening process. In fact, the direction of stress axis will deflect due to load induced by earthquake, vehicle, etc., and thus cause plastic deformation.

The reason why the Classic Plastic Theory cannot compute the plastic deformation induced by rotation of principal axis of stress is that it supposes the yield surface is a function of stress invariants [1]. When the principal stress axe rotates, the principal values are constant for stresses, and the stress invariants keep constant too. So, from the viewpoint of the Classic Plastic Theory, principal axis rotation is neutral loading and will not produce plastic deformation.

In order to calculate plastic deformation due to the rotation of principal axe of stress, domestic and foreign scholars have done a lot of work, mainly focusing on the following two aspects [2].

First, to establish directly the relation schema between the component of the general stress increment and general strain increment components [3, 4]. The essence of rotation of principal axes of stress is the existence of shear component in stress increment, which cannot be reflected in the Classic Plastic Theory. Matsouka transformed rotation of principal axes of stress into the change of component of the general stress increment, established the relationship between the stress increment and strain increment under the general coordinate system through experimentation, and got the plastic deformation caused by the rotation of principal axes of stress. But this is too complicated, especially for the three-dimensional case.

Second, to adopt the kinematic hardening model [5, 6], namely, the yield surface hardens with the rotation of principal axes of stress, but it is difficult to give hardening rule of yield surface when the stress path is complex.

This chapter aims to set up the generalized plastic potential theory considering the effect of variation of principal value and principal vector of stress, and put forward stress–strain relations based on generalized plastic potential theory, which presents a simple and useful elastoplastic method to compute rotation of principal axis of stress [7–10].

Y. Liu and Y. Zheng, *Plastic Mechanics of Geomaterial*, Springer Geophysics,
https://doi.org/10.1007/978-981-13-3753-6_8

8.1 Decomposition of the Stress Increment

Currently, the plastic mechanic for geomaterial generally holds that the deflection of principal axis of stress can be neglected. That is to say, there is only change in principal value of stress, not in principal vector. In this case, the principal axis of stress, stress increment and strain increment, is the same as that of strain in the Classic Plastic Theory. In fact, there is part of stress increment, which would make the rotation of principal axis of stress. In this section, general stress increment is divided into two parts according to matrix theory: Part I shares the principal axis with stress and is named the coaxial component; Part II makes rotation of the principal axis of stress and is called the rotating component. As a result, based on decomposition of the stress increment, a complex three-dimensional problem involving rotation of principal axis of stress is simplified into a three-dimensional coaxial problem, and three rotations around one principal axis of stress without change of principal value of stress, and the computational difficulty is reduced significantly.

8.1.1 Decomposition of Two-Dimensional Stress Increment

Let principal values of stress be σ_1 and σ_2, and corresponding principal vectors N_1, N_2 on $x0y$ plane (Fig. 8.1), then the stress σ can be expressed as

$$\boldsymbol{\sigma} = \begin{bmatrix} N_1 & N_2 \end{bmatrix} \begin{bmatrix} \sigma_1 & 0 \\ 0 & \sigma_2 \end{bmatrix} \begin{bmatrix} N_1 \\ N_2 \end{bmatrix} = \boldsymbol{T}_1 \wedge \boldsymbol{T}_1^{\mathrm{T}} \quad (8.1.1)$$

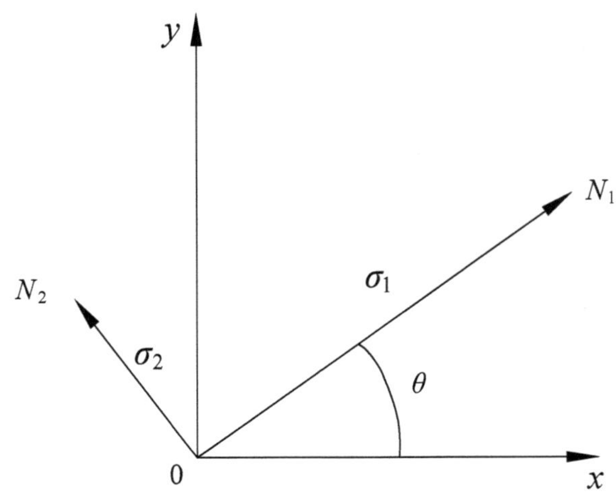

Fig. 8.1 2D general stress

Let θ be the angle between axis N_1 and axis x, then

$$\boldsymbol{T}_1 = \begin{bmatrix} \cos\theta & -\sin\theta \\ \sin\theta & \cos\theta \end{bmatrix},\ \boldsymbol{T}_1^{\mathrm{T}} = \begin{bmatrix} \cos\theta & \sin\theta \\ -\sin\theta & \cos\theta \end{bmatrix} \tag{8.1.2}$$

1. **Coaxial part of stress increment**

Stress increment of the coaxial part, namely, the stress increment that only induces the changing of principal values of stress, not principal vector. In this case, the direction matrix is a constant matrix in Eq. (8.1.1), and only the elements σ_1, σ_2 change in the diagonal matrix $\wedge$.

$$\mathrm{d}\boldsymbol{\sigma}_{\mathrm{c}} = \mathrm{d}\left(\boldsymbol{T}_1 \wedge \boldsymbol{T}_1^{\mathrm{T}}\right) = \boldsymbol{T}_1(\mathrm{d}\wedge)\boldsymbol{T}_1^{\mathrm{T}} = \boldsymbol{T}_1 \begin{bmatrix} \mathrm{d}\sigma_1 & 0 \\ 0 & \mathrm{d}\sigma_2 \end{bmatrix} \boldsymbol{T}_1^{\mathrm{T}} \tag{8.1.3}$$

From the above expressions, we can obtain the characteristic of the coaxial stress increment that the diagonal elements are nonzero and vice-diagonal elements are zero under the principal stress coordinate system. This is the coaxial part of stress increment which only brings about the variation of the principal value ($\mathrm{d}\sigma_1$, $\mathrm{d}\sigma_2$) of stress, and it is not considered for the rotation for the principal axis of stress.

2. **Rotational part of stress increment**

The rotational part of stress increment $\mathrm{d}\boldsymbol{\sigma}_{\mathrm{r}}$, namely, the stress increment, leads to the rotation of principal axis of stress and no change in principal value of stress. In this case, the diagonal matrix is a constant matrix in Eq. (8.1.1), and $\boldsymbol{T}_1$, $\boldsymbol{T}_1^{\mathrm{T}}$ changes, then

$$\mathrm{d}\boldsymbol{\sigma}_{\mathrm{r}} = \mathrm{d}\left(\boldsymbol{T}_1 \wedge \boldsymbol{T}_1^{\mathrm{T}}\right) = \mathrm{d}\boldsymbol{T}_1 \wedge \boldsymbol{T}_1^{\mathrm{T}} + \boldsymbol{T}_1 \wedge \boldsymbol{T}_1^{\mathrm{T}} \tag{8.1.4}$$

$\mathrm{d}\boldsymbol{\sigma}_{\mathrm{r}}$ is a symmetric tensor. Differentiating $\boldsymbol{T}_1$, $\boldsymbol{T}_1^{\mathrm{T}}$ in Eq. (8.1.2), respectively, and we get

$$\mathrm{d}\boldsymbol{T}_1 = \begin{bmatrix} -\sin\theta & -\cos\theta \\ \cos\theta & -\sin\theta \end{bmatrix}\mathrm{d}\theta,\ \mathrm{d}\boldsymbol{T}_1^{\mathrm{T}} = \begin{bmatrix} -\sin\theta & \cos\theta \\ -\cos\theta & -\sin\theta \end{bmatrix}\mathrm{d}\theta \tag{8.1.5}$$

Combining Eqs. (8.1.2) and (8.1.5), we get

$$\boldsymbol{T}_1^{\mathrm{T}}\mathrm{d}\boldsymbol{T}_1 = \begin{bmatrix} 0 & -1 \\ 1 & 0 \end{bmatrix}\mathrm{d}\theta,\ \mathrm{d}\boldsymbol{T}_1^{\mathrm{T}}\boldsymbol{T} = \begin{bmatrix} 0 & 1 \\ -1 & 0 \end{bmatrix}\mathrm{d}\theta \tag{8.1.6}$$

The above $\mathrm{d}\boldsymbol{\sigma}_\mathrm{r}$ is expressed in the general stress space. If $\mathrm{d}\boldsymbol{\sigma}_\mathrm{r}$ is converted into principal stress space from general stress space, and $\boldsymbol{T}_1^\mathrm{T}, \boldsymbol{T}_1 = \boldsymbol{I}$, then

$$\begin{aligned}
\boldsymbol{T}_1^\mathrm{T}\mathrm{d}\boldsymbol{\sigma}_\mathrm{r}\boldsymbol{T}_1 &= \boldsymbol{T}_1^\mathrm{T}(\mathrm{d}\boldsymbol{T}_1 \wedge \boldsymbol{T}_1^\mathrm{T} + \boldsymbol{T}_1 \wedge \mathrm{d}\boldsymbol{T}_1^\mathrm{T})\boldsymbol{T}_1 \\
&= (\boldsymbol{T}_1^\mathrm{T}\mathrm{d}\boldsymbol{T}_1) \wedge (\boldsymbol{T}_1^\mathrm{T}\boldsymbol{T}_1) + (\boldsymbol{T}_1^\mathrm{T}\boldsymbol{T}_1) \wedge (\mathrm{d}\boldsymbol{T}_1^\mathrm{T}\boldsymbol{T}_1) \\
&= \begin{bmatrix} 0 & -1 \\ 1 & 0 \end{bmatrix}\mathrm{d}\theta \begin{bmatrix} \sigma_1 & 0 \\ 0 & \sigma_2 \end{bmatrix}\boldsymbol{I} + \boldsymbol{I}\begin{bmatrix} \sigma_1 & 0 \\ 0 & \sigma_2 \end{bmatrix}\begin{bmatrix} 0 & 1 \\ -1 & 0 \end{bmatrix}\mathrm{d}\theta \\
&= \begin{bmatrix} 0 & \mathrm{d}\theta(\sigma_1 - \sigma_2) \\ \mathrm{d}\theta(\sigma_1 - \sigma_2) & 0 \end{bmatrix}
\end{aligned} \tag{8.1.7}$$

The above expression indicates that the rotational part of stress increment $\mathrm{d}\boldsymbol{\sigma}_\mathrm{r}$ results in the rotation of principal axis of stress at an angle increment of $\mathrm{d}\theta$, which establishes the relationship between stress increment of rotation and rotational angle increment of principal axis of stress. In the principal stress space, the diagonal elements are zero and vice-diagonal elements are equal, and vice-diagonal elements equal multiplication of $\mathrm{d}\theta$ and the difference of two principal values of stress.

3. **Stress increment analysis**

From the above expressions, we know that stress increment can be divided into coaxial component $\mathrm{d}\boldsymbol{\sigma}_\mathrm{c}$ and rotational component $\mathrm{d}\boldsymbol{\sigma}_\mathrm{r}$, namely

$$\mathrm{d}\boldsymbol{\sigma} = \mathrm{d}\boldsymbol{\sigma}_\mathrm{c} + \mathrm{d}\boldsymbol{\sigma}_\mathrm{r} = \boldsymbol{T}_1\begin{bmatrix} K_1 & 0 \\ 0 & K_3 \end{bmatrix}\boldsymbol{T}_1^\mathrm{T} + \boldsymbol{T}_1\begin{bmatrix} 0 & K_2 \\ K_2 & 0 \end{bmatrix}\boldsymbol{T}_1^\mathrm{T} \tag{8.1.8}$$

where $K_1 = \mathrm{d}\sigma_1$, $K_2 = \mathrm{d}\sigma_2$, $K_3 = \mathrm{d}\theta(\sigma_1 - \sigma_2)$.

8.1.2 Decomposition of Three-Dimensional Stress Increment

Let the three principal values of stress $\boldsymbol{\sigma}$ be $\sigma_1, \sigma_2, \sigma_3$, and the corresponding unit principal vectors are N_1, N_2, N_3, then

$$\boldsymbol{\sigma} = \begin{bmatrix} N_1 & N_2 & N_3 \end{bmatrix}\begin{bmatrix} \sigma_1 & 0 & 0 \\ 0 & \sigma_2 & 0 \\ 0 & 0 & \sigma_3 \end{bmatrix}\begin{bmatrix} N_1 \\ N_2 \\ N_3 \end{bmatrix} = \boldsymbol{T} \wedge \boldsymbol{T}^\mathrm{T} \tag{8.1.9}$$

Converting the stress increment $\mathrm{d}\boldsymbol{\sigma}$ into the principal stress space, we get

$$\boldsymbol{T}^\mathrm{T}\mathrm{d}\boldsymbol{\sigma}\boldsymbol{T} = \begin{bmatrix} M_1 & A_1 & C_1 \\ A_1 & M_2 & B_1 \\ C_1 & B_1 & M_3 \end{bmatrix} \tag{8.1.10}$$

Coaxial part of Stress increments $\mathrm{d}\boldsymbol{\sigma}_{\mathrm{c}}$ and the rotational parts $\mathrm{d}\boldsymbol{\sigma}_{\mathrm{r}}$ will be

$$\mathrm{d}\boldsymbol{\sigma}_{\mathrm{c}} = \boldsymbol{T}\begin{bmatrix} M_1 & 0 & 0 \\ 0 & M_2 & 0 \\ 0 & 0 & M_3 \end{bmatrix}\boldsymbol{T}^{\mathrm{T}} \tag{8.1.11}$$

where $M_1 = \mathrm{d}\sigma_1, M_2 = \mathrm{d}\sigma_2, M_3 = \mathrm{d}\sigma_3$.

$$\begin{aligned} \mathrm{d}\boldsymbol{\sigma}_{\mathrm{r}} &= \mathrm{d}\boldsymbol{\sigma}_{\mathrm{r}1} + \mathrm{d}\boldsymbol{\sigma}_{\mathrm{r}2} + \mathrm{d}\boldsymbol{\sigma}_{\mathrm{r}3} \\ &= \boldsymbol{T}\begin{bmatrix} 0 & A_1 & 0 \\ A_1 & 0 & 0 \\ 0 & 0 & 0 \end{bmatrix}\boldsymbol{T}^{\mathrm{T}} + \boldsymbol{T}\begin{bmatrix} 0 & 0 & 0 \\ 0 & 0 & B_1 \\ 0 & B_1 & 0 \end{bmatrix}\boldsymbol{T}^{\mathrm{T}} + \boldsymbol{T}\begin{bmatrix} 0 & 0 & C_1 \\ 0 & 0 & 0 \\ C_1 & 0 & 0 \end{bmatrix}\boldsymbol{T}^{\mathrm{T}} \end{aligned} \tag{8.1.12}$$

$$\mathrm{d}\boldsymbol{\sigma}_{\mathrm{r}1} = \boldsymbol{T}\begin{bmatrix} 0 & A_1 & 0 \\ A_1 & 0 & 0 \\ 0 & 0 & 0 \end{bmatrix}\boldsymbol{T}^{\mathrm{T}} \quad A_1 = \mathrm{d}\theta_1(\sigma_1 - \sigma_2) = \mathrm{d}\sigma_{12}$$

$$\mathrm{d}\boldsymbol{\sigma}_{\mathrm{r}2} = \boldsymbol{T}\begin{bmatrix} 0 & 0 & 0 \\ 0 & 0 & B_1 \\ 0 & B_1 & 0 \end{bmatrix}\boldsymbol{T}^{\mathrm{T}} \quad B_1 = \mathrm{d}\theta_2(\sigma_2 - \sigma_3) = \mathrm{d}\sigma_{23}$$

$$\mathrm{d}\boldsymbol{\sigma}_{\mathrm{r}3} = \boldsymbol{T}\begin{bmatrix} 0 & 0 & C_1 \\ 0 & 0 & 0 \\ C_1 & 0 & 0 \end{bmatrix}\boldsymbol{T}^{\mathrm{T}} \quad C_1 = \mathrm{d}\theta_3(\sigma_1 - \sigma_3) = \mathrm{d}\sigma_{13}$$

where $\mathrm{d}\theta_1$, $\mathrm{d}\theta_2$, $\mathrm{d}\theta_3$ in the above expressions indicate rotational angle increment of principal axis of stress induced by the rotational increment of stress $\mathrm{d}\boldsymbol{\sigma}_{\mathrm{r}1}$, $\mathrm{d}\boldsymbol{\sigma}_{\mathrm{r}2}$, $\mathrm{d}\boldsymbol{\sigma}_{\mathrm{r}3}$ around the third, first, and second principal stress axis, respectively.

Then, the total stress increment can be expressed as

$$\begin{aligned} \mathrm{d}\boldsymbol{\sigma} &= \mathrm{d}\boldsymbol{\sigma}_{\mathrm{c}} + \mathrm{d}\boldsymbol{\sigma}_{\mathrm{r}} = \mathrm{d}\boldsymbol{\sigma}_{\mathrm{c}} + \mathrm{d}\boldsymbol{\sigma}_{\mathrm{r}1} + \mathrm{d}\boldsymbol{\sigma}_{\mathrm{r}2} + \mathrm{d}\boldsymbol{\sigma}_{\mathrm{r}3} \\ &= \boldsymbol{T}\begin{bmatrix} \mathrm{d}\sigma_1 & \mathrm{d}\theta_1(\sigma_1 - \sigma_2) & \mathrm{d}\theta_3(\sigma_1 - \sigma_3) \\ \mathrm{d}\theta_1(\sigma_1 - \sigma_2) & \mathrm{d}\sigma_2 & \mathrm{d}\theta_2(\sigma_2 - \sigma_3) \\ \mathrm{d}\theta_3(\sigma_1 - \sigma_3) & \mathrm{d}\theta_2(\sigma_2 - \sigma_3) & \mathrm{d}\sigma_3 \end{bmatrix}\boldsymbol{T}^{\mathrm{T}} \end{aligned} \tag{8.1.13}$$

8.2 Generalized Plastic Potential Theory Considering Rotation of Principal Stress Axis

Based on what we have learned in the previous sections, the stress increment can be decomposed into two parts which can cause plastic deformation, so plastic strain increment and the total strain increment can be written as

$$d\boldsymbol{\varepsilon} = d\boldsymbol{\varepsilon}^{e} + d\boldsymbol{\varepsilon}^{p} \tag{8.2.1}$$

$$d\boldsymbol{\varepsilon}^{p} = d\boldsymbol{\varepsilon}_{c}^{p} + d\boldsymbol{\varepsilon}_{r}^{p} = d\boldsymbol{\varepsilon}_{c}^{p} + d\boldsymbol{\varepsilon}_{r1}^{p} + d\boldsymbol{\varepsilon}_{r2}^{p} + d\boldsymbol{\varepsilon}_{r3}^{p} \tag{8.2.2}$$

where $d\boldsymbol{\varepsilon}_{c}^{p}$ is the plastic strain increment caused by coaxial increment of stress $d\sigma_{c}$; $d\boldsymbol{\varepsilon}_{r}^{p}$ is the plastic strain increment caused by rotational increment of stress $d\boldsymbol{\sigma}_{c}$; $d\boldsymbol{\varepsilon}_{r1}^{p}$, $d\boldsymbol{\varepsilon}_{r2}^{p}$, $d\boldsymbol{\varepsilon}_{r3}^{p}$ are plastic strain increments caused by rotational increment of stress $d\boldsymbol{\sigma}_{r1}$, $d\boldsymbol{\sigma}_{r2}$, $d\boldsymbol{\sigma}_{r3}$, respectively.

The coaxial increment of plastic strain can be calculated by the generalized plastic theory without considering the rotation of principal axis of stress.

$$d\boldsymbol{\varepsilon}_{c}^{p} = \sum_{k=1}^{3} d\lambda_{k} \frac{\partial Q_{k}}{\partial \boldsymbol{\sigma}} \tag{8.2.3}$$

where Q_{k}, $d\lambda_{k}$ are three linearly independent plastic potential and the corresponding plastic factor.

Based on what we have concluded from the soil tests, the rotational increment of stress $d\boldsymbol{\sigma}_{r}$ can cause plastic strain increment with six linearly independent components (three increments of normal strain, and three increments of shear strain) in the space of principal stress, so six linear independent functions of plastic potential are needed. Potential functions are selected randomly, but the selected plastic potential functions are linearly independent. For instance, six stress components can be expressed as six potential functions, and the corresponding plastic factor is the relevant increment of plastic strain. So generalized flow rule considering stress increment of rotation can be written as

$$d\boldsymbol{\varepsilon}_{ijr}^{p} = \sum_{k=1}^{6} d\lambda_{kr} \frac{\partial Q_{kr}}{\partial \sigma_{ij}} \tag{8.2.4}$$

where $d\lambda_{kr}$ stands for six plastic factors; Q_{kr} stands for six plastic potential functions, such as $Q_{1r} = \sigma_{1}$, $Q_{2r} = \sigma_{2}$, $Q_{3r} = \sigma_{3}$, $Q_{4r} = \sigma_{12}$, $Q_{5r} = \sigma_{13}$, $Q_{6r} = \sigma_{23}$.

The rotational increment of stress $d\boldsymbol{\sigma}_{r}$ can be expressed as three rotational components of the stress increment involving the rotation around only one principal axis of stress, respectively, namely, $d\boldsymbol{\sigma}_{r1}$, $d\boldsymbol{\sigma}_{r2}$, $d\boldsymbol{\sigma}_{r3}$, and $d\boldsymbol{\sigma}_{r}$ can be expressed as $d\boldsymbol{\sigma}_{r1} + d\boldsymbol{\sigma}_{r2} + d\boldsymbol{\sigma}_{r3}$. The rotational increment of stress $d\boldsymbol{\sigma}_{ri}$ could cause the plastic components of strain increment in four independent directions. So generalized flow rule considering rotational increment of stress can be written in the following form:

$$\begin{cases} \mathrm{d}\varepsilon_{ijr1}^{\mathrm{p}} = \mathrm{d}\lambda_{11\mathrm{r}1}\frac{\partial Q_{1\mathrm{r}}}{\partial \sigma_{ij}} + \mathrm{d}\lambda_{22\mathrm{r}1}\frac{\partial Q_{2\mathrm{r}}}{\partial \sigma_{ij}} + \mathrm{d}\lambda_{33\mathrm{r}1}\frac{\partial Q_{3\mathrm{r}}}{\partial \sigma_{ij}} + \mathrm{d}\lambda_{12\mathrm{r}1}\frac{\partial Q_{4\mathrm{r}}}{\partial \sigma_{ij}} \\ \mathrm{d}\varepsilon_{ijr2}^{\mathrm{p}} = \mathrm{d}\lambda_{11\mathrm{r}2}\frac{\partial Q_{1\mathrm{r}}}{\partial \sigma_{ij}} + \mathrm{d}\lambda_{22\mathrm{r}2}\frac{\partial Q_{2\mathrm{r}}}{\partial \sigma_{ij}} + \mathrm{d}\lambda_{33\mathrm{r}2}\frac{\partial Q_{3\mathrm{r}}}{\partial \sigma_{ij}} + \mathrm{d}\lambda_{23\mathrm{r}2}\frac{\partial Q_{6\mathrm{r}}}{\partial \sigma_{ij}} \\ \mathrm{d}\varepsilon_{ijr3}^{\mathrm{p}} = \mathrm{d}\lambda_{11\mathrm{r}3}\frac{\partial Q_{1\mathrm{r}}}{\partial \sigma_{ij}} + \mathrm{d}\lambda_{22\mathrm{r}3}\frac{\partial Q_{2\mathrm{r}}}{\partial \sigma_{ij}} + \mathrm{d}\lambda_{33\mathrm{r}3}\frac{\partial Q_{3\mathrm{r}}}{\partial \sigma_{ij}} + \mathrm{d}\lambda_{13\mathrm{r}3}\frac{\partial Q_{5\mathrm{r}}}{\partial \sigma_{ij}} \end{cases} \tag{8.2.5}$$

Equations (8.2.4) and (8.2.5) are named generalized plastic potential theory or flow rule for the rotational part of stress. Combining Eq. (8.2.4) with Eq. (8.2.5), we get the generalized plastic potential theory including rotation of principal axis of stress.

$$\mathrm{d}\varepsilon_{ij}^{\mathrm{p}} = \sum_{k=1}^{3} \mathrm{d}\lambda_{\mathrm{k}}\frac{\partial Q_{\mathrm{k}}}{\partial \sigma_{ij}} + \sum_{k=1}^{6} \mathrm{d}\lambda_{\mathrm{kr}}\frac{\partial Q_{\mathrm{kr}}}{\partial \sigma_{ij}} \tag{8.2.6}$$

8.3 Complete Stress Increment Expression of Elastoplastic Stress–Strain Relationship for Geomaterial

In order to compute the plastic deformation, we generally establish the relationship between plastic strain increment $\mathrm{d}\varepsilon_{\mathrm{v}}^{\mathrm{p}}$, $\mathrm{d}\varepsilon_{\mathrm{s}}^{\mathrm{p}}$ and stress increment $\mathrm{d}p$, $\mathrm{d}q$.

$$\begin{cases} \mathrm{d}\varepsilon_{\mathrm{v}}^{\mathrm{p}} = A\mathrm{d}p + B\mathrm{d}q \\ \mathrm{d}\varepsilon_{\mathrm{s}}^{\mathrm{p}} = C\mathrm{d}p + D\mathrm{d}q \end{cases} \tag{8.3.1}$$

However, stress increment expression is incomplete in Eq. (8.3.1), because there are not only $\mathrm{d}p$, $\mathrm{d}q$, but also Lode angle increment $\mathrm{d}\theta_{\sigma}$ and rotational angle increment of principal axis of stress $\mathrm{d}\theta$. Another way is that plastic strain increment and stress increment in Eq. (8.3.1) are generalized as increment for strain invariants $\mathrm{d}\varepsilon_{\mathrm{v}}^{\mathrm{p}'}$, $\mathrm{d}\varepsilon_{s}^{\mathrm{p}'}$ and increment of stress invariant $\mathrm{d}p'$, $\mathrm{d}q'$, then

$$\begin{cases} \mathrm{d}\varepsilon_{\mathrm{v}}^{\mathrm{p}'} = A\mathrm{d}p' + B\mathrm{d}q' \\ \mathrm{d}\varepsilon_{\mathrm{s}}^{\mathrm{p}'} = C\mathrm{d}p' + D\mathrm{d}q' \end{cases} \tag{8.3.2}$$

where increment of strain invariants and increment of stress invariants in the general stress space are

$$\begin{cases} \mathrm{d}\varepsilon_{\mathrm{v}}^{\mathrm{p}'} = \mathrm{d}\varepsilon_{11}^{\mathrm{p}} + \mathrm{d}\varepsilon_{22}^{\mathrm{p}} + \mathrm{d}\varepsilon_{33}^{\mathrm{p}} \\ \mathrm{d}\varepsilon_{\mathrm{s}}^{\mathrm{p}'} = \frac{\sqrt{2}}{3}[\left(\mathrm{d}\varepsilon_{11}^{\mathrm{p}} - \mathrm{d}\varepsilon_{22}^{\mathrm{p}}\right)^2 + \left(\mathrm{d}\varepsilon_{22}^{\mathrm{p}} - \mathrm{d}\varepsilon_{33}^{\mathrm{p}}\right)^2 + \left(\mathrm{d}\varepsilon_{11}^{\mathrm{p}} - \mathrm{d}\varepsilon_{33}^{\mathrm{p}}\right)^2 \\ \quad + \frac{3}{2}\left(\mathrm{d}\varepsilon_{12}^{\mathrm{p}2} + \mathrm{d}\varepsilon_{13}^{\mathrm{p}2} + \mathrm{d}\varepsilon_{23}^{\mathrm{p}2}\right)]^{\frac{1}{2}} \\ \quad = \frac{\sqrt{2}}{3}\sqrt{3\left(\mathrm{d}\varepsilon_{12}^{\mathrm{p}2} + \mathrm{d}\varepsilon_{13}^{\mathrm{p}2} + \mathrm{d}\varepsilon_{23}^{\mathrm{p}2}\right) - \mathrm{d}\varepsilon_{\mathrm{v}}^{\mathrm{p}2} + \frac{3}{2}(\mathrm{d}\varepsilon_{12}^{\mathrm{p}2} + \mathrm{d}\varepsilon_{13}^{\mathrm{p}2} + \mathrm{d}\varepsilon_{23}^{\mathrm{p}2})} \end{cases} \tag{8.3.3}$$

$\mathrm{d}q'$ in Eq. (8.3.2) contains not only $\mathrm{d}q$ but also $\mathrm{d}\theta_\sigma$ and $\mathrm{d}\theta$, which is the full expression of stress increment.

In the principal stress space, they will be

$$\begin{cases} \mathrm{d}p' = \frac{1}{3}(\mathrm{d}\sigma_1 + \mathrm{d}\sigma_2 + \mathrm{d}\sigma_3) \\ \mathrm{d}q' = \frac{1}{\sqrt{2}}[(\mathrm{d}\sigma_1 - \mathrm{d}\sigma_2)^2 + (\mathrm{d}\sigma_2 - \mathrm{d}\sigma_3)^2 + (\mathrm{d}\sigma_1 - \mathrm{d}\sigma_3)^2 + 6\left(\mathrm{d}\sigma_{12}^2 + \mathrm{d}\sigma_{13}^2 + \mathrm{d}\sigma_{23}^2\right)]^{\frac{1}{2}} \end{cases} \tag{8.3.4}$$

When the rotation of principal axis of stress is not considered, there is $\mathrm{d}\sigma_{12} = \mathrm{d}\sigma_{13} = \mathrm{d}\sigma_{23} = 0$. Furthermore, when the change of Lode angle of stress is ignored, there is $\mathrm{d}\sigma_2 = \mathrm{d}\sigma_3$, namely, conventional triaxial test conditions. Then there is

$$\begin{cases} \mathrm{d}p' = \frac{1}{3}(\mathrm{d}\sigma_1 + \mathrm{d}\sigma_2 + \mathrm{d}\sigma_3) = \frac{1}{3}(\mathrm{d}\sigma_1 + 2\mathrm{d}\sigma_3) = \mathrm{d}p \\ \mathrm{d}q' = \frac{1}{\sqrt{2}}\sqrt{(\mathrm{d}\sigma_1 - \mathrm{d}\sigma_2)^2 + (\mathrm{d}\sigma_2 - \mathrm{d}\sigma_3)^2 + (\mathrm{d}\sigma_1 - \mathrm{d}\sigma_3)^2 + 6(0^2 + 0^2 + 0^2)} \\ \quad = \frac{1}{\sqrt{2}}\sqrt{2(\mathrm{d}\sigma_1 - \mathrm{d}\sigma_3)^2} \\ \quad = \mathrm{d}\sigma_1 - \mathrm{d}\sigma_3 \\ \quad = \mathrm{d}q \end{cases} \tag{8.3.5}$$

Similarly, under conventional triaxial test condition,

$$\mathrm{d}\varepsilon_{\mathrm{v}}^{\mathrm{p}'} = \mathrm{d}\varepsilon_{\mathrm{v}}^{\mathrm{p}},\ \mathrm{d}\varepsilon_{\mathrm{s}}^{\mathrm{p}'} = \mathrm{d}\varepsilon_{\mathrm{s}}^{\mathrm{p}} \tag{8.3. 6}$$

So Eq. (8.3.2) is changed into (8.3.1) under conventional triaxial test condition.

When only the change of stress Lode angle $\mathrm{d}\theta_\sigma$ is considered, there are $\mathrm{d}\sigma_{12} = \mathrm{d}\sigma_{13} = \mathrm{d}\sigma_{23} = 0$, $\mathrm{d}p = 0$, $\mathrm{d}q = 0$. According to the relationships between three principal stress values σ_1, σ_2, σ_3 and p, q, θ_σ in the principal stress space,

$$\begin{bmatrix} \sigma_1 \\ \sigma_2 \\ \sigma_3 \end{bmatrix} = \frac{2}{3}q\begin{bmatrix} \sin(\theta_\sigma + \frac{2}{3}\pi) \\ \sin(\theta_\sigma) \\ \sin(\theta_\sigma - \frac{2}{3}\pi) \end{bmatrix} + \begin{bmatrix} p \\ p \\ p \end{bmatrix} \tag{8.3.7}$$

When pure stress Lode angle changes, the above expression can be differentiated into

$$\begin{bmatrix} d\sigma_1 \\ d\sigma_2 \\ d\sigma_3 \end{bmatrix} = \frac{2}{3}q \begin{bmatrix} \cos(\theta_\sigma + \frac{2}{3}\pi) \\ \cos\theta_\sigma \\ \cos(\theta_\sigma - \frac{2}{3}\pi) \end{bmatrix} d\theta_\sigma \tag{8.3.8}$$

When pure stress Lode angle changes, dp', dq' are calculated, respectively, as follows:

$$\begin{aligned} dp' &= \frac{1}{3}(d\sigma_1 + d\sigma_2 + d\sigma_3) \\ &= \frac{1}{3}\frac{2}{3}q d\theta_\sigma[\cos(\theta_\sigma + \frac{2}{3}\pi) + \cos\theta_\sigma + \cos(\theta_\sigma - \frac{2}{3}\pi)] \\ &= 0 \end{aligned} \tag{8.3.9}$$

$$\begin{aligned} dq' &= \frac{1}{\sqrt{2}}\sqrt{(d\sigma_1 - d\sigma_2)^2 + (d\sigma_2 - d\sigma_3)^2 + (d\sigma_1 - d\sigma_3)^2} \\ &= \frac{1}{\sqrt{2}}\frac{2}{3}q d|d\theta_\sigma|\{[\cos(\theta_\sigma + \frac{2}{3}\pi) - \cos\theta_\sigma]^2 + [\cos\theta_\sigma - \cos(\theta_\sigma - \frac{2}{3}\pi)]^2 \\ &+ [\cos(\theta_\sigma + \tfrac{2}{3}\pi) - \cos(\theta_\sigma - \tfrac{2}{3}\pi)]^2\}^{\frac{1}{2}} \end{aligned}$$

where

$$[\cos(\theta_\sigma + \frac{2}{3}\pi) - \cos\theta_\sigma]^2 = \frac{3}{4}\left(1 + 2\cos^2\theta_\sigma + \sqrt{3}\sin 2\theta_\sigma\right)$$

$$[\cos\theta_\sigma - \cos(\theta_\sigma - \frac{2}{3}\pi)]^2 = \frac{3}{4}\left(1 + 2\cos^2\theta_\sigma - \sqrt{3}\sin 2\theta_\sigma\right)$$

$$[\cos(\theta_\sigma + \frac{2}{3}\pi) - \cos(\theta_\sigma - \frac{2}{3}\pi)]^2 = 3\sin^2\theta_\sigma$$

Then there is

$$\begin{aligned} dq' &= \frac{\sqrt{2}}{3}q|d\theta_\sigma|\sqrt{\frac{3}{4}\left(1 + 2\cos^2\theta_\sigma + \sqrt{3}\sin 2\theta_\sigma + 1 + 2\cos^2\theta_\sigma - \sqrt{3}\sin 2\theta_\sigma + 4\sin^2\theta_\sigma\right)} \\ &= q|d\theta_\sigma| \end{aligned} \tag{8.3.10}$$

Obviously, the direction of dq' differs from dq in Eq. (8.3.10). The direction of dq is in the direction of q, and the direction of $q|d\theta_\sigma|$ is perpendicular to the direction of q.

When considering the stress increment $d\boldsymbol{\sigma}_{r1}$ involving the rotation only around the third principal stress axis, there is $d\boldsymbol{\sigma}_{r2} = d\boldsymbol{\sigma}_{r3} = d\boldsymbol{\sigma}_{c} = 0$. The corresponding increment of stress invariants is

$$\begin{cases} dp' = \frac{1}{3}(d\sigma_{r1} + d\sigma_{r2} + d\sigma_{r3}) = 0 \\ dq' = \frac{1}{\sqrt{2}}\sqrt{6d\theta_1^2(\sigma_1 - \sigma_2)^2} = \sqrt{3}|\sigma_1 - \sigma_2||d\theta_1| \end{cases} \tag{8.3.11}$$

Similarly, when only $d\boldsymbol{\sigma}_{r2}$ and $d\boldsymbol{\sigma}_{r3}$ are considered, there are

$$\begin{cases} dp' = 0 \\ dq' = \sqrt{3}|\sigma_2 - \sigma_3||d\theta_2| \end{cases} \tag{8.3.12}$$

and

$$\begin{cases} dp' = 0 \\ dq' = \sqrt{3}|\sigma_1 - \sigma_3||d\theta_3| \end{cases} \tag{8.3.13}$$

Substituting these expressions into Eq. (8.3.2), we get plastic deformation caused by change of stress Lode angle and rotation of principal stress axis.

The plastic deformation for change of stress Lode angle will be

$$\begin{cases} d\varepsilon_v^{p'} = Adp' + Bdq' = A.0 + Bdq' = Bq|d\theta_\sigma| \\ d\varepsilon_s^{p'} = Cdp' + Ddq' = C.0 + Ddq' = Dq|d\theta_\sigma| \end{cases} \tag{8.3.14}$$

That for the rotation around the third principal stress axis $d\boldsymbol{\sigma}_{r1}$ is

$$\begin{cases} d\varepsilon_v^{p'} = Adp' + Bdq' = A.0 + Bdq' = B\sqrt{3}|\sigma_1 - \sigma_2||d\theta_1| \\ d\varepsilon_s^{p'} = Cdp' + Ddq' = C.0 + Ddq' = D\sqrt{3}|\sigma_1 - \sigma_3||d\theta_1| \end{cases} \tag{8.3.15}$$

That for the rotation around the first principal stress axis $d\boldsymbol{\sigma}_{r2}$ is

$$\begin{cases} d\varepsilon_v^{p'} = B\sqrt{3}|\sigma_2 - \sigma_3||d\theta_2| \\ d\varepsilon_s^{p'} = D\sqrt{3}|\sigma_2 - \sigma_3||d\theta_2| \end{cases} \tag{8.3.16}$$

That for the rotation around the second principal stress axis $d\boldsymbol{\sigma}_{r3}$ is

$$\begin{cases} d\varepsilon_v^{p'} = B\sqrt{3}|\sigma_1 - \sigma_3||d\theta_3| \\ d\varepsilon_s^{p'} = D\sqrt{3}|\sigma_1 - \sigma_3||d\theta_3| \end{cases} \tag{8.3.17}$$

As you can see from the above formula for the plastic deformation Eqs. (8.3.14)–(8.3.17), the essence for the plastic deformation brought into by change of stress Lode angle and the rotation of principal stress axis is that the shear deformation and dilatancy caused by the generalized shear division of stress increment.

8.4 Plastic Deformation Caused by Coaxial Stress Increment

There are many elastoplastic models to calculate the volumetric component and generalized shear component of plastic strain in the *p*-*q* plane. That is to say, it can be easy to determine the plastic coefficient *A*, *B*, *C*, *D* in Eq. (8.3.1) according to the study subject by choosing an appropriate. The key is the determination of plastic coefficients to compute plastic deformation caused by the variation of stress Lode angle in the coaxial stress increment.

The abovementioned direction of dq' caused by the change of stress Lode angle is perpendicular to dq. Therefore, *B*, *D* cannot be directly adopted to compute the plastic deformation by change of stress Lode angle.

In plane π, the yield locus is made up of six 60° fan area. Each sector is the same for the yield locus. The following is to discuss the plastic coefficient for pure change of stress Lode angle in the area of −30° to 30° (see Fig. 8.2).

It starts from the same point of stress state *A* to two points *B* and *C* in the same yield surface along the stress path 1 (stress path of pure *q* change) or path 2 (stress path of pure change of stress Lode angle). According to the concept of yield surface, the plastic deformation should be the same for these two stress paths, that is

$$\mathrm{d}\varepsilon_{\mathrm{s}1}^{\mathrm{p}} = B\mathrm{d}q = \mathrm{d}\varepsilon_{\mathrm{s}2}^{\mathrm{p}} = n_1 B\mathrm{d}q' = n_1 Bq\mathrm{d}\theta_\sigma \tag{8.4.1}$$

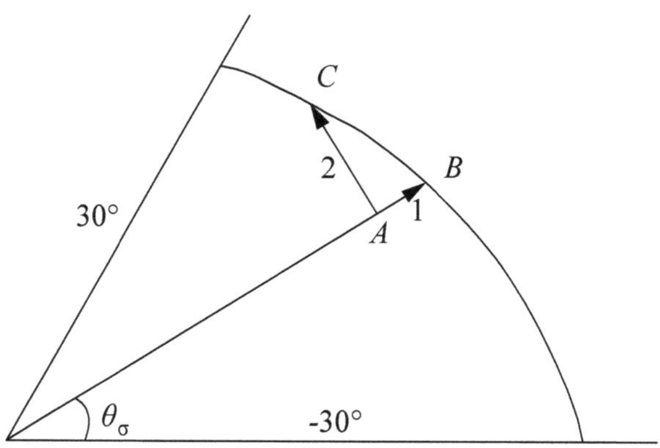

Fig. 8.2 Yield locus in the π plane (−30° to 30°)

From Fig. 8.2, dq is the reduction of q for the yield surface with the increase of stress Lode angle. The shape function is $g(\theta_\sigma)$ for the yield surface in plane π, namely

$$q = q_0 g(\theta_\sigma) \tag{8.4.2}$$

where q_0 is the value of q at $\theta_\sigma = 30°$.

Then

$$\mathrm{d}q = q_0 g'(\theta_\sigma)\mathrm{d}\theta_\sigma \tag{8.4.3}$$

Combine Eq. (8.4.1) with Eq. (8.4.3), and we get

$$\mathrm{d}\varepsilon_{s1}^{p} = B\mathrm{d}q = Bq_0 g'(\theta_\sigma)\mathrm{d}\theta_\sigma = \mathrm{d}\varepsilon_{s2}^{p} = n_1 Bq\mathrm{d}\theta_\sigma = n_1 Bq_0 g(\theta_\sigma)\mathrm{d}\theta_\sigma$$

where n_1 is the correction coefficient of the plastic coefficient for plastic deformation caused by variation of stress Lode angle.
and

$$Bq_0 g'(\theta_\sigma)\mathrm{d}\theta_\sigma = n_1 Bq_0 g(\theta_\sigma)\mathrm{d}\theta_\sigma$$

$$n_1 = \frac{g'(\theta_\sigma)}{g(\theta_\sigma)} \tag{8.4.4}$$

According to Eq. (8.3.2), we can obtain the plastic deformation for variation of stress Lode angle.

$$\begin{cases} \mathrm{d}\varepsilon_{v}^{p'} = n_1 Bq|\mathrm{d}\theta_\sigma| \\ \mathrm{d}\varepsilon_{s}^{p'} = n_1 Dq|\mathrm{d}\theta_\sigma| \end{cases} \tag{8.4.5}$$

Thus, plastic deformation could be calculated by Eq. (8.3.2) for the change of all factors (dp, dq, dθ_σ) of the coaxial stress increment.

$$\begin{cases} \mathrm{d}\varepsilon_{v}^{p'} = A\mathrm{d}p + B\mathrm{d}q + n_1 Bq|\mathrm{d}\theta_\sigma| \\ \mathrm{d}\varepsilon_{s}^{p'} = C\mathrm{d}p + B\mathrm{d}q + n_1 Dq|\mathrm{d}\theta_\sigma| \end{cases} \tag{8.4.6}$$

Under conventional triaxial cases,

$$\mathrm{d}\varepsilon_{v}^{p'} = \mathrm{d}\varepsilon_{v}^{p},\ \mathrm{d}\varepsilon_{s}^{p'} = \mathrm{d}\varepsilon_{s}^{p}$$

If we can determine the Lode angle of plastic strain increment at this time, the components of plastic strain increment in principal stress space will be obtained, and then it can be turned into the general stress space.

Take the Lode angle of plastic strain increment as follows:

$$\theta_{\varepsilon_c^p} = w\theta_\sigma + (1 - w)\theta_{d\sigma} \tag{8.4.7}$$

where $\theta_{\varepsilon_c^p}$, θ_σ, $\theta_{d\sigma}$ are Lode angles for coaxial plastic strain increment, stress, and stress increment, respectively; w is the proportional coefficient, taken as

$$w = \frac{q}{Mp}$$

The plastic strain increment in principal stress space will be

$$d\varepsilon_c^p = \begin{bmatrix} d\varepsilon_1^p \\ d\varepsilon_2^p \\ d\varepsilon_3^p \end{bmatrix} = \begin{bmatrix} \frac{1}{3}\sin(\theta_{\varepsilon_c^p} + \frac{2}{3}\pi) \\ \frac{1}{3}\sin(\theta_{\varepsilon_c^p}) \\ \frac{1}{3}\sin(\theta_{\varepsilon_c^p} - \frac{2}{3}\pi) \end{bmatrix} \begin{bmatrix} d\varepsilon_v^p \\ d\varepsilon_s^p \end{bmatrix} \tag{8.4.8}$$

Its tensor expression in principal stress space is

$$d\varepsilon_c^p = \begin{bmatrix} d\varepsilon_1^p & 0 & 0 \\ 0 & d\varepsilon_2^p & 0 \\ 0 & 0 & d\varepsilon_3^p \end{bmatrix} \tag{8.4.9}$$

The coaxial plastic deformation in general stress space can be obtained by the similar transformation as Eq. (8.1.1), which can be applied in the similar cases to come.

8.5 Plastic Deformation Caused by the Rotational Increment of Stress

Similarly, dq' is orthogonal with dq for rotation of principal stress axes; therefore, it also needs to calculate the correction coefficient n_2 of plastic coefficient for the rotation of principal stress axes.

To obtain the plastic strain caused by the rotational increment of stress, the plastic coefficients must be known for $d\lambda_{kr}$ or $d\lambda_{11r1}$, $d\lambda_{22r1}$, $d\lambda_{33r1}$, $d\lambda_{12r1}$, etc. In order to get these coefficients, $d\varepsilon_c^{p'}$ and $d\varepsilon_s^{p'}$ are needed in the first. Equations (8.3.15)–(8.3.17) give the corresponding formulas on condition that the plastic coefficients B and D are known. The following are the methods to seek B and D.

In general, the method to obtain plastic coefficient is to establish yield surface. But for the difficulty under the rotation of principal stress axis, no one has tried in this respect successfully. However, the plastic coefficients can be got by way of test fitting or empirical formula obtained from experimentation.

8.5.1 Plastic Deformation Caused by Stress Increment $d\sigma_{r1}$

Plastic deformation caused by stress increment $d\boldsymbol{\sigma}_{r1}$, which induces the rotation around the third principal stress axes, is provided below.

In 1987, Matsouka proposed that the relationship was hyperbolic between the main shear stress ratio ($\frac{\tau_{xy}}{\sigma_x}$ or $\frac{\tau_{xy}}{\sigma_y}$) and shear strain under the rotation of principal stress axis in plane strain (Figs. 8.3 and 8.4).

$$\gamma_{xy} = \frac{1}{G_0} \frac{\left(\frac{\tau_{xy}}{\sigma_x}\right)_f \left(\frac{\tau_{xy}}{\sigma_x}\right)}{\left(\frac{\tau_{xy}}{\sigma_x}\right)_f - \left(\frac{\tau_{xy}}{\sigma_x}\right)} \tag{8.5.1}$$

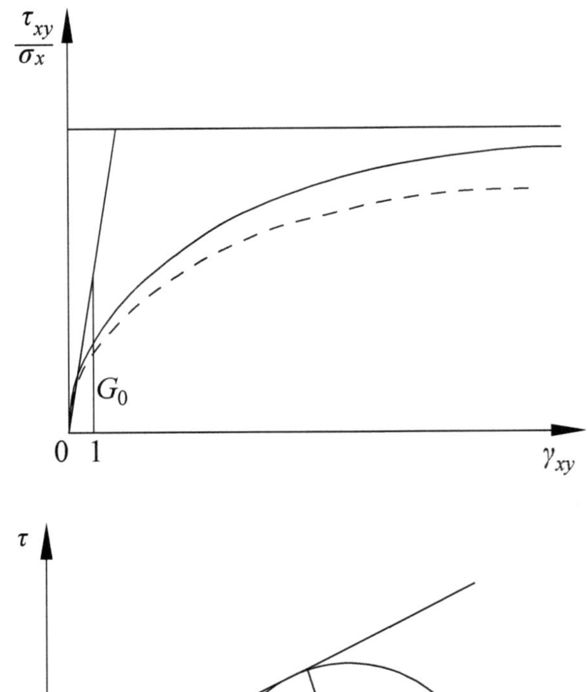

Fig. 8.3 The hyperbolic relationship between ratios of shearing stress and normal stress with shear strain

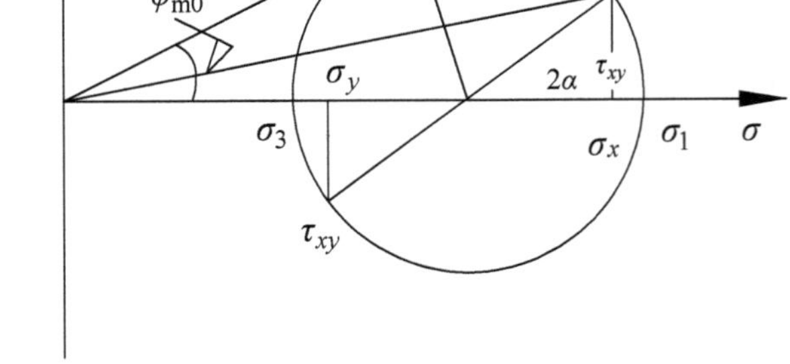

Fig. 8.4 The shear stress and normal stress in Mohr stress circle

or

$$\gamma_{xy} = \frac{1}{G_0} \frac{\left(\frac{\tau_{xy}}{\sigma_y}\right)_{\mathrm{f}} \left(\frac{\tau_{xy}}{\sigma_y}\right)}{\left(\frac{\tau_{xy}}{\sigma_y}\right)_{\mathrm{f}} - \left(\frac{\tau_{xy}}{\sigma_y}\right)} \tag{8.5.2}$$

where $\left(\frac{\tau_{xy}}{\sigma_x}\right)_{\mathrm{f}}$, $\left(\frac{\tau_{xy}}{\sigma_y}\right)_{\mathrm{f}}$ are shear stress ratios at failure; G_0 is tangent modulus at initial point of the curve of γ_{xy} and $\frac{\tau_{xy}}{\sigma_x}$ (or $\frac{\tau_{xy}}{\sigma_y}$) (Fig. 8.3).

From Fig. 8.4, we get

$$\sin\varphi_{\mathrm{m}0} = \frac{R}{\frac{\sigma_x+\sigma_y}{2}}$$

$$\sin 2\alpha = \frac{\tau_{xy}}{R}$$

$$\cos 2\alpha = \frac{\frac{\sigma_x-\sigma_y}{2}}{R}$$

and we obtain

$$\left.\begin{matrix}\left(\dfrac{\tau_{xy}}{\sigma_x}\right) \\ \left(\dfrac{\tau_{xy}}{\sigma_y}\right)\end{matrix}\right\} = \frac{\sin\varphi_{\mathrm{m}0} \cdot \sin 2\theta}{1 \pm \sin\varphi_{\mathrm{m}0} \cdot \cos 2\theta} \tag{8.5.3}$$

$$\left.\begin{matrix}\left(\dfrac{\tau_{xy}}{\sigma_x}\right)_{\mathrm{f}} \\ \left(\dfrac{\tau_{xy}}{\sigma_y}\right)_{\mathrm{f}}\end{matrix}\right\} = \frac{\sin\varphi_{\mathrm{m}0} \cdot \sin 2\theta}{1 \pm \sin\varphi_{\mathrm{m}0} \cdot \cos 2\theta} \tag{8.5.4}$$

where $\varphi_{\mathrm{m}0}$ is the critical internal friction angle; θ is the angle between the arbitrary plane and the principal stress plane.

$$\sin\varphi_{\mathrm{m}0} = \frac{\frac{\sigma_1-\sigma_3}{2}}{\frac{\sigma_1+\sigma_3}{2}} = \frac{\sigma_1 - \sigma_3}{\sigma_1 + \sigma_3} = \frac{\sqrt{3}q\cos\theta_\sigma}{3p - q\sin\theta_\sigma}$$

Substituting Eqs. (8.5.3), (8.5.4) into Eqs. (8.5.1) and (8.5.2), (8.5.1) can be written as

$$\gamma_{xy} = \frac{1}{G_0} \frac{\sin\varphi \sin\varphi_{\mathrm{m}0} \sin 2\theta}{\sin\varphi - \sin\varphi_{\mathrm{m}0}} \tag{8.5.5}$$

Differentiating both sides of Eq. (8.5.5) when only θ changes, we get shear strain increment caused by principal stress rotation.

$$\mathrm{d}\gamma_{xy} = \frac{2}{G_0}\frac{\sin\varphi\sin\varphi_{\mathrm{m0}}\sin 2\theta}{\sin\varphi - \sin\varphi_{\mathrm{m0}}}\mathrm{d}\theta \tag{8.5.6}$$

Rotation of principal stress axis would cause that stress and strain do not coaxes. Let the angel be δ between σ_1-axis and $\mathrm{d}\varepsilon_1$-axis, so the difference of angle between Mohr circle of strain increment and Mohr circle of stress increment is $90° - 2\delta$. 2θ in Eq. (8.5.6) can be replaced by $[2\theta - (90° - 2\delta)]$, then Eq. (8.5.6) can be expressed as follows:

$$\mathrm{d}\gamma_{xy} = \frac{2}{G_0}\frac{\sin\varphi\sin\varphi_{\mathrm{m0}}\sin 2(\theta + \delta)}{\sin\varphi - \sin\varphi_{\mathrm{m0}}}\mathrm{d}\theta \tag{8.5.7}$$

Based on the minimum energy principle, Rowe proposed the relationship of stress ratio-dilatancy according to specific energy theory.

$$R_1 = \frac{\sigma_1}{\sigma_3} = K \cdot \left(-\frac{\mathrm{d}\varepsilon_3}{\mathrm{d}\varepsilon_1}\right) \tag{8.5.8}$$

where $K = \tan^2(45° + \frac{\varphi_{\mathrm{f}}}{2})$; φ_{f} is equivalent internal friction angle.

From Eq. (8.5.7), we get

$$\mathrm{d}\gamma_{xy} = (\mathrm{d}\varepsilon_1 - \mathrm{d}\varepsilon_3)\sin 2(\theta + \delta) = \left[1 + \left(\frac{R_1}{K}\right)\right]\sin 2(\theta + \delta)\mathrm{d}\varepsilon_1 \tag{8.5.9}$$

$$\left.\begin{array}{l}\mathrm{d}\varepsilon_x\\ \mathrm{d}\varepsilon_y\end{array}\right\} = \frac{\mathrm{d}\varepsilon_1 + \mathrm{d}\varepsilon_3}{2} \pm \frac{\mathrm{d}\varepsilon_1 - \mathrm{d}\varepsilon_3}{2}\cos 2(\theta + \delta)$$

$$= \left[\frac{1 - \left(\frac{R_1}{K}\right)}{2} \pm \frac{1 + \left(\frac{R_1}{K}\right)}{2}\cos 2(\theta + \delta)\right]\mathrm{d}\varepsilon_1 \tag{8.5.10}$$

Principal strain increments are got from Eqs. (8.5.8) and (8.5.9).

$$\mathrm{d}\varepsilon_1 = \frac{K}{K + R_1}\frac{2}{G_0}\frac{\sin\varphi\sin\varphi_{\mathrm{m0}}\sin 2(\theta + \delta)}{\sin\varphi - \sin\varphi_{\mathrm{m0}}}\mathrm{d}\theta \tag{8.5.11}$$

$$\mathrm{d}\varepsilon_3 = \frac{K - R_1}{K + R_1}\frac{2}{G_0}\frac{\sin\varphi\sin\varphi_{\mathrm{m0}}}{\sin\varphi - \sin\varphi_{\mathrm{m0}}}\mathrm{d}\theta \tag{8.5.12}$$

Plastic volumetric strain $\mathrm{d}\varepsilon_{\mathrm{v}}^{\mathrm{p}}$ is obtained by summing up the above expressions (the elastic volumetric strain is ignored).

$$
\begin{aligned}
\mathrm{d}\varepsilon_{\mathrm{v}}^{\mathrm{p}} &= \frac{K - R_1}{K + R_1}\frac{2}{G_0}\frac{\sin\varphi\sin\varphi_{\mathrm{m0}}}{\sin\varphi - \sin\varphi_{\mathrm{m0}}}\mathrm{d}\theta \\
&= \frac{K - R_1}{K + R_1}\frac{2}{G_0}\frac{\sin\varphi\sin\varphi_{\mathrm{m0}}}{\sin\varphi - \sin\varphi_{\mathrm{m0}}}\frac{\mathrm{d}q'}{\sqrt{3}(\sigma_1 - \sigma_3)} \\
&= n_2 B\mathrm{d}q'
\end{aligned} \tag{8.5.13}
$$

Similarly, according to the model, the plastic strain increment can be obtained for pure change of q.

$$
\begin{aligned}
\mathrm{d}\varepsilon_{\mathrm{vs}}^{\mathrm{p}} &= \frac{K - R_1}{K + R_1}\frac{1}{G_0}\frac{\sin^2\varphi\sin\varphi_{\mathrm{m0}}}{(\sin\varphi - \sin\varphi_{\mathrm{m0}})^2}\frac{3p}{(3p - q\sin\theta_\sigma)q}\mathrm{d}q \\
&= B\mathrm{d}q
\end{aligned} \tag{8.5.14}
$$

Let $\mathrm{d}q' = \mathrm{d}q$, then

$$
n_2 = \frac{\mathrm{d}\varepsilon_{\mathrm{vr}}^{\mathrm{p}}}{\mathrm{d}\varepsilon_{\mathrm{vs}}^{\mathrm{p}}} = \frac{2(\sin\varphi - \sin\varphi_{\mathrm{m0}})(3p - q\sin\theta_\sigma)q}{3p\sin\varphi\sqrt{3}(\sigma_1 - \sigma_3)} \tag{8.5.15}
$$

The plastic deformation caused by rotational component of stress $\mathrm{d}\boldsymbol{\sigma}_{\mathrm{r1}}$ can be calculated.

$$
\begin{cases}
\mathrm{d}\varepsilon_{\mathrm{v}}^{\mathrm{p}'} = A\mathrm{d}p' + B\mathrm{d}q' = A.0 + B\mathrm{d}q' = n_2 B\sqrt{3}|\sigma_1 - \sigma_2||\mathrm{d}\theta_1| \\
\mathrm{d}\varepsilon_{\mathrm{s}}^{\mathrm{p}'} = C\mathrm{d}p' + D\mathrm{d}q' = C.0 + D\mathrm{d}q' = n_2 D\sqrt{3}|\sigma_1 - \sigma_3||\mathrm{d}\theta_1|
\end{cases} \tag{8.5.16}
$$

From Eq. (8.2.5), we know it is not enough to know only plastic volumetric deformation and shear deformation caused by $\mathrm{d}\boldsymbol{\sigma}_{\mathrm{r1}}$. In order to get $\mathrm{d}\lambda_{11\mathrm{r1}}$, $\mathrm{d}\lambda_{22\mathrm{r1}}$, $\mathrm{d}\lambda_{33\mathrm{r1}}$, $\mathrm{d}\lambda_{12\mathrm{r1}}$ and plastic deformation increment $\mathrm{d}\varepsilon_{11\mathrm{r1}}^{\mathrm{p}}$, $\mathrm{d}\varepsilon_{22\mathrm{r1}}^{\mathrm{p}}$, $\mathrm{d}\varepsilon_{33\mathrm{r1}}^{\mathrm{p}}$, $\mathrm{d}\varepsilon_{12\mathrm{r1}}^{\mathrm{p}}$ in the direction of ε_1, ε_2, ε_3 caused by $\mathrm{d}\boldsymbol{\sigma}_{\mathrm{r1}}$, we need to introduce other assumptions.

From Eqs. (8.3.15) to (8.3.16), we know the effect of rotation of principal stress axis can be attributed to shear deformation and dilatancy caused by generalized shear component $\mathrm{d}q'$ in stress increment, which is the indication of soil dilatancy. Based on this, we can assume that the plastic flow in the direction of principal stress axis when rotation of principal stress axis is similar to the relationship of stress ratio-dilatancy proposed by Rowe.

$$
\begin{cases}
\dfrac{\mathrm{d}\varepsilon_{11\mathrm{r1}}^{\mathrm{p}}}{\mathrm{d}\varepsilon_{22\mathrm{r1}}^{\mathrm{p}}} = R_1\dfrac{\sigma_1}{\sigma_2} \\
\dfrac{\mathrm{d}\varepsilon_{33\mathrm{r1}}^{\mathrm{p}}}{\mathrm{d}\varepsilon_{11\mathrm{r1}}^{\mathrm{p}}} = \dfrac{0.3\sigma_3}{R_1\sigma_2}
\end{cases} \tag{8.5.17}
$$

where $R_1 = \tan^2\left(45^\circ + \frac{\varphi}{2}\right)$

Hollow torsional shear experiments show that plastic strain increment perpendicular to the rotation plane is smaller than plastic flow in the two principal vectors of rotating plane. By fitting the experimental results, the proportion of factor 0.3 is reasonable in Eq. (8.5.17).

And

$$d\varepsilon_{\text{vr1}}^{\text{p}} = d\varepsilon_{11\text{r1}}^{\text{p}} + d\varepsilon_{22\text{r1}}^{\text{p}} + d\varepsilon_{33\text{r1}}^{\text{p}} \tag{8.5.18}$$

In the above expression, $d\varepsilon_{11\text{r1}}^{\text{p}}$, $d\varepsilon_{22\text{r1}}^{\text{p}}$ and $d\varepsilon_{33\text{r1}}^{\text{p}}$ are the plastic strain increments in the principal vectors caused by $d\boldsymbol{\sigma}_{\text{r1}}$.

From the above two expressions, we get

$$\begin{cases} d\varepsilon_{11\text{r1}}^{\text{p}} = E_{\text{r1}}^{1}|d\theta_1|,\ d\varepsilon_{22\text{r1}}^{\text{p}} = E_{\text{r2}}^{2}|d\theta_1|,\ d\varepsilon_{33\text{r1}}^{\text{p}} = E_{\text{r3}}^{3}|d\theta_1| \\ E_{\text{r1}}^{1} = R_1\sigma_1 K_{\text{f1}},\ E_{\text{r1}}^{2} = \sigma_2 K_{\text{f1}},\ E_{\text{r1}}^{3} = 0.3\sigma_3 K_{\text{f1}} \\ K_{\text{f1}} = \frac{n_2 B\sqrt{3}|\sigma_1-\sigma_2|}{R_1\sigma_1+\sigma_2+0.3\sigma_3} \end{cases} \tag{8.5.19}$$

From Eqs. (8.5.16) and (8.5.19), we get

$$\begin{aligned} d\varepsilon_{12\text{r1}}^{\text{p}} &= \sqrt{(9D^2 + 2B^2)n_2^2(\sigma_1 - \sigma_2)^2 - 2\left(E_{\text{r1}}^{12} + E_{\text{r1}}^{22} + E_{\text{r1}}^{32}\right)}d\theta_1 \\ &= E_{\text{r1}}^{4} d\theta_1 \end{aligned} \tag{8.5.20}$$

The plastic strain increment caused by $d\boldsymbol{\sigma}_{\text{r1}}$ in principal stress space can be expressed as

$$d\boldsymbol{\varepsilon}_{\text{r1}}^{\text{p}} = \begin{bmatrix} d\varepsilon_{11\text{r1}}^{\text{p}} & d\varepsilon_{12\text{r1}}^{\text{p}} & 0 \\ d\varepsilon_{12\text{r1}}^{\text{p}} & d\varepsilon_{22\text{r1}}^{\text{p}} & 0 \\ 0 & 0 & d\varepsilon_{33\text{r1}}^{\text{p}} \end{bmatrix} \tag{8.5.21}$$

8.5.2 *Plastic Deformation Caused by Rotational Stress Increment* $d\sigma_{\text{r2}}$ *and* $d\sigma_{\text{r3}}$

Similarly, plastic deformation can be obtained for $d\boldsymbol{\sigma}_{\text{r2}}$ and $d\boldsymbol{\sigma}_{\text{r3}}$, inducing rotation around the first and second principal stress axes.

$$\begin{cases} d\varepsilon_{11r2}^{p} = E_{r2}^{1}d\theta_2,\ d\varepsilon_{22r2}^{p} = E_{r2}^{2}d\theta_2,\ d\varepsilon_{33r2}^{p} = E_{r2}^{3}d\theta_2 \\ d\varepsilon_{23r2}^{p} = E_{r2}^{4}d\theta_2 \\ E_{r2}^{1} = 0.3\sigma_1 K_{f2},\ E_{r2}^{2} = R_1\sigma_2 K_{f2},\ E_{r2}^{3} = \sigma_3 K_{f2} \\ E_{r2}^{4} = \sqrt{\left(9D^2 + 2B^2\right)n_2^2(\sigma_2 - \sigma_3)^3 - 2\left(E_{r2}^{12} + E_{r2}^{22} + E_{r2}^{32}\right)} \\ E_{f2} = \frac{n_2 B\sqrt{3}|\sigma_2 - \sigma_3}{0.3\sigma_1 + R\sigma_2 + \sigma_3} \end{cases} \tag{8.5.22}$$

$$\begin{cases} d\varepsilon_{11r3}^{p} = E_{r3}^{1}d\theta_3,\ d\varepsilon_{22r3}^{p} = E_{r3}^{2}d\theta_3,\ d\varepsilon_{33r3}^{p} = E_{r3}^{3}d\theta_3 \\ d\varepsilon_{13r3}^{p} = E_{r3}^{4}d\theta_3 \\ E_{r3}^{1} = R_1\sigma_1 K_{f3},\ E_{r3}^{2} = 0.3\sigma_2 K_{f3},\ E_{r3}^{3} = \sigma_3 K_{f3} \\ E_{r4}^{3} = \sqrt{\left(9D^2 + 2B^2\right)n_2^2(\sigma_1 - \sigma_3)^2 - 2\left(E_{r3}^{12} + E_{r3}^{22} + E_{r3}^{32}\right)} \\ K_{f3} = n_2 B\sqrt{3}|\sigma_1 - \sigma_3| \Big/ (R_1\sigma_1 + 0.3\sigma_2 + \sigma_3) \end{cases} \tag{8.5.23}$$

From Eqs. (8.5.19) to (8.5.23), the plastic coefficients corresponding with the rotation of principal stress axis in Eqs. (8.2.5) and (8.2.6) are got.

$$\begin{cases} d\lambda_{11r1} = d\varepsilon_{11r1}^{p},\ d\lambda_{22r1} = d\varepsilon_{22r1}^{p},\ d\lambda_{33r1} = d\varepsilon_{33r1}^{p},\ d\lambda_{12r1} = d\varepsilon_{12r1}^{p} \\ d\lambda_{11r2} = d\varepsilon_{11r2}^{p},\ d\lambda_{22r2} = d\varepsilon_{22r2}^{p},\ d\lambda_{33r2} = d\varepsilon_{33r2}^{p},\ d\lambda_{23r2} d\varepsilon_{23r2}^{p} \\ d\lambda_{11r3} = d\varepsilon_{11r3}^{p},\ d\lambda_{22r3} = d\varepsilon_{22r3}^{p},\ d\lambda_{33r3} = d\varepsilon_{33r3}^{p},\ d\lambda_{13r3} = d\varepsilon_{13r3}^{p} \end{cases} \tag{8.5.24}$$

$$\begin{cases} d\lambda_{1r} = d\varepsilon_{11r1}^{p} + d\varepsilon_{11r2}^{p} + d\varepsilon_{11r3}^{p} \\ d\lambda_{2r} = d\varepsilon_{22r1}^{p} + d\varepsilon_{22r2}^{p} + d\varepsilon_{22r3}^{p} \\ d\lambda_{3r} = d\varepsilon_{33r1}^{p} + d\varepsilon_{33r2}^{p} + d\varepsilon_{33r3}^{p} \\ d\lambda_{4r} = d\varepsilon_{12r1}^{p} \\ d\lambda_{5r} = d\varepsilon_{13r3}^{p} \\ d\lambda_{6r} = d\varepsilon_{23r2}^{p} \end{cases} \tag{8.5.25}$$

In principal stress space, the plastic deformation caused by $d\boldsymbol{\sigma}_{r2}$ and $d\boldsymbol{\sigma}_{r3}$ can be expressed as follows:

$$d\boldsymbol{\varepsilon}_{r2}^{p} = \begin{bmatrix} d\varepsilon_{11r2}^{p} & 0 & 0 \\ 0 & d\varepsilon_{22r2}^{p} & d\varepsilon_{23r2}^{p} \\ 0 & d\varepsilon_{23r2}^{p} & d\varepsilon_{33r2}^{p} \end{bmatrix} \tag{8.5.26}$$

$$\mathrm{d}\boldsymbol{\varepsilon}^{\mathrm{p}}_{\mathrm{r3}} = \begin{bmatrix} \mathrm{d}\varepsilon^{\mathrm{p}}_{11\mathrm{r3}} & 0 & \mathrm{d}\varepsilon^{\mathrm{p}}_{13\mathrm{r3}} \\ 0 & \mathrm{d}\varepsilon^{\mathrm{p}}_{22\mathrm{r3}} & 0 \\ \mathrm{d}\varepsilon^{\mathrm{p}}_{13\mathrm{r3}} & 0 & \mathrm{d}\varepsilon^{\mathrm{p}}_{33\mathrm{r3}} \end{bmatrix} \tag{8.5.27}$$

8.6 Elastoplastic Stress–Strain Relations Considering Rotation of Principal Stress Axis

The stress–strain relation is built for considering rotation of principal stress axis in the following chapter. In the first, the elastic–plastic compliance matrix is presented, and then the elastic–plastic stiffness matrix is obtained.

8.6.1 Elastic Compliance Matrix

The elastic compliance matrix is obtained based on Hook's law.

$$\begin{bmatrix} \mathrm{d}\varepsilon^{\mathrm{e}}_{1} \\ \mathrm{d}\varepsilon^{\mathrm{e}}_{2} \\ \mathrm{d}\varepsilon^{\mathrm{e}}_{3} \\ \mathrm{d}\varepsilon^{\mathrm{e}}_{12} \\ \mathrm{d}\varepsilon^{\mathrm{e}}_{23} \\ \mathrm{d}\varepsilon^{\mathrm{e}}_{13} \end{bmatrix} = \frac{1}{E} \begin{bmatrix} 1 & -\mu & -\mu & 0 & 0 & 0 \\ -\mu & 1 & -\mu & 0 & 0 & 0 \\ -\mu & -\mu & 1 & 0 & 0 & 0 \\ 0 & 0 & 0 & A & 0 & 0 \\ 0 & 0 & 0 & 0 & A & 0 \\ 0 & 0 & 0 & 0 & 0 & A \end{bmatrix} \begin{bmatrix} \mathrm{d}\sigma_1 \\ \mathrm{d}\sigma_2 \\ \mathrm{d}\sigma_3 \\ \mathrm{d}\sigma_{12} \\ \mathrm{d}\sigma_{23} \\ \mathrm{d}\sigma_{13} \end{bmatrix}$$
$$= [C_{\mathrm{e}}]\begin{bmatrix} \mathrm{d}\sigma_1 & \mathrm{d}\sigma_2 & \mathrm{d}\sigma_3 & \mathrm{d}\sigma_{12} & \mathrm{d}\sigma_{23} & \mathrm{d}\sigma_{13} \end{bmatrix}^{\mathrm{T}} \tag{8.6.1}$$

where $A = 2(1 + \mu)$.

8.6.2 Coaxial Plastic Compliance Matrix

Combine Eq. (8.4.6) with Eq. (8.4.8), and the coaxial plastic compliance matrix can be got.

$$
\begin{aligned}
\mathrm{d}\boldsymbol{\varepsilon}_{\mathrm{c}}^{\mathrm{p}} = \begin{bmatrix} \mathrm{d}\varepsilon_1^{\mathrm{p}} \\ \mathrm{d}\varepsilon_2^{\mathrm{p}} \\ \mathrm{d}\varepsilon_3^{\mathrm{p}} \end{bmatrix} &= \begin{bmatrix} \frac{1}{3}\sin(\theta_{\varepsilon_{\mathrm{c}}^{\mathrm{p}}} + \frac{2}{3}\pi) \\ \frac{1}{3}\sin(\theta_{\varepsilon_{\mathrm{c}}^{\mathrm{p}}}) \\ \frac{1}{3}\sin(\theta_{\varepsilon_{\mathrm{c}}^{\mathrm{p}}} - \frac{2}{3}\pi) \end{bmatrix} \begin{bmatrix} A & B & n_1 B \\ C & D & n_2 D \end{bmatrix} \begin{bmatrix} \mathrm{d}p \\ \mathrm{d}q \\ \mathrm{d}\theta_\sigma \end{bmatrix} \\
&= \begin{bmatrix} \frac{1}{3}\sin(\theta_{\varepsilon_{\mathrm{c}}^{\mathrm{p}}} + \frac{2}{3}\pi) \\ \frac{1}{3}\sin(\theta_{\varepsilon_{\mathrm{c}}^{\mathrm{p}}}) \\ \frac{1}{3}\sin(\theta_{\varepsilon_{\mathrm{c}}^{\mathrm{p}}} - \frac{2}{3}\pi) \end{bmatrix} \begin{bmatrix} A & B & n_1 B \\ C & D & n_2 D \end{bmatrix} \begin{bmatrix} \frac{1}{3} & \frac{1}{3} & \frac{1}{3} \\ \frac{3(\sigma_1 - p)}{2q} & \frac{3(\sigma_2 - p)}{2q} & \frac{3(\sigma_3 - p)}{2q} \\ K_{\mathrm{e}}(\sigma_2 - \sigma_3) & K_{\mathrm{e}}(\sigma_1 - \sigma_3) & K_{\mathrm{e}}(\sigma_1 - \sigma_2) \end{bmatrix} \begin{bmatrix} \mathrm{d}\sigma_1 \\ \mathrm{d}\sigma_2 \\ \mathrm{d}\sigma_3 \end{bmatrix} \\
&= \begin{bmatrix} C_{11} & C_{12} & C_{13} \\ C_{21} & C_{22} & C_{23} \\ C_{31} & C_{32} & C_{33} \end{bmatrix} \begin{bmatrix} \mathrm{d}\sigma_1 \\ \mathrm{d}\sigma_2 \\ \mathrm{d}\sigma_3 \end{bmatrix} \\
&= [C_{\mathrm{cp}}] \begin{bmatrix} \mathrm{d}\sigma_1 \\ \mathrm{d}\sigma_2 \\ \mathrm{d}\sigma_3 \end{bmatrix}
\end{aligned} \tag{8.6.2}
$$

where $K_{\mathrm{e}} = \frac{2}{\sqrt{3}(\sigma_1 - \sigma_3)^2 (1 + \tan^2 \theta_\sigma)}$.

The complete form of coaxial plastic compliance matrix in principal stress space is

$$
[C_{\mathrm{cp}}] = \begin{bmatrix} C_{11} & C_{12} & C_{13} & 0 & 0 & 0 \\ C_{21} & C_{22} & C_{23} & 0 & 0 & 0 \\ C_{31} & C_{32} & C_{33} & 0 & 0 & 0 \\ 0 & 0 & 0 & 0 & 0 & 0 \\ 0 & 0 & 0 & 0 & 0 & 0 \\ 0 & 0 & 0 & 0 & 0 & 0 \end{bmatrix} \tag{8.6.3}
$$

8.6.3 Rotating Plastic Compliance Matrix

From Sect. 8.5, we know that rotating plastic stress–strain relations is

$$\begin{bmatrix} d\varepsilon_{11}^{p} \\ d\varepsilon_{22}^{p} \\ d\varepsilon_{33}^{p} \\ d\varepsilon_{12}^{p} \\ d\varepsilon_{23}^{p} \\ d\varepsilon_{13}^{p} \end{bmatrix} = \begin{bmatrix} 0\,0\,0 & \frac{E_{r1}^{1}}{|\sigma_1-\sigma_2|} & \frac{E_{r2}^{1}}{|\sigma_2-\sigma_3|} & \frac{E_{r3}^{1}}{|\sigma_1-\sigma_3|} \\ 0\,0\,0 & \frac{E_{r1}^{2}}{|\sigma_1-\sigma_2|} & \frac{E_{r2}^{2}}{|\sigma_2-\sigma_3|} & \frac{E_{r3}^{2}}{|\sigma_1-\sigma_3|} \\ 0\,0\,0 & \frac{E_{r1}^{3}}{|\sigma_1-\sigma_2|} & \frac{E_{r2}^{3}}{|\sigma_2-\sigma_3|} & \frac{E_{r3}^{3}}{|\sigma_1-\sigma_3|} \\ 0\,0\,0 & \frac{E_{r1}^{4}}{|\sigma_1-\sigma_2|} & 0 & 0 \\ 0\,0\,0 & 0 & \frac{E_{r2}^{4}}{|\sigma_2-\sigma_3|} & 0 \\ 0\,0\,0 & 0 & 0 & \frac{E_{r3}^{4}}{|\sigma_1-\sigma_3|} \end{bmatrix} \begin{bmatrix} d\sigma_1 \\ d\sigma_2 \\ d\sigma_3 \\ d\sigma_{12} \\ d\sigma_{23} \\ d\sigma_{13} \end{bmatrix}$$

$$= \left[C_{rp}\right]\left[d\sigma_1\ d\sigma_2\ d\sigma_3\ d\sigma_{12}\ d\sigma_{23}\ d\sigma_{13}\right]^{T} \tag{8.6.4}$$

where $[C_{rp}]$ is the rotating plastic compliance matrix.

If we add up the three compliance matrixes in Eqs. (8.6.1), (8.6.3) and (8.6.4), the total elastoplastic compliance matrix in the principal stress space is obtained.

$$\left[C_{ep}^{'}\right] = [C_e] + \left[C_{cp}\right] + \left[C_{rp}\right] \tag{8.6.5}$$

Let the three principal vectors be

$$T_1 = \left[L_1\ L_2\ L_3\right]^{T},\ T_2 = \left[M_1\ M_2\ M_3\right]^{T},\ T_3 = \left[N_1\ N_2\ N_3\right]^{T}$$

Then elastoplastic compliance matrix containing principal stress rotation in general stress space is

$$\left[C_{ep}\right] = [T_A]\left[C_{ep}^{'}\right][T_A]^{T} \tag{8.6.6}$$

where

$$[T_A] = \begin{bmatrix} L_1^2 & M_1^2 & N_1^2 & 2L_1M_1 & 2M_1N_1 & 2L_1N_1 \\ L_2^2 & M_2^2 & N_2^2 & 2L_2M_2 & 2M_2N_2 & 2L_2N_2 \\ L_3^2 & M_3^2 & N_3^2 & 2L_3M_3 & 2M_3N_3 & 2L_3N_3 \\ L_1L_2 & M_1M_3 & N_1N_3 & L_1M_2+L_2M_1 & M_1N_2+M_2N_1 & L_1N_2+L_2N_1 \\ L_2L_3 & M_2M_3 & N_2N_3 & L_2M_3+L_3M_2 & M_2N_3+M_3N_2 & L_2N_3+L_3N_2 \\ L_1L_3 & M_1M_3 & N_1N_3 & L_1M_3+L_3M_1 & M_1N_3+M_3N_1 & L_1N_3+L_3N_1 \end{bmatrix}$$

If we inverse $[C_{ep}]$, elastoplastic stiffness matrix $[D_{ep}]$ will be got containing rotation of principal stress axis in general stress space.

8.7 Example

The example is a plane strain problem with displacement constrained on both sides and the bottom, applying uniformly distributing load q_0 and local load q_1 on the top (Fig. 8.5). The example gives the distribution of stress and deformation when applying uniform load (including gravity) or local load.

The vertical displacement is illustrated in Figs. 8.6 and 8.7 for considering the rotation of principal stress axis or not when applying uniform load. Obviously, they are the same.

Figure 8.8 shows vertical displacements not considering the rotation of principal stress axis under local and uniform load. Figure 8.9 shows vertical displacements considering the rotation of principal stress axis under local and uniform load.

These results lead to the following conclusions:

1. Soil deformation is equal settlement under uniform load (including gravity). The effect of the rotation of principal stress axis is small, so the results are essentially the same when considering the rotation of principal stress axis or not (Figs. 8.6 and 8.7). Then the effect of the rotation of principal stress axis can be ignored.
2. Under the local and uniform load, principal stress axis rotates apparently, which has greater effects on vertical displacement and principal vector of σ_1. The maximum rotation angle is 31.9°. The effects may reach 20% to the maximum vertical

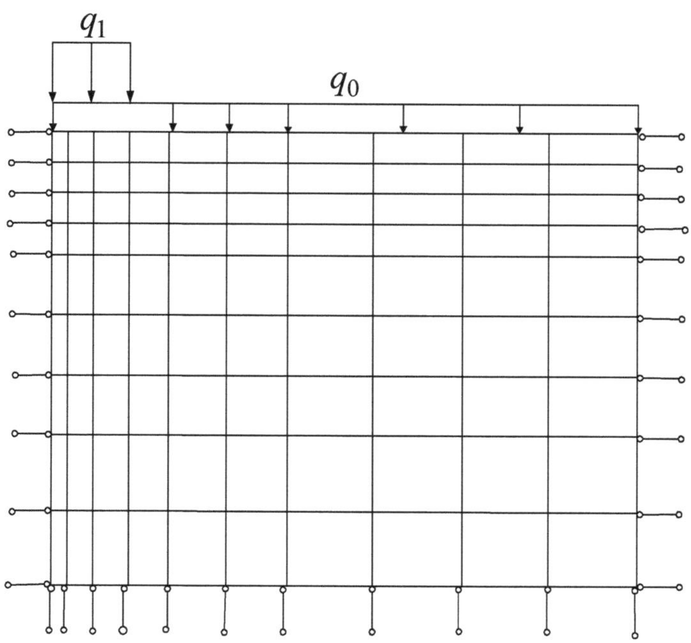

Fig. 8.5 Schematic diagram of element generation and boundary conditions

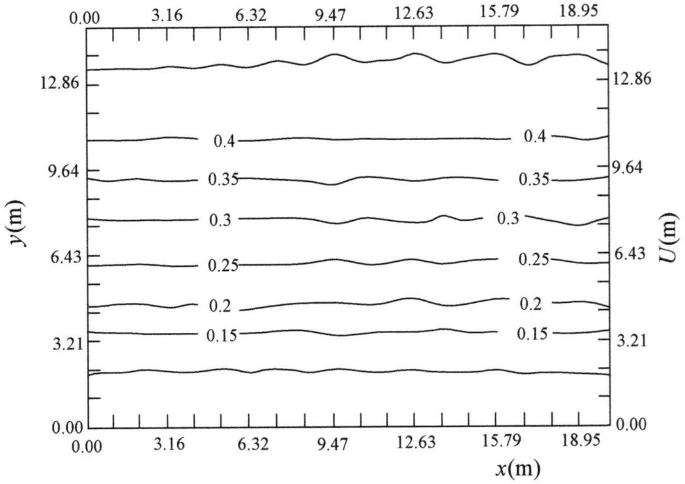

Fig. 8.6 Vertical displacement contours considering the rotation of principal stress axis when applying uniform load

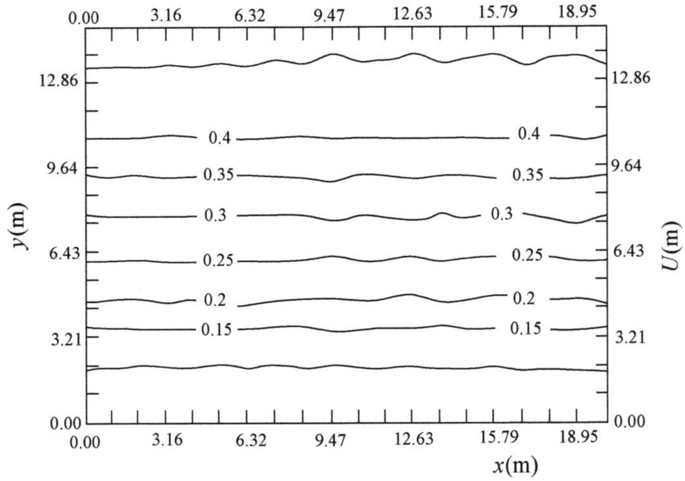

Fig. 8.7 Vertical displacement contours not considering the rotation of principal stress axis when applying uniform load

displacement at the relevant zone. So the effects of the rotation of principal stress axis should be considered under irregularly distributed load.

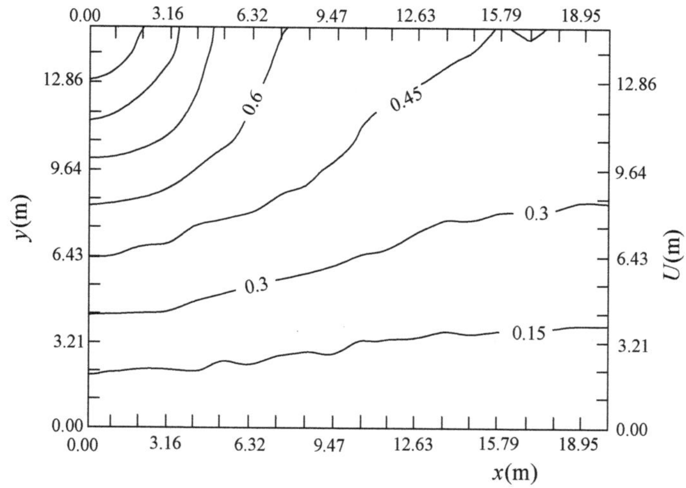

Fig. 8.8 Vertical displacement contours not considering the rotation of principal stress axis when applying local load and uniform load

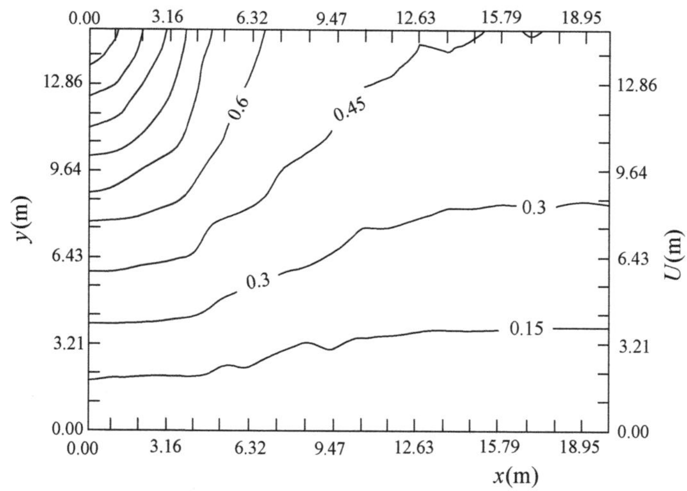

Fig. 8.9 Vertical displacement contours considering the rotation of principal stress axis when applying local load and uniform load

Questions

1. What are the characteristics of the stress increment of principal stress rotation?
2. What is the mechanism of the influence of the change of stress Lode angle and rotation of principal stress axes?
3. What is the condition under which the effect needs to be considered for the rotation of principal stress axes?

References

1. Liu YX (1997) The general stress strain relation of soils involving the rotation of principal axes of stress. Doctoral dissertation, Chongqing, Logistic Engineering University
2. Liu YX, Zheng YR (2000) Research development of soils constitutive relation involving the rotation of principal axes of stress. Adv Mech 30(4):597–604
3. Matsuoka H, Sakakihara K (1987) A constitutive model for sands and clays evaluating principal stress rotation. Soils Found 27(4):73–88
4. Matsuoka H, Suzuki Y (1990) A constitutive model for soils evaluating principal stress rotation and its application to some deformation problems. Soils Found 30(1):142–154
5. Nakai T, Fujii J, Taki H (1991) Kinematic of an isotropic hardening model for sand. In: Proceedings of 3rd international conference on constitutive laws for engineering materials, pp 36–45
6. Nakai T, Hoshikawa T (1991) Kinematic hardening models for clay in three-dimensional stresses. Comput Methods Adv Geomech 655–660
7. Liu YX, Zheng YR, Chen ZH (1998) The general stress strain relation of soils involving the rotation of principal stress axes. Appl Math Mech 19(5):407–413
8. Liu YX, Zheng YR (1998) A new method to analyze the influence of principal axe rotation of stress on the stress-strain relation of soils. Chin J Geotech Eng 20(2):45–47
9. Liu YX, Zheng YR (1999) Study on the effect of lode angle variation of stress. J Hydraul Eng 8:6–10
10. Liu YX, Zheng YR (2000) Generalized plastic potential theory involving the rotation of principal axes of stress. Chin Q Mech 21(1):129–133

Chapter 9
The Dynamic Constitutive Model of Geomaterial

Geomaterial dynamic constitutive model (or dynamic stress–strain relationship) is the foundation for understanding the dynamic behaviors of geomaterial or geomaterial structure, which is under the dynamic loads (such as earthquake loads, wave loads, traffic loads, wind loads, etc.), and also the prerequisite of dynamic and stability analysis by numerical simulation methods (such as finite element method, boundary element method). So far, a variety of dynamic constitutive models have been developed for geomaterial. First, the dynamic stress–strain characteristics of geomaterial will be briefly introduced in this chapter. Second, some important models are introduced in detailed.

9.1 Basic Characteristics of Dynamic Stress–Strain Relationship of Geomaterial

The geomaterial deformation under dynamic loads can be divided into two parts: elastic deformation and plastic deformation. When the dynamic load is small, there exists mainly elastic deformation, while when the dynamic load increases, the plastic deformation emerges and develops gradually. Therefore, in small strain amplitude cases, geomaterial will show the approximate elastic characteristics, such as the vibration of foundation under the working machine, etc. This small strain characteristic controls the spread speed of wavelet in geomaterial, which is the main factor to determine the dynamic response of ground base and foundation. However, when the dynamic strain amplitude increases, such as earthquake, blasting and construction, dynamic loads, it will cause the change of geomaterial structure, the residual deformation and strength loss of soil. The dynamic characteristics will be clearly different from small strain amplitude. At this time, the law of strength and deformation, for saturated sand (including the light saturated clay, etc.) needs to be studied, and the phenomenon of the sudden strength loses due to structural damage and rapid growth

Y. Liu and Y. Zheng, *Plastic Mechanics of Geomaterial*, Springer Geophysics,
https://doi.org/10.1007/978-981-13-3753-6_9

of pore water pressure (the so-called vibration liquefaction phenomenon) must also be considered. Therefore, for the problems of dynamic performance of geomaterial under dynamic load, the two different strain amplitudes must be distinguished for small strain and large strain at first.

For small strain (strain amplitude is less than 10^{-4} for axial strain or shear strain), the variation of shear modulus and damping ratio are generally regarded as the necessary parameters for the dynamic analysis of building foundation, machinery foundation, or dam. However, in large strain amplitude circumstances, not only the shear modulus and damping ratio variation but also the parameters for strength and plastic deformation are needed.

The stress–strain relationship has four basic characteristics cyclic loading, namely the hysteresis, nonlinear, deformation accumulating, and strain rate dependence.

1. **Hysteresis**

The hysteresis of geomaterial is shown by the hysteresis curve for dynamic load. From Fig. 9.1, we can see that the stain response is later than the applied stress. The maximum stain does not appear at the maximum stress. It appears that the stress is reducing. It is the demonstration of the hysteresis for geomaterial at dynamic load.

2. **Nonlinear**

The nonlinear of dynamic stress–stress relation is illustrated by the backbone curve in Fig. 9.2.

The stress–strain relationship curves will form a hysteresis loop for the cyclic shear stress in one cycle. Different hysteresis loops will be formed for shear stress cycles of different amplitudes. The track of each maximum stress and strain for hysteresis loops can be called skeleton curves of stress–strain. Backbone curve shows the relationship between maximum shear stress and maximum shear strain, and the relationship can reflect the nonlinear.

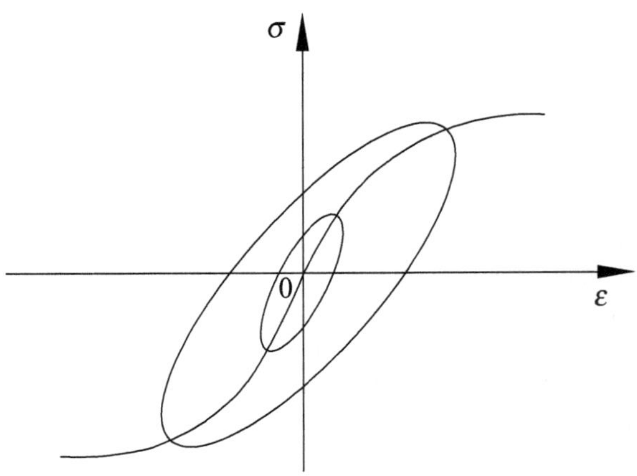

Fig. 9.1 The hysteresis curve

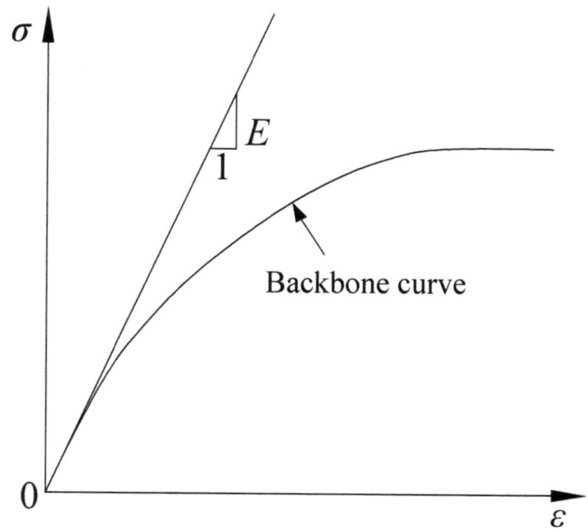

Fig. 9.2 Backbone curve

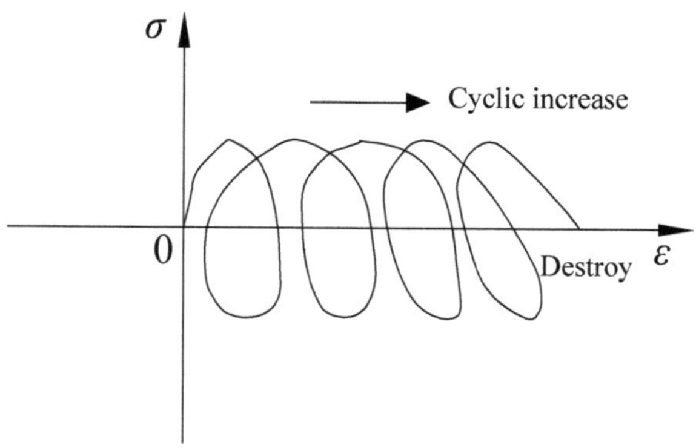

Fig. 9.3 Deformation accumulating

3. **Deformation accumulating**

In addition, because the irrecoverable deformation of soil will be produced under unloading process, this part of the deformation will accumulate gradually in cyclic load. Even the amplitude of loading is invariable, the deformation is increasing with the increasing number of load cycles, and the center of hysteresis loops moves constantly in one direction. The change of hysteresis loops reflects the accumulation effect of dynamic load, and it produces the plastic strain, namely the unrecoverable structural damage (in Fig. 9.3). The cumulative effect of deformation includes the influence to stress and strain, too.

4. **Strain rate dependence**

Geomaterial shows different behaviors under high strain rate loading. Test data using a Split-Hopkinson Pressure Bar (SHPB) [1, 2] shows that the density of soil and the

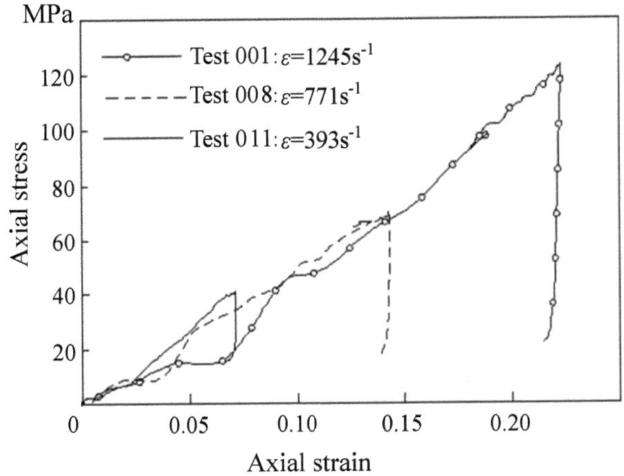

Fig. 9.4 Split-Hopkinson pressure bar test data

shock velocity are increasing with moisture content increasing. The confined axial stress–strain curves of the soil specimens from SHPB tests at three different strain rates are presented in Fig. 9.4 [3]. The strength increases with the higher strain rate.

9.2 Empirical Model for Dynamic Stress–Strain Relationship

Based on a lot of experiments, empirical models are presented for dynamic stress –strain relationship of geomaterial.

The morphological backbone curves were generally depicted by the expression of hyperbola (Fig. 9.5). The expressions are given below.

$$\tau = f(\gamma) = \frac{\gamma}{a + b\gamma} \tag{9.2.1}$$

where τ, γ, a, b represent the shear stress, shear strain, the initial shear modulus, and the maximum shear stress.

Obviously, $\frac{1}{a}$ is the slope of the backbone curve at the origin, denoted as

$$G\mathrm{max} = \frac{1}{a} \tag{9.2.2}$$

where $G_{\max}$ is the initial shear modulus.

$\frac{1}{b}$ is the intercept of vertical axis for the horizontal asymptote of skeleton curve, denoted as

$$\tau_{\mathrm{f}} = \frac{1}{b} \tag{9.2.3}$$

Fig. 9.5 Backbone curve and the construction method of hysteresis loop

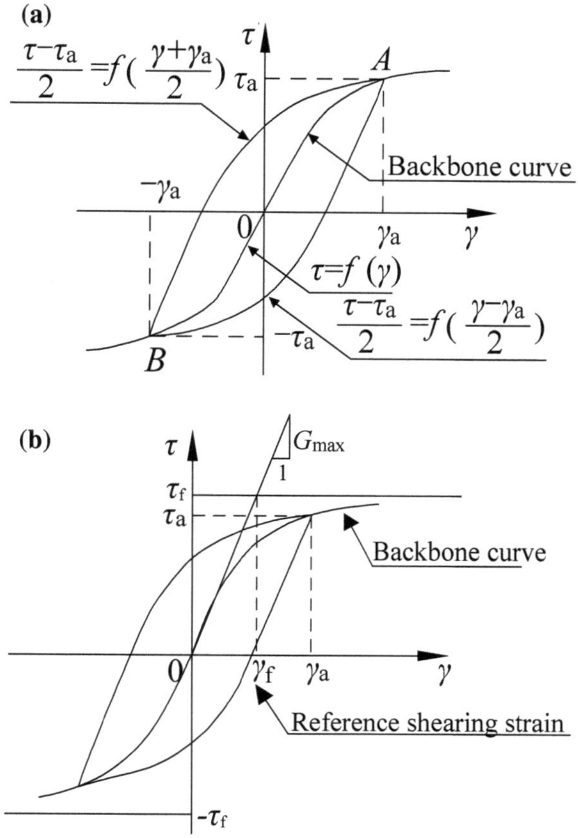

where τ_f is the failure shear stress.

$$\gamma_r = \frac{a}{b} = \frac{\tau_f}{G_{max}} \tag{9.2.4}$$

where γ_r is the reference shear strain.

Then Eq. (9.2.1) can be expressed as

$$\tau = f(\gamma) = \frac{G_{max}\gamma}{1 + \frac{\gamma}{\gamma_r}} \tag{9.2.5}$$

The loading–unloading curve is constructed by the Masing rule, through the coordinate translation of origin, rotation of 180°, and the backbone curve is enlarged and a hysteresis loop is formed.

Through origin translate to (γ_a, τ_a), and shrinking to a half, the backbone curve will transform into the unloading curve.

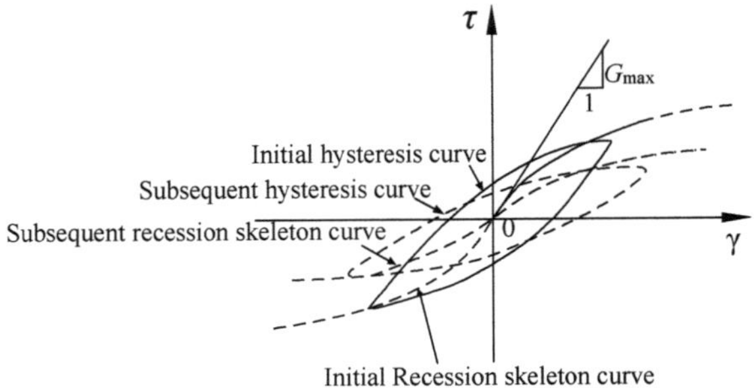

Fig. 9.6 The Construction method for backbone curve and subsequent hysteresis loop

$$\frac{\tau - \tau_a}{2} = f(\frac{\gamma - \gamma_a}{2}) \tag{9.2.6}$$

where γ_a, τ_a are the location of reverse point at peak.

The unloading curve will be

$$\tau = \tau_a + \frac{G_{max}(\gamma - \gamma_a)}{1 - \frac{\gamma - \gamma_a}{2\gamma_r}} \tag{9.2.7}$$

Through origin translate to $(-\gamma_a, -\tau_a)$, and shrinking to a half, the backbone curve will transform into the loading curve.

$$\frac{\tau + \tau_a}{2} = f(\frac{\gamma + \gamma_a}{2})$$

$$\tau = -\tau_a + \frac{G_{max}(\gamma + \gamma_a)}{1 - \frac{\gamma + \gamma_a}{2\gamma_r}} \tag{9.2.8}$$

According to the test results of reciprocating load for the saturated sand, the relationship was proposed by Matasovic and Vucetic [4] for initial hysteresis loop and any subsequent hysteresis loops (Fig. 9.6). The subsequent hysteresis loop is constructed by recession backbone curve and Masing rules. Reciprocating decline of soil properties can be obtained by reduction of vertical coordinate value of the initial backbone curve. The initial backbone curve is expressed as

$$\tau = \frac{G_{max}\gamma}{1 + \psi\left(\frac{\gamma}{\gamma_r}\right)^s} \tag{9.2.9}$$

where ψ, s are the experimental parameters. For sands, $\psi = 1.0 \sim 2.0$, $s = 0.65 \sim 1.0$.

For cohesionless or less cohesion soil, the recession of backbone curve may be regarded as the results of increase of pore pressure [5]. The recession is expressed as

$$G^*_{max} = G_{max}(1 - u)^n, \ \tau^*_{ult} = \tau_{ult}(1 - u^{\mu}) \tag{9.2.10}$$

The reference dynamic shear strain in Fig. 9.6 can be represented as

$$\gamma_r^* = \frac{\tau_{ult}^*}{G_{max}^*} = \frac{\tau_{ult}(1 - u^\mu)}{G_{max}(1 - u^n)} = \gamma_r \frac{1 - u^\mu}{(1 - u)^n} \tag{9.2.11}$$

The recession of subsequent backbone curve can be represented as

$$\tau = \frac{G_{max}^* \gamma}{1 + \psi\left(\frac{\gamma}{\gamma_r^*}\right)^s} \tag{9.2.12}$$

Masing rule is used to build the instantaneous loading–unloading curve and hysteresis loop.

9.3 Equivalent Dynamic Linear Viscoelastic Model of Soil

Soil is supposed as a viscoelastic body in equivalent dynamic linear viscoelastic model of geomaterial. The dynamic nonlinearity and hysteresis are simulated by using the equivalent shear modulus G and equivalent damping ratio λ, and the equivalent shear modulus and damping ratio are expressed as the function of amplitude of dynamic strain. These models have clear concept, easy application, but the deformation accumulation effect cannot be reflected reasonably.

9.3.1 Viscoelastic Model

To viscoelastic model, geomaterial is regarded as in parallel of a linear elastic element and a cohesive element (Fig. 9.7).

$$\boldsymbol{\sigma} = \boldsymbol{\sigma}^{e} + \boldsymbol{\sigma}^{v} \tag{9.3.1}$$

where $\boldsymbol{\sigma}$, $\boldsymbol{\sigma}^e$, $\boldsymbol{\sigma}^v$ are the total stress, elastic stress (stress of linear elastic element), and viscous stress (that of cohesive element).

The elastic stress $\boldsymbol{\sigma}^e$ is calculated by Hooke's Law.

$$\boldsymbol{\sigma}^{e} = [D_e]\boldsymbol{\varepsilon} \tag{9.3.2}$$

where

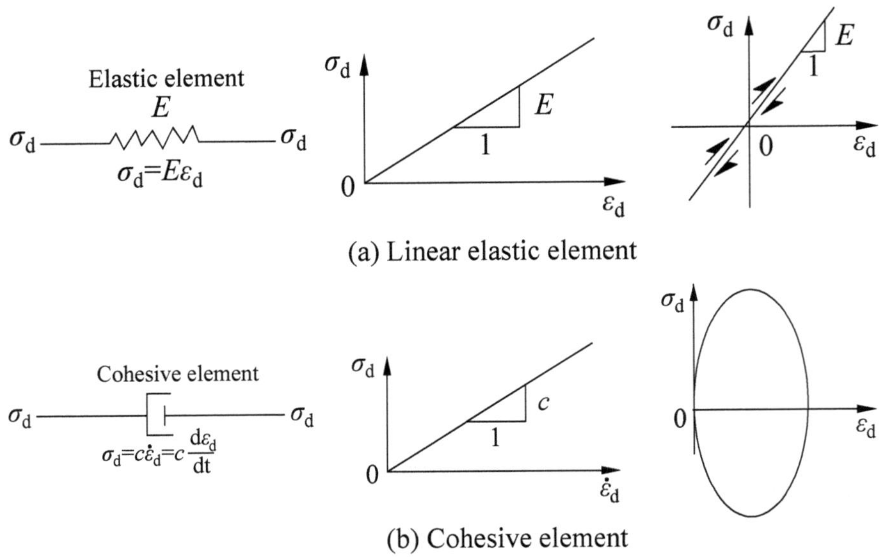

Fig. 9.7 Viscoelastic element model and its mechanical response

$$[D_e] = \frac{E}{(1+\nu)(1-2\nu)}\begin{bmatrix} 1-\nu & \nu & \nu & 0 & 0 & 0 \\ \nu & 1-\nu & \nu & 0 & 0 & 0 \\ \nu & \nu & 1-\nu & 0 & 0 & 0 \\ 0 & 0 & 0 & 1-2\nu & 0 & 0 \\ 0 & 0 & 0 & 0 & 1-2\nu & 0 \\ 0 & 0 & 0 & 0 & 0 & 1-2\nu \end{bmatrix}$$

where E is the elastic modulus, and ν is Poisson's Ratio.

The viscous stress is calculated as

$$\boldsymbol{\sigma}^{\mathrm{v}} = c\dot{\boldsymbol{\varepsilon}} = c\frac{\mathrm{d}\varepsilon}{\mathrm{d}t} \tag{9.3.3}$$

where c is viscous coefficient.

The total stress can be represented as

$$\boldsymbol{\sigma} = \boldsymbol{\sigma}^{\mathrm{e}} + \boldsymbol{\sigma}^{\mathrm{v}} = [D]\boldsymbol{\varepsilon} + c\frac{\mathrm{d}\boldsymbol{\varepsilon}}{\mathrm{d}t} \tag{9.3.4}$$

For one-dimensional case

$$\boldsymbol{\sigma} = E\boldsymbol{\varepsilon} + c\frac{\mathrm{d}\boldsymbol{\varepsilon}}{\mathrm{d}t} \tag{9.3.5}$$

Take stress for change of sine function.

$$\boldsymbol{\sigma} = \sigma_{\mathrm{a}} \sin wt \tag{9.3.6}$$

where σ_{a}, w, t are dynamic stress amplitude, frequency, and time, respectively.

To solve the differential Eq. (9.3.5), we will get

$$\boldsymbol{\varepsilon} = \varepsilon_{\mathrm{a}} \sin(wt - \delta) \tag{9.3.7a}$$

where ε_{a}, δ are dynamic strain amplitude and phase angle, respectively, (phase difference of dynamic strain lagging behind the dynamic stress).

$$\varepsilon_{\mathrm{a}} = \frac{\sigma_{\mathrm{a}}}{\sqrt{E^2 + (c\omega)^2}} \tag{9.3.7b}$$

$$\delta = \text{arc tan}\,\frac{c\omega}{E} \tag{9.3.7c}$$

Take stress for change of the cosine function.

$$\boldsymbol{\sigma} = \sigma_{\mathrm{a}} \cos wt \tag{9.3.8}$$

The dynamic strain will be

$$\boldsymbol{\varepsilon} = \varepsilon_{\mathrm{a}} \cos(wt - \delta) \tag{9.3.9}$$

If the dynamic stress is applied in the complex form

$$\boldsymbol{\sigma} = \sigma_{\mathrm{a}} e^{\mathrm{iwt}} = \sigma_{\mathrm{a}}(\cos\ wt + i\ \sin\ wt) \tag{9.3.10}$$

Then the dynamic strain will be obtained.

$$\varepsilon = \varepsilon_a e^{\mathrm{i(wt-\delta)}} = \sigma_{\mathrm{a}}[\cos(wt - \delta) + i\ \sin(wt - \delta)] \tag{9.3.11}$$

Obviously, the real part and imaginary part of dynamic strain only relate to the real part and imaginary part of the dynamic stress. And dynamic stress–strain relationship can also be represented as

$$\frac{\boldsymbol{\sigma}}{\boldsymbol{\varepsilon}} = \frac{\sigma_{\mathrm{a}}}{\varepsilon_{\mathrm{a}}} e^{\mathrm{i}\delta} = \frac{\sigma_{\mathrm{a}}}{\varepsilon_{\mathrm{a}}}(\cos\delta + i \sin\delta) \tag{9.3.12}$$

Let

$$E^* = \frac{\boldsymbol{\sigma}}{\boldsymbol{\varepsilon}} \tag{9.3.13a}$$

$$E = \frac{\sigma_{\mathrm{a}}}{\varepsilon_{\mathrm{a}}} \cos\delta \tag{9.3.13b}$$

$$E' = \frac{\sigma_{\mathrm{a}}}{\varepsilon_{\mathrm{a}}} \sin\delta \tag{9.3.13c}$$

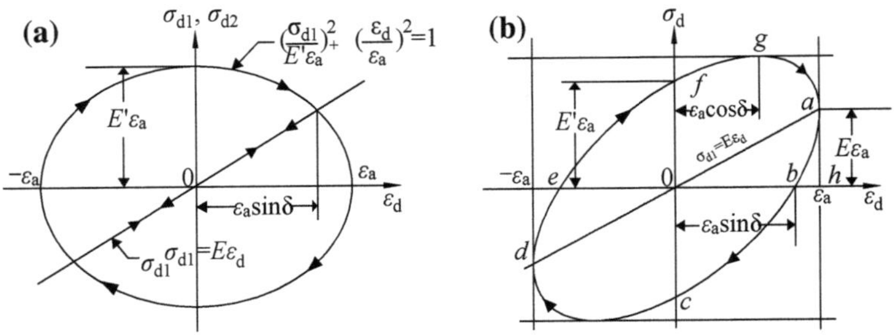

Fig. 9.8 Decomposition of elastic and viscous component for viscoelastic model

where E^*, E, E' are called complex modulus, dynamic elastic modulus, and loss modulus, respectively.

And

$$E^* = E + iE' \tag{9.3.13d}$$

$$\tan\delta = \frac{E'}{E} = \eta \tag{9.3.14}$$

where η is called loss coefficient, a parameter represents the energy loss and the damping characteristics.

Simultaneous Eqs. (9.3.6) and (9.3.7a), and we eliminate t, it will be

$$\left(\frac{\boldsymbol{\sigma}}{\sigma_a}\right)^2 - 2\cos\delta\frac{\boldsymbol{\sigma}}{\sigma_a}\frac{\boldsymbol{\varepsilon}}{\varepsilon_a} + \left(\frac{\boldsymbol{\varepsilon}}{\varepsilon_a}\right)^2 - \sin^2\delta = 0 \tag{9.3.15}$$

It can also be represented as

$$\boldsymbol{\sigma} = E\boldsymbol{\varepsilon} \pm E'\sqrt{\varepsilon_{\mathrm{a}}^2 - \boldsymbol{\varepsilon}^2} \tag{9.3.16}$$

It can be decomposed into two parts.

$$\boldsymbol{\sigma} = \boldsymbol{\sigma}_1 + \boldsymbol{\sigma}_2 \tag{9.3.17a}$$

$$\boldsymbol{\sigma}_1 = E\boldsymbol{\varepsilon} \tag{9.3.17b}$$

$$\left(\frac{\boldsymbol{\sigma}_2}{E'\varepsilon_a}\right)^2 + \left(\frac{\boldsymbol{\varepsilon}}{\varepsilon_a}\right)^2 = 1 \tag{9.3.17c}$$

The first part (Eq. 9.3.17a) is expressed as a line with the slope of E, the second part is expressed as an ellipse with semimajor axis $E'\varepsilon_{\mathrm{a}}$, and semiminor axis ε_{a} (Fig. 9.8a). Total stress–strain relationship is an inclined ellipse (Fig. 9.8b).

The total stress and strain is an inclined ellipse. Its intersection point with the vertical axis is $(0, \pm E'\varepsilon_{\mathrm{a}})$. The loss modulus E' can describe the flat degree of inclined ellipse. The larger loss modulus E', the less flatter of the inclined ellipse, the greater the energy losses and damping in the loading–unloading cycle. The smaller loss modulus E', the more flatter of the inclined ellipse, the smaller the energy losses and damping in the loading–unloading cycle.

The energy loss in one cycle is generally used to describe the viscoelastic damping characteristics. The energy loss of one cycle is equal to the area of the hysteresis loop.

$$\Delta W = \oint \sigma \mathrm{d}\boldsymbol{\varepsilon} = E'\pi\varepsilon_a^2 \tag{9.3.18a}$$

Energy loss of a cycle ΔW can be used as a measure of material damping characteristics, but it is a function of dynamic strain amplitude ε_{a}, not a reasonable quantitative characterization of performance indicators. The ratio of energy loss and elastic energy storage of one cycle is adopted as the quantitative index of damping characteristics.

$$\frac{\Delta W}{W} = \frac{E'\pi\varepsilon_{\mathrm{a}}^2}{\frac{1}{2}E\varepsilon_{\mathrm{a}}^2} = 2\pi\frac{E'}{E} \tag{9.3.18b}$$

Combining the definition of dissipative coefficient η,

$$\eta = \frac{1}{2\pi}\frac{\Delta W}{W} = \frac{E'}{E} = \tan\delta \tag{9.3.18c}$$

Equation (9.3.18c) reveals the relationship between the energy dissipation expressed in hysteresis loop area and the damping described by phase angle.

From Fig. 9.9, the most simple calculation method can be obtained for dissipative coefficient.

$$\eta = \frac{\text{stress at 0 strain}}{\text{stress at maximum strain}} \tag{9.3.18d}$$

The significance of this equation is the most convenient method for the calculation of the dissipative coefficient, even if the linear viscoelastic theory is no longer valid.

9.3.2 Parameter for Viscoelastic Model

Generally, two dynamic parameters are needed for the engineering computation for viscoelastic model, shear modulus G, and damping ratio λ. Its definition is shown in Fig. 9.10.

The secant modulus of backbone curve is used to describe the nonlinear characteristic of dynamic deformation.

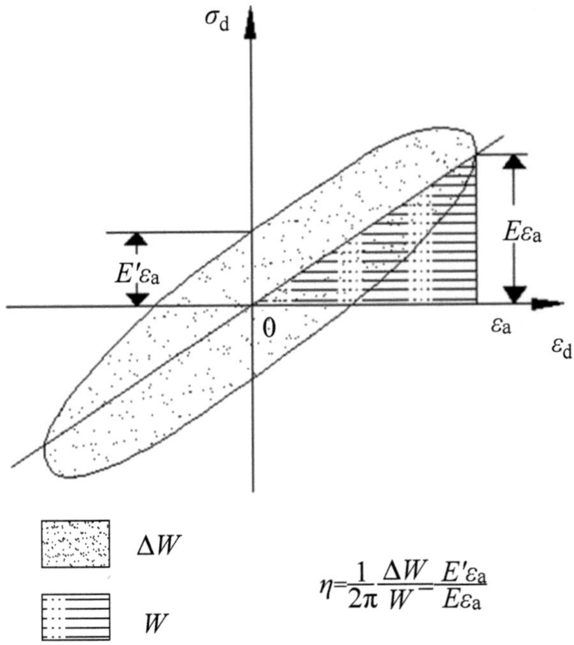

Fig. 9.9 The definition of dissipative coefficient η

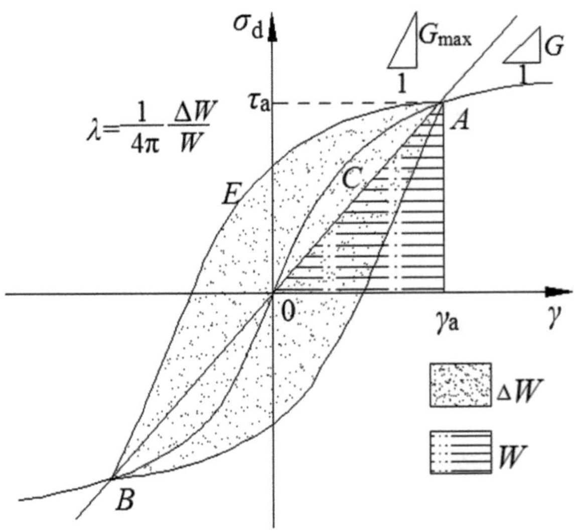

Fig. 9.10 Parameters for linear viscoelastic model

$$G = \frac{\tau_a}{\gamma_a} = \frac{f(\gamma_a)}{\gamma_a} \tag{9.3.19a}$$

Elastic modulus

$$E = 2(1 + \upsilon)G \tag{9.3.19b}$$

Damping ratio λ

$$\lambda = \frac{1}{4\pi}\frac{\Delta W}{W} = \frac{\eta}{2} \tag{9.3.20}$$

$$\lambda = \frac{\tan\delta}{2} = \frac{c\omega}{E} \tag{9.3.21}$$

$$c = \frac{E}{\omega}\lambda \tag{9.3.22}$$

Of course, dynamic characteristic is nonlinear for the actual geomaterial. The nonlinear model should be adopted. The mechanical parameters can also be determined by the function of strain amplitude.

9.4 Viscoelastoplastic Dynamic Constitutive Model of Geomaterial

The viscoelastic model can reflect the hysteresis of geomaterial, but it cannot reflect nonlinearity and cumulative plastic deformation. A better way is to use the viscoelastic plasticity model, which can well simulate the four mechanical characteristics.

9.4.1 Framework for Viscoelastic Plasticity Model

Viscoelastoplastic model regards geomaterial as the parallel connection of elastoplastic element with viscous element (Fig. 9.11).

$$\boldsymbol{\sigma} = \boldsymbol{\sigma}^{\mathrm{ep}} + \boldsymbol{\sigma}^{\mathrm{v}} \tag{9.4.1}$$

where $\boldsymbol{\sigma}$, $\boldsymbol{\sigma}^{\mathrm{ep}}$, $\boldsymbol{\sigma}^{\mathrm{v}}$ are the total stress, elastic–plastic stress, and viscous stress, respectively.

9.4.2 The Computation of Viscoelastic Part

The computation of viscoelastic part is based on Newton's law of viscosity.

$$\boldsymbol{\sigma}^{\mathrm{v}} = [C]\dot{\boldsymbol{\varepsilon}} \tag{9.4.2}$$

Fig. 9.11 Viscoelastic plasticity element model and its mechanical response

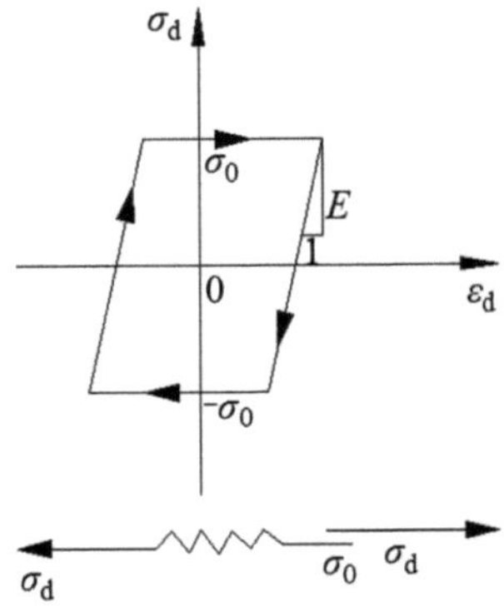

(a) Ideal elastoplastic model

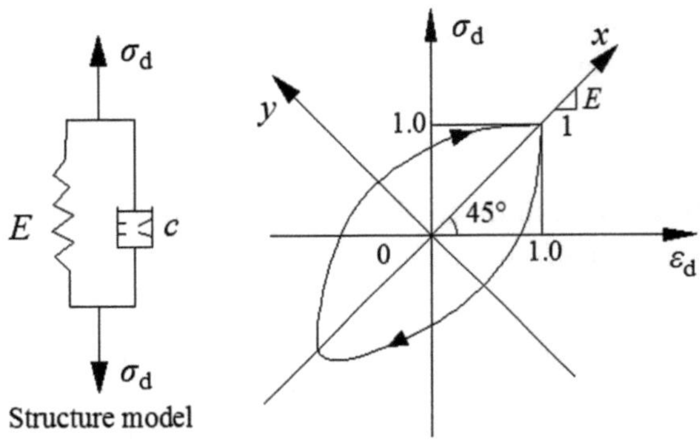

(b) Viscous elastic model

where $[C]$ is viscous matrix which is different with the different viscoelastic model, and the simplest one is to use a scalar.

9.4.3 The Computation of Elastoplastic Part

Elastic–plastic deformation is expressed as

$$\boldsymbol{\varepsilon} = \boldsymbol{\varepsilon}^{\mathrm{e}} + \boldsymbol{\varepsilon}^{\mathrm{p}} \tag{9.4.3}$$

where $\boldsymbol{\varepsilon}$, $\boldsymbol{\varepsilon}^{\mathrm{e}}$, $\boldsymbol{\varepsilon}^{\mathrm{p}}$ are total strain, elastic strain, and plastic strain.

The elastic strain is calculated by Hooke's Law.

$$\boldsymbol{\varepsilon}^{\mathrm{e}} = [D]^{-1}\boldsymbol{\sigma} \tag{9.4.4a}$$

where $[D]$ is elastic stiffness matrix.

$$\mathrm{d}\boldsymbol{\varepsilon}^{\mathrm{e}} = [D]^{-1}\mathrm{d}\boldsymbol{\sigma} \tag{9.4.4b}$$

The computation of plastic deformation is based on the plastic theory.

$$\boldsymbol{\varepsilon}^{\mathrm{p}} = \left[D_{\mathrm{p}}\right]^{-1}\boldsymbol{\sigma} \tag{9.4.5}$$

where $[D_{\mathrm{p}}]$ is plastic stiffness matrix.

At present, the plastic model is generally more complex for the calculation of dynamic deformation, with too many parameters, and it is not convenient to the application. The dynamic deformation mechanism is also not completely clear for geomaterial, so the mixed hardening theory is suggested to calculate the dynamic plastic deformation.

$$\mathrm{d}\boldsymbol{\varepsilon}^{\mathrm{p}} = \mathrm{d}\lambda\frac{\partial f}{\partial \boldsymbol{\sigma}} = \frac{1}{A}\frac{\partial f}{\partial \boldsymbol{\sigma}}\mathrm{d}\boldsymbol{\sigma}\frac{\partial f}{\partial \boldsymbol{\sigma}} \tag{9.4.6}$$

The mixed hardening yield function is defined as

$$f(\boldsymbol{\sigma}, H_{\alpha}) = F[\boldsymbol{\sigma} - \alpha(\boldsymbol{\varepsilon}^{\mathrm{p}})] - k(\boldsymbol{\varepsilon}^{\mathrm{p}}) = 0 \tag{9.4.7}$$

where $\alpha(\boldsymbol{\varepsilon}^{\mathrm{p}})$, $k(\boldsymbol{\varepsilon}^{\mathrm{p}})$ describe the motion of center and size change of the yield surface.

The hardening modulus will be

$$\mathrm{d}f(\boldsymbol{\sigma}, H_{\alpha}) = \mathrm{d}F[\boldsymbol{\sigma} - \alpha(\boldsymbol{\varepsilon}^{\mathrm{p}})] - \mathrm{d}k(\boldsymbol{\varepsilon}^{\mathrm{p}}) = 0$$

$$\frac{\partial F[\boldsymbol{\sigma} - \alpha(\boldsymbol{\varepsilon}^{\mathrm{p}})]}{\partial \boldsymbol{\sigma}}\mathrm{d}\boldsymbol{\sigma} + \frac{\partial F[\boldsymbol{\sigma} - \alpha(\boldsymbol{\varepsilon}^{\mathrm{p}})]}{\partial \alpha}\mathrm{d}\alpha - \frac{\mathrm{d}k(\boldsymbol{\varepsilon}^{\mathrm{p}})}{\mathrm{d}\boldsymbol{\varepsilon}^{\mathrm{p}}}\mathrm{d}\boldsymbol{\varepsilon}^{\mathrm{p}} = 0$$

$$\frac{\partial F}{\partial \boldsymbol{\sigma}}\mathrm{d}\boldsymbol{\sigma} - \frac{\partial F}{\partial \alpha}\frac{\mathrm{d}\alpha}{\mathrm{d}\boldsymbol{\varepsilon}^{\mathrm{p}}}\mathrm{d}\boldsymbol{\varepsilon}^{\mathrm{p}} - \frac{\mathrm{d}k}{\mathrm{d}\boldsymbol{\varepsilon}^{\mathrm{p}}}\mathrm{d}\boldsymbol{\varepsilon}^{\mathrm{p}} = 0$$

$$\frac{\partial F}{\partial \boldsymbol{\sigma}}\mathrm{d}\boldsymbol{\sigma} - \frac{\partial F}{\partial \alpha}\frac{\mathrm{d}\alpha}{\mathrm{d}\boldsymbol{\varepsilon}^{\mathrm{p}}}\frac{1}{A}\frac{\partial f}{\partial \boldsymbol{\sigma}}\mathrm{d}\boldsymbol{\sigma}\frac{\partial f}{\partial \boldsymbol{\sigma}} - \frac{\mathrm{d}k}{\mathrm{d}\boldsymbol{\varepsilon}^{\mathrm{p}}}\frac{1}{A}\frac{\partial f}{\partial \boldsymbol{\sigma}}\mathrm{d}\boldsymbol{\sigma}\frac{\partial f}{\partial \boldsymbol{\sigma}} = 0$$

$$\frac{\partial F}{\partial \boldsymbol{\sigma}}\mathrm{d}\boldsymbol{\sigma} - \frac{\partial F}{\partial \alpha}\frac{\mathrm{d}\alpha}{\mathrm{d}\boldsymbol{\varepsilon}^{\mathrm{p}}}\frac{1}{A}\frac{\partial F}{\partial \boldsymbol{\sigma}}\mathrm{d}\boldsymbol{\sigma}\frac{\partial F}{\partial \boldsymbol{\sigma}} - \frac{\mathrm{d}k}{\mathrm{d}\boldsymbol{\varepsilon}^{\mathrm{p}}}\frac{1}{A}\frac{\partial F}{\partial \boldsymbol{\sigma}}\mathrm{d}\boldsymbol{\sigma}\frac{\partial F}{\partial \boldsymbol{\sigma}} = 0$$

where A is the plastic modulus.

$$\frac{1}{A} = \frac{\frac{\partial F}{\partial \boldsymbol{\sigma}}\mathrm{d}\boldsymbol{\sigma}}{\frac{\partial F}{\partial \alpha}\frac{\mathrm{d}\alpha}{\mathrm{d}\boldsymbol{\varepsilon}^{\mathrm{p}}}\frac{\partial F}{\partial \boldsymbol{\sigma}}\mathrm{d}\boldsymbol{\sigma}\frac{\partial F}{\partial \boldsymbol{\sigma}} + \frac{\mathrm{d}k}{\mathrm{d}\boldsymbol{\varepsilon}^{\mathrm{p}}}\frac{\partial F}{\partial \boldsymbol{\sigma}}\mathrm{d}\boldsymbol{\sigma}\frac{\partial F}{\partial \boldsymbol{\sigma}}} = \frac{1}{\frac{\partial F}{\partial \alpha}\frac{\mathrm{d}\alpha}{\mathrm{d}\boldsymbol{\varepsilon}^{\mathrm{p}}}\frac{\partial F}{\partial \boldsymbol{\sigma}} + \frac{\mathrm{d}k}{\mathrm{d}\boldsymbol{\varepsilon}^{\mathrm{p}}}\frac{\partial F}{\partial \boldsymbol{\sigma}}} \tag{9.4.8}$$

The plastic strain would be

$$\mathrm{d}\boldsymbol{\varepsilon}^{\mathrm{p}} = \frac{1}{\frac{\partial F}{\partial \alpha}\frac{\mathrm{d}\alpha}{\mathrm{d}\boldsymbol{\varepsilon}^{\mathrm{p}}}\frac{\partial F}{\partial \sigma} + \frac{\mathrm{d}k}{\mathrm{d}\boldsymbol{\varepsilon}^{\mathrm{p}}}\frac{\partial F}{\partial \sigma}} \frac{\partial F}{\partial \boldsymbol{\sigma}}\mathrm{d}\boldsymbol{\sigma}\frac{\partial F}{\partial \boldsymbol{\sigma}}$$

$$\mathrm{d}\boldsymbol{\varepsilon}^{\mathrm{p}} = \frac{\frac{\partial F}{\partial \sigma}\frac{\partial F}{\partial \sigma}\mathrm{d}\boldsymbol{\sigma}}{\frac{\partial F}{\partial \alpha}\frac{\mathrm{d}\alpha}{\mathrm{d}\boldsymbol{\varepsilon}^{\mathrm{p}}}\frac{\partial F}{\partial \sigma} + \frac{\mathrm{d}k}{\mathrm{d}\boldsymbol{\varepsilon}^{\mathrm{p}}}\frac{\partial F}{\partial \sigma}} = [D_{\mathrm{p}}]^{-1}\mathrm{d}\boldsymbol{\sigma} \tag{9.4.9}$$

9.4.4 The Computation of Total Part

The total deformation will be

$$\mathrm{d}\boldsymbol{\varepsilon} = \mathrm{d}\boldsymbol{\varepsilon}^{\mathrm{e}} + \mathrm{d}\boldsymbol{\varepsilon}^{\mathrm{p}} = ([D]^{-1} + [D_{\mathrm{p}}]^{-1})\mathrm{d}\boldsymbol{\sigma} = [D_{\mathrm{ep}}]^{-1}\mathrm{d}\boldsymbol{\sigma} \tag{9.4.10}$$

$$\mathrm{d}\boldsymbol{\sigma} = \mathrm{d}\boldsymbol{\sigma}^{\mathrm{ep}} + \mathrm{d}\boldsymbol{\sigma}^{\mathrm{v}} = [D_{\mathrm{ep}}]\mathrm{d}\boldsymbol{\varepsilon} + c\ddot{\boldsymbol{\varepsilon}} \tag{9.4.11}$$

The model parameters are determined on the basis of the related model.
It is calculated by the viscoelastic theory if the stress is within the yield surface.

$$f(\boldsymbol{\sigma}, H_{\alpha}) < 0 \tag{9.4.12}$$

$$\mathrm{d}\boldsymbol{\sigma} = \mathrm{d}\boldsymbol{\sigma}^{\mathrm{e}} + \mathrm{d}\boldsymbol{\sigma}^{\mathrm{v}} = [D]\mathrm{d}\boldsymbol{\varepsilon} + [C]\ddot{\boldsymbol{\varepsilon}} \tag{9.4.13}$$

Related to time, the numerical difference method should be used for the numerical computation.

9.5 Viscoplastic Cap Models

Viscoplasticity is defined as a rate-dependent (as opposed to inviscid means rate independent) plasticity model and may be applied to the soil constitutive laws to account for the strain rate effect. A variety of viscoplastic formulations for soils have been proposed in the literature. The formulation of viscoplasticity based on Perzyna's theory [6] is the most well-known formulation, where viscous behavior is modeled with a time-rate flow rule. The flow rule is assumed to be associative and the viscoplastic potential is identical or at least proportional to the yield surface [7]. After the transition into rate-independent plasticity, this identity becomes essential although it has no great significance in viscoplasticity. The viscoplastic formulation has the following advantages:

1. The generality of the viscous flow rule offers the capability of simulating time-dependent material behavior over a wide range of loading.
2. The extension of an inviscid cap model for viscoplasticity is relatively straightforward [7].

Another viscoplastic formulation of the Duvant–Lions type has been advocated by Simo et al. [8]. The viscous behavior is constructed directly based on the difference between solutions for inviscid and the viscoplastic formulations. The main advantage is its ease in numerical implementation, only a stress update needs to be added in an inviscid formulation in order to obtain the corresponding viscoplastic solution.

The viscoplastic cap model is an effective material model to simulate soil behavior under high strain rate loading. Tong [7] applied viscoplastic cap model in LS-DYNA to simulate a series of explosions in soil. Comparisons with experimental results, the simulations of soil ejecta, crater, and explosive clouds from landmine-explosion tests are reasonably good.

Two types of viscoplastic cap models [7] are proposed based on Perzyna's theory and Duvant–Lions' theory. The plastic yield functions are patterned on the generalized two-invariant cap model. Numerical algorithm is presented. The performance of viscoplastic cap model is examined using a hypothetical uniaxial strain test and compared against experimental data under rapid loading.

In the viscoplasticity model, the total strain rate vector $\dot{\boldsymbol{\varepsilon}}$ is decomposed into an elastic component $\dot{\boldsymbol{\varepsilon}}^{\mathrm{e}}$ and a viscoplastic component $\dot{\boldsymbol{\varepsilon}}^{\mathrm{vp}}$.

$$\dot{\boldsymbol{\varepsilon}} = \dot{\boldsymbol{\varepsilon}}^{\mathrm{e}} + \dot{\boldsymbol{\varepsilon}}^{\mathrm{vp}} \tag{9.5.1}$$

The elastic component is expressed as

$$\dot{\boldsymbol{\sigma}} = [C]\dot{\boldsymbol{\varepsilon}}^{\mathrm{e}} \tag{9.5.2}$$

where $\dot{\boldsymbol{\sigma}}$ is the stress rate vector, and $[C]$ is an elastic constitutive matrix.

For the viscoplastic component, it is different from other types.

9.5.1 The Perzyna-Type Viscoplastic Cap Model

The viscoplastic strain rate vector is assumed to be delayed with time and is expressed as follows when assuming associated flow rule.

$$\dot{\boldsymbol{\varepsilon}}^{\mathrm{vp}} = \eta < \varphi(f) > \frac{\partial f}{\partial \boldsymbol{\sigma}} \tag{9.5.3}$$

where η is a material constant called fluidity parameter; the notion$< >$refers to the ramp function defined by $< x >= \frac{x+|x|}{2}$; f is plastic yield function; $\varphi(f)$ is dimensionless viscous flow function and commonly expressed in the form of

$$\varphi(f) = (\frac{f}{f_0})^{\mathrm{N}} \tag{9.5.4}$$

where N is an exponent, and f_0 is a normalizing constant with the same units as f.

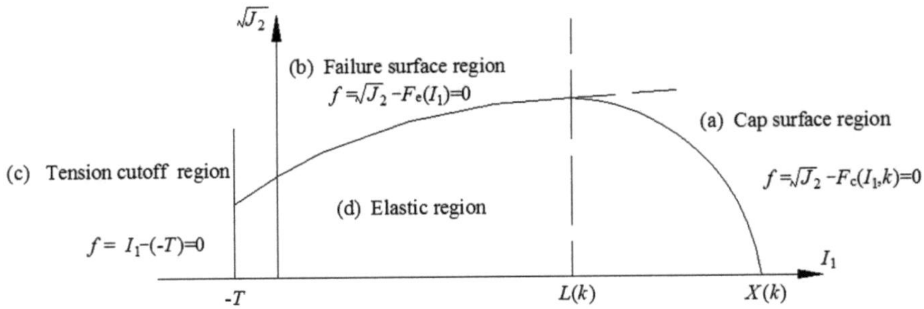

Fig. 9.12 Static yield surface for cap model

1. **Static yield functions**

The plastic yield function f is patterned in the inviscid cap model which is formulated in terms of the first stress invariant I_1 and the second deviator stress invariant J_2 as shown in Fig. 9.12. The static yield surface is divided into three regions:

(1) when $I_1 \geq L$, the cap surface region $f = \sqrt{J_2} - F_c(I_1, k) = 0$
(2) when $L > I_1 > -T$, the failure surface region $f = \sqrt{J_2} - F_e(I_1) = 0$
(3) when $I_1 \leq -T$, the tension cutoff region $f = I_1 + T = 0$.

(a) Cap surface portion: the cap surface is a hardening surface in the shape of an ellipse quadrant in the stress space of I_1 and J_2. It is generally defined by

$$f(I_1,\ \sqrt{J_2},\ k) = \sqrt{J_2} - F_c(I_1, k) = \sqrt{J_2} - \frac{1}{R}\sqrt{[X(k) - L(k)]^2 - [I_1 - L(k)]^2} = 0 \qquad (9.5.5)$$

where $F_c(I_1, k)$ is the loading function for cap envelope; R is a material parameter; k is a hardening parameter related to the actual viscoplastic volumetric change ε_v^{vp}.

$$\varepsilon_v^{vp}[X(k)] = W\{1 - \exp[-D[X(k) - X_0]]\} \qquad (9.5.6)$$

$$X(k) = k + RF_e(k) \qquad (9.5.7)$$

where $F_e(k)$ is the loading function.

$L(k)$ is the value of I_1 at the location of the start of cap and is defined by

$$L(k) = \begin{cases} k & if\ k > 0 \\ 0 & if\ k \leq 0 \end{cases} \qquad (9.5.8)$$

The cap surface may be expressed alternatively as

$$f(I_1,\ \sqrt{J_2},\ k) = \sqrt{\frac{(I_1 - L)^2}{R^2} + J_2} - \frac{l - X}{R} = 0 \qquad (9.5.9)$$

(b) Failure surface portion: the failure surface is a nonhardening, modified Drucker–Prager form with a yield function defined as

$$f(I_1,\ \sqrt{J_2}) = \sqrt{J_2} - F_e(I_1) = \sqrt{J_2} - \left[\alpha - \gamma \exp(-\beta I_1) + \theta I_1\right] = 0 \quad (9.5.10)$$

where α, β, γ, and θ are material parameters.

(c) Tension cutoff portion: the tension cutoff surface is defined by

$$f(I_1) = I_1 - (-T) = 0 \quad (9.5.11)$$

where $-T$ is tension cutoff value.

9.5.2 *Solution Algorithms*

The strain rate in Eqs. (9.5.1) and (9.5.2) is integrated over a time step t from t to $t+\Delta t$, to yield the incremental strains and stresses.

$$\Delta\boldsymbol{\varepsilon} = \Delta\boldsymbol{\varepsilon}^{\mathrm{e}} + \Delta\boldsymbol{\varepsilon}^{\mathrm{vp}} \quad (9.5.12)$$

$$\Delta\boldsymbol{\sigma} = C\Delta\boldsymbol{\varepsilon}^{\mathrm{e}} = C(\Delta\boldsymbol{\varepsilon} - \Delta\boldsymbol{\varepsilon}^{\mathrm{vp}}) \quad (9.5.13)$$

where $\Delta\boldsymbol{\varepsilon}$ is the total incremental strain vector; $\Delta\boldsymbol{\varepsilon}^{\mathrm{e}}$ is the elastic incremental strain vector; $\Delta\boldsymbol{\varepsilon}^{\mathrm{vp}}$ is the viscoplastic incremental strain vector; $\Delta\boldsymbol{\sigma}$ is the incremental stress vector.

Based on the Euler method, the viscoplastic incremental strain vector $\Delta\boldsymbol{\varepsilon}$ can be approximated as

$$\Delta\boldsymbol{\varepsilon} = [(1-\chi)\dot{\boldsymbol{\varepsilon}}_{\mathrm{t}}^{\mathrm{vp}} + \chi\dot{\boldsymbol{\varepsilon}}_{\mathrm{t+\Delta t}}^{\mathrm{vp}}]\Delta t \quad (9.5.14)$$

where χ is an adjustable integration parameter, $0 \le \chi \le 1$.

The integration scheme is explicit if $\chi = 0$ and fully implicit if $\chi = 1$. This solution algorithm is conditionally stable when $\chi \le 0.5$ and unconditionally stable when $\chi > 0.5$. The fully implicit integration scheme, $\chi = 1$, is used here in the numerical algorithm just for simplification.

In the full implicit integration scheme, the viscoplastic flow (Eq. 9.5.14) is only determined by the gradient of the yield surface at time $t+\Delta t$. Thus, $\Delta\boldsymbol{\varepsilon}^{\mathrm{vp}}$ may be rewritten as

$$\Delta\boldsymbol{\varepsilon}^{\mathrm{vp}} = \dot{\boldsymbol{\varepsilon}}^{\mathrm{vp}}\Delta t = \eta < \varphi(f) > \Delta t\frac{\partial f}{\partial \boldsymbol{\sigma}} \quad (9.5.15)$$

If a plastic multiplier $\Delta\lambda$ is introduced.

$$\Delta\lambda = \eta < \varphi(f) > \Delta t \tag{9.5.16}$$

Then Eq. (9.5.16) may be rewritten as

$$\Delta\boldsymbol{\varepsilon}^{\mathrm{vp}} = \Delta\lambda\frac{\partial f}{\partial\boldsymbol{\sigma}} \tag{9.5.17}$$

This viscoplastic problem can be solved under the condition that the residual ρ, defined in Eq. (9.5.18), is reduced to zero during a local iteration.

$$\rho = \frac{\Delta\lambda}{\eta\Delta t} - \varphi(f) \to 0 \tag{9.5.18}$$

Substituting Eq. (9.5.17) into Eq. (9.5.14) yields

$$\Delta\boldsymbol{\sigma} = [C](\Delta\boldsymbol{\varepsilon} - \Delta\lambda\frac{\partial f}{\partial\boldsymbol{\sigma}}) \tag{9.5.19}$$

To compute $\Delta\lambda$, a local Newton–Raphson iteration process is applied. Note that the yield function takes the general form $f=f(\boldsymbol{\sigma}, k)$. Differentiate Eq. (9.5.19) during iteration i.

$$\delta\boldsymbol{\sigma} = C(\delta\boldsymbol{\varepsilon} - \delta\lambda\frac{\partial f}{\partial\boldsymbol{\sigma}} - \Delta\lambda^{(i)}\frac{\partial^2 f}{\partial\boldsymbol{\sigma}^2} - \Delta\lambda^{(i)}\frac{\partial^2 f}{\partial\boldsymbol{\sigma}\partial\lambda}\delta\lambda) \tag{9.5.20}$$

where $\delta\boldsymbol{\sigma}, \delta\boldsymbol{\varepsilon}, \delta\lambda$ are the iterative improvements of $\Delta\boldsymbol{\sigma}, \Delta\boldsymbol{\varepsilon}, \Delta\lambda$, respectively, within the local iteration process.

Equation (9.5.20) may be expressed alternatively as

$$\delta\boldsymbol{\sigma} = H[\delta\boldsymbol{\varepsilon} - (\frac{\partial f}{\partial\boldsymbol{\sigma}} + \Delta\lambda^{(i)}\frac{\partial^2 f}{\partial\boldsymbol{\sigma}\partial\lambda})\delta\lambda] \tag{9.5.21}$$

With a pseudoelastic stiffness matrix H

$$H = \left\{[C]^{-1} + \Delta\lambda^{(i)}\frac{\partial^2 f}{\partial\boldsymbol{\sigma}^2}\right\}^{-1} \tag{9.5.22}$$

By differentiation of Eq. (9.5.18), the Newton–Raphson process at iteration i is expressed as

$$\rho^{(i)} = (\frac{1}{\eta\Delta t} - \frac{\partial\varphi}{\partial\lambda})\delta\lambda - (\frac{\partial\varphi}{\partial\lambda})^{\mathrm{T}}\delta\boldsymbol{\sigma} \tag{9.5.23}$$

Substituting Eq. (9.5.20) into Eq. (9.5.23)

$$\delta\lambda = \frac{1}{\xi}[(\frac{\partial\varphi}{\partial\boldsymbol{\sigma}})^{\mathrm{T}}H\delta\boldsymbol{\varepsilon} + \rho^{(i)}] \tag{9.5.24}$$

With

$$\xi = (\frac{\partial \varphi}{\partial \boldsymbol{\sigma}})^{\mathrm{T}} H[\frac{\partial f}{\partial \boldsymbol{\sigma}} + \Delta\lambda^{(i)} \frac{\partial^2 f}{\partial \boldsymbol{\sigma} \partial \lambda}] + \frac{1}{\eta \Delta t} - \frac{\partial \varphi}{\partial \lambda} \quad (9.5.25)$$

If a local iteration is applied, the iterative strain increment $\delta\boldsymbol{\varepsilon}$ will turn to a fixed total strain increment $\Delta\boldsymbol{\varepsilon}$ during a global iteration.

9.5.3 The Duvant–Lions Type Viscoplastic Cap Model

The viscoplastic strain rate vector and hardening parameter are, respectively, defined as

$$\dot{\boldsymbol{\varepsilon}}^{\mathrm{vp}} = \frac{1}{\tau}[C]^{-1}(\boldsymbol{\sigma} - \bar{\boldsymbol{\sigma}}) \quad (9.5.26)$$

$$\dot{k} = \frac{1}{\tau}(k - \bar{k}) \quad (9.5.27)$$

where τ is a material constant called the relaxation time; the pair $(\bar{\boldsymbol{\sigma}}, \ \bar{k})$ are the stress and hardening parameter of the inviscid material (a bar is used to denote the variable of the inviscid plastic model), which can be viewed as a projection of the current stress on the current yield surface; k and $\dot{k}$ are hardening parameter and its differential with respect to time.

It can be seen from Eq. (9.5.26) that the viscoplastic strain rate is simply defined by the difference between the true stresses and the stresses obtained by the inviscid model which is quite different from that of the Perzyna type (Eq. 9.5.12).

1. **Static yield functions**

The Duvant–Lions-type cap model plastic yield surface function f is the same with the Perzyna's type.

2. **Solution algorithms**

The strain rate in Eqs. (9.5.26) and (9.5.27) is integrated over a time step t from t to $t+\Delta t$, to yield the incremental strains and stresses.

$$\Delta\boldsymbol{\varepsilon} = \Delta\boldsymbol{\varepsilon}^{\mathrm{e}} + \Delta\boldsymbol{\varepsilon}^{\mathrm{vp}} \quad (9.5.28)$$

$$\Delta\boldsymbol{\sigma} = [C]\Delta\boldsymbol{\varepsilon}^{\mathrm{e}} = [C](\Delta\boldsymbol{\varepsilon} - \Delta\boldsymbol{\varepsilon}^{\mathrm{vp}}) \quad (9.5.29)$$

where $\Delta\boldsymbol{\varepsilon}$ is the total incremental strain vector; $\Delta\boldsymbol{\varepsilon}^{\mathrm{e}}$ is the elastic incremental strain vector; $\Delta\boldsymbol{\varepsilon}^{\mathrm{vp}}$ is the viscoplastic incremental strain vector. $\Delta\boldsymbol{\sigma}$ is the incremental stress vector.

Based on the Euler method, the viscoplastic incremental strain vector $\Delta\boldsymbol{\varepsilon}^{\mathrm{vp}}$ can be approximated as

$$\Delta\boldsymbol{\varepsilon} = [(1-\chi)\dot{\boldsymbol{\varepsilon}}_{\mathrm{t}}^{\mathrm{vp}} + \chi\dot{\boldsymbol{\varepsilon}}_{\mathrm{t+\Delta t}}^{\mathrm{vp}}]\Delta t \tag{9.5.30}$$

where χ is an adjustable integration parameter, $0 \le \chi \le 1$. The integration scheme is explicit if $\chi = 0$ and fully implicit if $\chi = 1$. This solution algorithm is conditionally stable when $\chi \le 0.5$ and unconditionally stable when $\chi > 0.5$. The fully implicit integration scheme, $\chi = 1$, is used here in the numerical algorithm just for simplification.

Integrate Eq. (9.5.26) over a time step t.

$$\Delta\boldsymbol{\varepsilon}^{\mathrm{vp}} = \frac{\Delta t}{\tau}[C]^{-1}(\boldsymbol{\sigma}_{n+1} - \bar{\boldsymbol{\sigma}}_{n+1}) \tag{9.5.31}$$

Substitute Eq. (9.5.31) into Eq. (9.5.29).

$$\Delta\boldsymbol{\sigma} = \boldsymbol{\sigma}_{n+1} - \bar{\boldsymbol{\sigma}}_{n+1} = [C]\Delta\boldsymbol{\varepsilon} - \frac{\Delta t}{\tau}(\boldsymbol{\sigma}_{n+1} - \bar{\boldsymbol{\sigma}}_{n+1}) \tag{9.5.32}$$

By solving $\Delta\boldsymbol{\sigma}_{\mathrm{n+1}}$ from Eq. (9.5.33), one obtains

$$\boldsymbol{\sigma}_{n+1} = \frac{(\boldsymbol{\sigma}_n + C\Delta\boldsymbol{\varepsilon}) + \frac{\Delta t}{\tau}\bar{\boldsymbol{\sigma}}_{n+1}}{1 + \frac{\Delta t}{\tau}} \tag{9.5.33}$$

where $\Delta\boldsymbol{\sigma}_{\mathrm{n}} + [C]\Delta\boldsymbol{\varepsilon}$ may be treated as an elastic trial stresses.

Similarly, the hardening parameter may be expressed as

$$k_{n+1} = \frac{k_n + \frac{\Delta t}{\tau}\bar{k}_{n+1}}{1 + \frac{\Delta t}{\tau}} \tag{9.5.34}$$

It is apparent that the Duvant–Lions' model is very easy to implement, since the viscoplastic solution is simply an update of the inviscid solution. The numerical implementation of the Duvant–Lions model is apparently easier compared with the Perzyna model, which requires many matrix operations.

9.5.4 Illustration Example

The simulated uniaxial strain test, presented by Kantona [9], has been used to prove the adequacy of this viscoplastic cap model under different loading/unloading strain rates.

A hypothetical uniaxial strain loading history: the axial strain of the soil under compression is increased at a constant rate ($\dot{\varepsilon}_1 = 0.03\%/\mathrm{s}$) for 1 s, held constant ($\dot{\varepsilon}_1 = 0.0$) for 4 s, unloaded at a constant rate ($\dot{\varepsilon}_1 = -0.015\%/\mathrm{s}$) for 0.5 s, and held constant afterwards (Fig. 9.13).

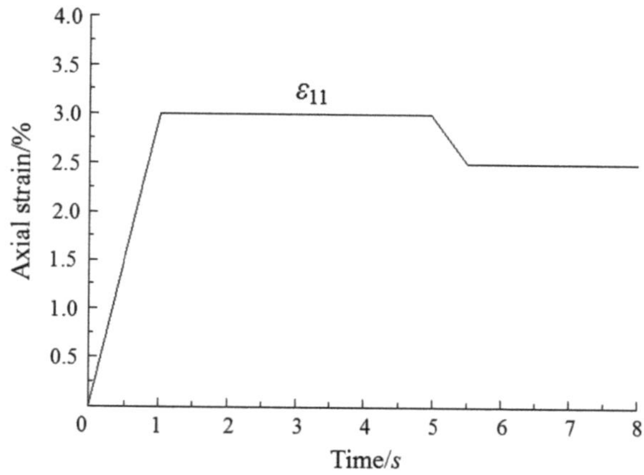

Fig. 9.13 Axial strain history for uniaxial strain test

The material parameters used for cap model are those for McCormick Ranch sand: K=66.7 ksi, G=40 ksi, α=0.25 ksi, β=0.67 ksi^{-1}, γ=0.18 ksi, θ=0.0, W=0.066, D=0.67 ksi^{-1}, R=2.5, X_0 =0.189 ksi, and T=0.0 ksi.

For the Perzyna model, the two parameters, N and f_0, were assumed to be 1.0 and 0.25 ksi based on experience data, respectively. Three values of the fluidity parameter (η=0.0035, 0.015, and 0.032) were examined similarly. According to Eq. (9.5.3), when η decreases, the viscoplastic strain decreases, and the stress is close to elastic, which implies the axial stress will increase. The stress response becomes purely elastic as $\eta \to 0$, and purely plastic as $\eta \to \infty$.

For the Duvant–Lions model, three values of the relaxation time (τ=1.0, 0.25, 0.125) were examined to illustrate its effects on the stress response. As shown in Fig. 9.14, the stress response increases as the relaxation time τ increases. According to Eqs. (9.5.26) and (9.5.31), when τ increases, the viscoplastic strain decreases, and the axial stress is close to elastic, which implies the stress response will increase. Although it is not plotted in Fig. 9.14, the stress responses will become purely elastic as $\tau \to \infty$, and purely plastic as $\tau \to 0$.

By comparing the stresses resulting from the two models in Fig. 9.14, it can be seen that each pair of the relaxation time and fluidity parameter yields nearly the same stresses. For instance, the axial stress history with τ=1.00 from using the Duvant–Lions model was very close to that with η=0.0035 from using the Perzyna model. Likewise, stresses obtained from using the Duvant–Lions model with τ=0.25 and 0.125 are nearly the same with those obtained from using the Perzyna model with η=0.015 and 0.032, respectively. The ratio of the three relaxation times is 8:2:1, while that of the fluidity parameters is approximately 1:2:9. Therefore, a certain relationship between τ and η may exist and the viscoplasticities of these two types may be equivalent for this example.

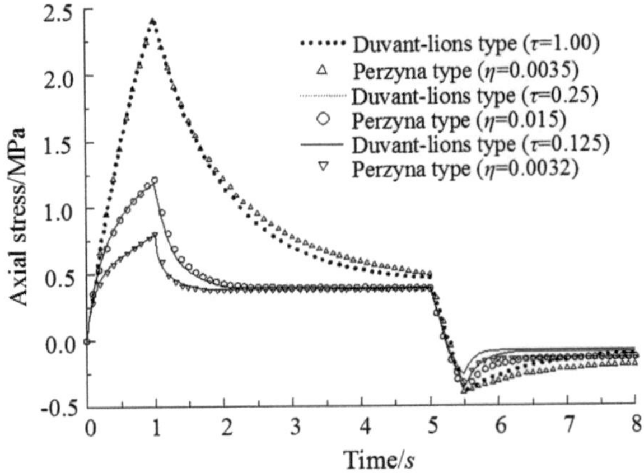

Fig. 9.14 Axial stresses for different values of τ and η

Questions

1. What is the basic mechanical characteristic for geomaterial?
2. What are the advantages and disadvantages of viscoelastic constitutive model for geomaterial?
3. How can you build viscoelastic–plastic constitutive model for geomaterial?

References

1. Bragov AM, Lomunov AK, Sergeichev IV, Proud W, Tsembelis K, Church PA (2005) Method for determining the main mechanics properties of soft soils at high strain rates (103–105 s^{-1}) and load amplitudes up to several gigapascals. Tech Phys Lett 31(6):530–531
2. Proud WG, Chapman DJ, Williamson DM, Sembelis KT, Addiss J, Bragov A, Lomunov A, Cullis IG, Church PD, Gould P, Porter D, Cogar JR, Borg J (2007) The dynamic compaction of sand and related porous systems. In: Shock compression of condensed matter, pp 1403–1408
3. An JX (2010) Soil behavior under blast loading. Doctoral dissertation, The University of Nebraska-Lincoln, USA
4. Matasovic N, Vucetic M (1993) Cyclic characterization of liquefiable sands. J Geotech Eng 11:1085–1121
5. Chen GX (2007) Geotechnical earthquake engineering (4–6 Empirical estimate of soil dynamic shear modulus and damping ratio). Science Press, Beijing, pp 150–166
6. Perzyna P (1966) Fundamental problems in viscoplasticity. Adv Appl Mech 9:243–377
7. Tong XL (2005) Finite element simulation of soil behaviors under high strain rate loading. Masteral dissertation, The University of Nebraska-Lincoln, USA
8. Simo JC, Ju JW, Pister KS, Taylor RL (1988) Assessment of cap model: consistent return algorithms and rate-dependent extension. J Eng Mech ASCE 114(2):191–218
9. Katona MG (1984) Evaluation of viscoplastic cap model. J Geotech Eng 110(8):1106–1125

Chapter 10
Limit Analysis for Geotechnical Engineering

10.1 Overview

Limit analysis is one of the most powerful aspects of the theory of plasticity due to its ability to easily predict approximate values for the collapse load in a very wide range of applications.

Limit analysis is based on the rigid-plastic model [1–3]. Under the action of load, the change is called limit state for part or all of the rigid-plastic body from the static equilibrium to the critical state of motion, and the corresponding load is called the limit load. Limit analysis is an analysis method for solving this kind of boundary value problem, its essence is to determine the stress field and velocity field by solving the static equation, motion equation and corresponding boundary conditions.

10.2 The Basic Equations for Limit Analysis

As an example, all of the equations for limit analysis are given below for two-dimensional problem of geotechnical engineering.

1. **Equilibrium equation**

$$\begin{cases} \frac{\partial \sigma_x}{\partial x} + \frac{\partial \tau_{xy}}{\partial y} = 0 \\ \frac{\partial \tau_{xy}}{\partial x} + \frac{\partial \sigma_y}{\partial y} - \gamma = 0 \end{cases} \tag{10.2.1}$$

2. **Yield criterion (M-C yield criterion)**

$$f = \sigma_1 - \sigma_2 \tan^2(45^\circ + \frac{\varphi}{2}) - 2c \tan(45^\circ + \frac{\varphi}{2}) = 0, \tag{10.2.2}$$

Y. Liu and Y. Zheng, *Plastic Mechanics of Geomaterial*, Springer Geophysics,
https://doi.org/10.1007/978-981-13-3753-6_10

where $\left.\begin{matrix}\sigma_1\\ \sigma_2\end{matrix}\right) = \frac{\sigma_x+\sigma_y}{2} \pm \sqrt{\left(\frac{\sigma_x-\sigma_y}{2}\right)^2 + \sigma_{xy}^2}$

3. **Geometric equation**

$$\begin{cases} \dot{\varepsilon}_x = \frac{\partial v_x}{\partial x} \\ \dot{\varepsilon}_z = \frac{\partial v_z}{\partial z} \\ \dot{\varepsilon}_{xz} = \frac{\partial v_x}{\partial z} + \frac{\partial v_z}{\partial x} \end{cases} \tag{10.2.3}$$

4. **Flow rule**

$$\begin{cases} \dot{\varepsilon}_x = \mathrm{d}\lambda \frac{\partial f}{\partial \sigma_x} \\ \dot{\varepsilon}_z = \mathrm{d}\lambda \frac{\partial f}{\partial \sigma_z} \\ \dot{\varepsilon}_{xz} = \mathrm{d}\lambda \frac{\partial f}{\partial \sigma_{xz}} \end{cases}, \tag{10.2.4}$$

where f is the yield surface, with no elastic strain in the rigid-plastic body.

Several kinds of commonly used methods are given below for limit analysis of geotechnical engineering.

10.3 The Characteristic Line Method in Limit Analysis

Strict limit analysis should solve the four classes equation in the previous section, but it is difficult. The simplified method is relaxed constraints.

10.3.1 Slip Line of Stress

Slip line of stress of limit analysis only considers the equilibrium equation and yield criterion.

For the geomaterial obeys M-C yield criterion or failure condition, the following equation can be obtained based on the relationship between the failure line and the limit stress circle in Fig. 10.1.

$$\begin{cases} \sigma_x = p - R\cos 2\theta \\ \sigma_y = p + R\cos 2\theta \\ \tau_{xy} = R\sin 2\theta \end{cases}, \tag{10.3.1}$$

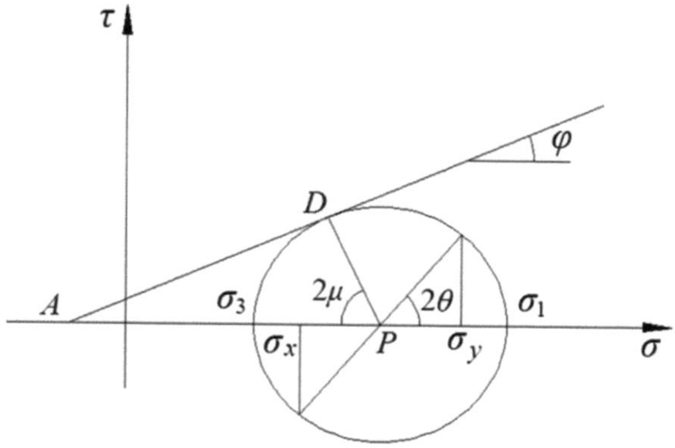

Fig. 10.1 The relationship between M-C yield surface and the Mohr circle limit of stress

where p, R are mean stress and radius of stress circle respectively, they are

$$p = \frac{1}{2}(\sigma_x + \sigma_y) = \frac{1}{2}(\sigma_1 + \sigma_3) \tag{10.3.2}$$

$$R = (p + \sigma_c)\sin\varphi, \tag{10.3.3}$$

where $\sigma_c = c\cot\varphi$.

Substituting Eq. (10.3.1) into equilibrium equation (Eq. 10.2.1), and performing the geometric operation, the limit equilibrium differential equation is obtained with the unknown quantity of p, θ.

$$\begin{cases} \frac{\partial p}{\partial y}(1 + \sin\varphi\cos 2\theta) + \frac{\partial p}{\partial x}\sin\varphi\sin 2\theta + 2R(-\frac{\partial \theta}{\partial y} + \frac{\partial \theta}{\partial x}\cos 2\theta) = 0 \\ \frac{\partial p}{\partial y}\sin\varphi\sin 2\theta + \frac{\partial p}{\partial x}(1 - \sin\varphi\cos 2\theta) + 2R(\frac{\partial \theta}{\partial y}\cos 2\theta + \frac{\partial \theta}{\partial x}\sin 2\theta) = \gamma \end{cases} \tag{10.3.4}$$

If p, θ can be calculated from Eq. (10.3.4), then the stress state σ_x, σ_y, τ_{xy} can be determined by Eq. (10.3.1). And the limit load p_u can be worked out according to the boundary conditions of the problem. Therefore, limit load problem is summed up as a mathematical problem solving limit equilibrium differential equations.

In mathematics, Eq. (10.3.4) is referred to as the first-order linear partial differential equation. It is difficult to be solved directly. It needs to use the method of characteristic line. This is because the solution for Eq. (10.3.4) can only be got along the special curve which is called the characteristic line, instead of along any line in plane domain x, y. So it is called the method of characteristic line. If the characteristic line is obtained through Eq. (10.3.4), and the solutions are obtained by integrating Eq. (10.3.4) along the characteristic line, it can be proved that Eq. (10.3.4) is a first-order linear partial differential equations of hyperbolic-type. It is accompanied with two groups of characteristic line (Fig. 10.2) with the following equation:

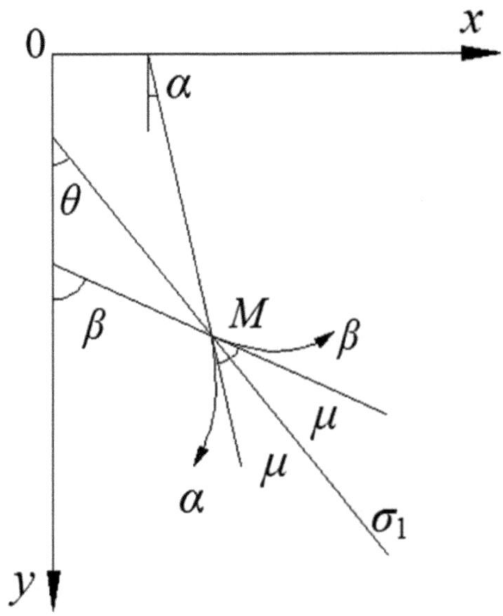

Fig. 10.2 The characteristic line of stress

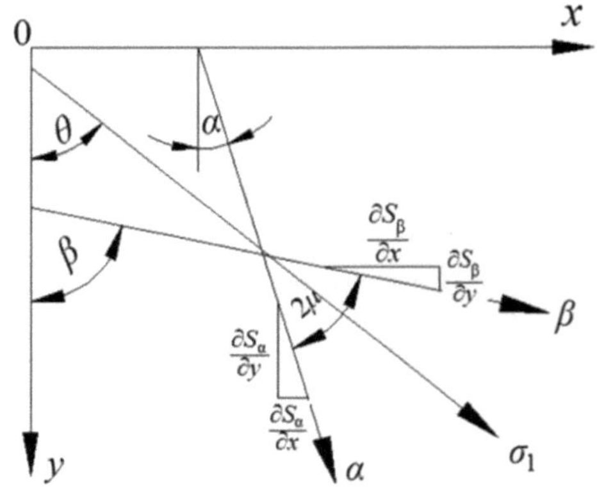

Fig. 10.3 Directional derivative along the slip line

$$\frac{\mathrm{d}x}{\mathrm{d}y} = \tan(\theta \mp \mu), \tag{10.3.5}$$

where $\mu = \frac{\pi}{4} - \frac{\varphi}{2}$.

The characteristic line in math is the slip lines in plastic mechanics. A solution of characteristic line is just that of slip line.

Slip line is used to solve Eq. (10.3.4). If curvilinear coordinates along α-slip line and β-slip line, as shown in Fig. 10.3, the directional derivative along S_α-slip line and S_β-slip line will be

$$\begin{cases} \frac{\partial}{\partial S_\alpha} = \cos(\theta-\mu)\frac{\partial}{\partial x} + \sin(\theta-\mu)\frac{\partial}{\partial y} \\ \frac{\partial}{\partial S_\beta} = \cos(\theta+\mu)\frac{\partial}{\partial x} + \sin(\theta+\mu)\frac{\partial}{\partial y} \end{cases} \tag{10.3.6a}$$

The following will be obtained:

$$\begin{cases} \sin 2\mu \frac{\partial}{\partial x} = \sin(\theta+\mu)\frac{\partial}{\partial S_\alpha} - \sin(\theta-\mu)\frac{\partial}{\partial S_\beta} \\ \sin 2\mu \frac{\partial}{\partial y} = -\cos(\theta+\mu)\frac{\partial}{\partial S_\alpha} - \cos(\theta-\mu)\frac{\partial}{\partial S_\beta} \end{cases} \tag{10.3.6b}$$

Equation (10.3.6a, 10.3.6b) will be suitable for any p, θ. Through substituting Eq. (10.3.6a) into Eq. (10.3.4) and using Eq. (10.3.6b), and the following will be obtained after simplifying.

Along α-slip line

$$\frac{\partial p}{\partial S_\alpha}\sin 2\mu - 2R\frac{\partial \theta}{\partial S_\alpha} = \gamma \sin(\theta-\mu) \tag{10.3.7a}$$

Along β-slip line

$$\frac{\partial p}{\partial S_\beta}\sin 2\mu + 2R\frac{\partial \theta}{\partial S_\beta} = -\gamma \sin(\theta+\mu) \tag{10.3.7b}$$

This is the differential equation of limit equilibrium along the α-slip line and β-slip line, it reflects the change law of p, θ along the α-slip line and β-slip line.

Equation (10.3.7a, 10.3.7b) is still a nonlinear partial differential equation, and it is difficult to integrate directly. So the numerical method is needed to solve it. Because of Eq. (10.3.7a, 10.3.7b) is the integral of p, θ along α-slip line and β-slip line respectively, there are

$$\frac{\partial}{\partial S_\alpha} = \frac{\mathrm{d}}{\mathrm{d}S_\alpha},\ \frac{\partial}{\partial S_\beta} = \frac{\mathrm{d}}{\mathrm{d}S_\beta}$$

And

$$\mathrm{d}S_\alpha = \mathrm{d}y\ \sec(\theta-\mu),\ \mathrm{d}S_\beta = \mathrm{d}y\ \sec(\theta+\mu),\ \sin 2\mu = \cos\phi$$

Substituting the above equations into Eq. (10.3.7a, 10.3.7b), we get

Along α-slip line

$$\mathrm{d}p - 2(p+\sigma_c)\tan\varphi \mathrm{d}\theta = \frac{\gamma \sin(\theta+\mu)\mathrm{d}y}{\cos\varphi\cos(\theta-\mu)} \tag{10.3.8a}$$

Along β-slip line

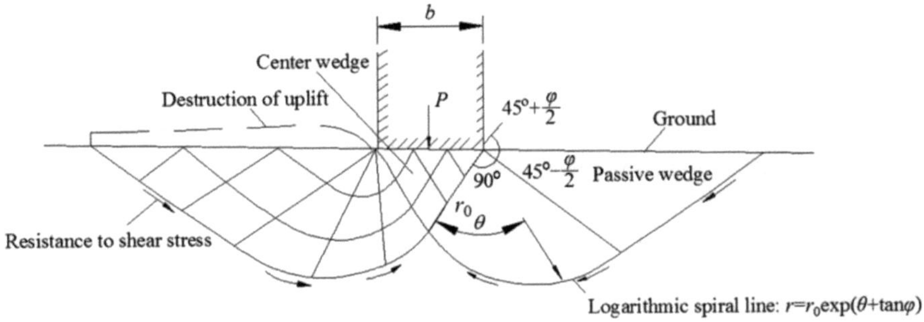

Fig. 10.4 The overall shear failure mode of Prandtl foundation

$$\mathrm{d}p + 2(p + \sigma_c)\tan\varphi \mathrm{d}\theta = -\frac{\gamma\sin(\theta - \mu)\mathrm{d}y}{\cos\varphi\cos(\theta + \mu)} \tag{10.3.8b}$$

This is the difference equation of geomaterial with weight for p, σ_c of the geomaterial along the α-slip line and β-slip line, respectively. The field distribution of slip line and ultimate load can be put forward by using the finite difference method or the characteristics of the slip line and solving all kinds of boundary value problem of geotechnical engineering with weight.

10.3.2 Prandtl Solution of Stress Characteristic Line for Ultimate Load of a Half-Infinite Plane Body

Prandtl uses the stress characteristic line method to solve the limit pressure of rigid die pressing into the massless semi-infinite rigid-plastic medium.

In Fig. 10.4, plastic limit equilibrium domain is divided into five parts. One is the center wedge below the base, also called active Rankine area. Within the central area, the major principal stress direction is vertical, and the direction of minor principal stress is horizontal. According to the theory of limit equilibrium, the included angle is $\frac{\pi}{4}+\frac{\varphi}{2}$ between the minor principal stress and fracture surface, which is known as the included angle between the two sides of the central area and the horizontal plane. And those that are adjacent to the central district are two radiation shear zone, also called the Prandtl area, and they are covered by a group of logarithmic spiral and a set of radiation straight line. It is a fan-shaped district with the boundary of a logarithmic spiral, and its central angle is a right angle. Two passive Rankine areas are adjacent to the other side of the two Prandtl areas. Its major principal stress direction is horizontal, and the minor principal stress direction is vertical, and the included angle is $\frac{\pi}{4} - \frac{\varphi}{2}$ between the fracture plane and the horizontal plane.

As shown in Fig. 10.5, the plastic zone under base with width of l can be divided into the active zone I, transition zone II and passive zone III according to the stress boundary condition and the movement trend. The specific boundary of each region

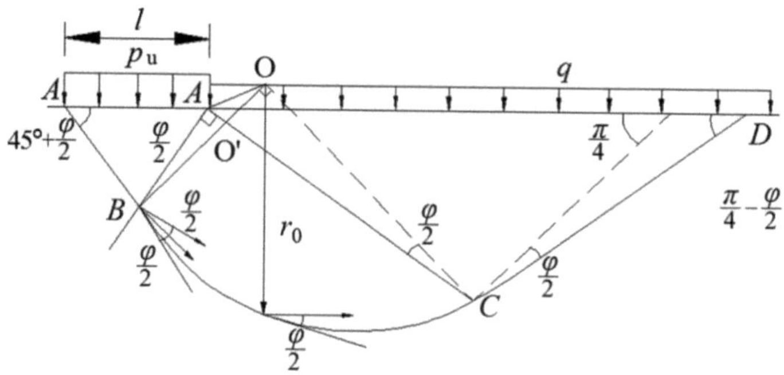

Fig. 10.5 Prandtl stress characteristic line for infinite body at half plane

is determined by the stress boundary conditions. The right half is analyzed for the symmetry of half-infinite body. First, the boundary condition is analyzed for zone III.

1. In zone III with the border *AD*, the boundary condition is $\sigma_n = q$, $\tau_n = 0$, $\theta = \pi$. Due to the effect of p_u on the surface of *AA*, soil below the surface *AD* is in a passive state, with a tendency to move up. Due to constant angle $\theta = \frac{\pi}{2}$, the slip line equation is

$$y = \cot[\frac{\pi}{2} + (\frac{\pi}{4} - \frac{\varphi}{2})]x + C$$

Then

$$p = q - (p + \sigma_c) \sin\varphi \cos 2(\pi - \frac{\pi}{2})$$

So

$$p = \frac{q + \sigma_c \sin\varphi}{1 - \sin\varphi} \tag{10.3.9}$$

According to the p, θ in zone III, the integral constants for β-slip line will be

$$C_\beta = (q + \sigma_c)\frac{e^{\pi \tan\varphi}}{(1 - \sin\varphi)} \tag{10.3.10}$$

2. In zone I, with the border *AAB*, the boundary condition is $\sigma_n = p_u$, $\tau_n = 0$, $\theta = \pi$. Due to the effect of p_u, soil below the surface *AA* is in an active state, with a tendency to move down. So $p_u = \sigma_1$, and $\theta = 0$. The slip line equation will be

$$y = \cot[(\frac{\pi}{4} - \frac{\varphi}{2})]x + C$$

The same procedure may be easily adapted to obtain

$$p = \frac{p_u - \sigma_c \sin\varphi}{1 + \sin\varphi} \tag{10.3.11}$$

After arranging, the integral constants for β-slip line will be

$$C_\beta = \frac{p_u + \sigma_c}{1 + \sin\varphi} \tag{10.3.12}$$

3. In transition zone II with the border ABC, because θ changes from 0 to $\frac{\pi}{2}$, the stress characteristic line should be logarithmic spiral in this zone, and $\angle BAC = \frac{\pi}{2}$. According to the stress condition of point B, it can be assumed that the polar equation of logarithmic spiral is

$$r = r_0 e^{(\theta - \frac{\pi}{4})\tan\frac{\varphi}{2}},$$

where θ here is the expansion angle of logarithmic spiral.

4. Determining the limit load p_u

According to the characteristic of the slip line, C_α, C_β should be identical for the same group of slip line. So Eqs. (10.3.10) and (10.3.12) are consistent, and we obtain

$$C_\beta = (q + \sigma_c)\frac{e^{\pi \tan\varphi}}{(1 - \sin\varphi)} = C_\beta = \frac{p_u + \sigma_c}{1 + \sin\varphi}$$

Then the limit load p_u can be put forward.

$$p_u = (q + \sigma_c)\tan^2\left(\frac{\pi}{4} + \frac{\varphi}{2}\right)e^{\pi \tan\frac{\varphi}{2}} - \sigma_c = qN_q + cN_c \tag{10.3.13}$$

where

$$N_q = \tan^2\left(\frac{\pi}{4} + \frac{\varphi}{2}\right)e^{\pi \tan\frac{\varphi}{2}}$$

$$N_c = (N_q - 1)\cot\varphi$$

where N_q, N_c are bearing capacity coefficient or ultimate load coefficient respectively related to the internal friction angle of geomaterial, which reflects the impact on the bearing capacity and the limit load for pressure intensity q and cohesion c.

10.3.3 Velocity Slip Line Field for Plane Strain Problem

The solution of the velocity slip line is based on the geometric equations and flow rule, ignoring the equilibrium condition.

Limit load solution of the slip line field has been adopted for nearly one century, but the most used one is stress characteristic line, and there are still many issues to be dealt with in solving the velocity field and velocity slip line. For example, the associated flow rule is always adopted in solving the velocity field, and the stress characteristic line will be the same as the velocity slip line. In fact, geomaterial does not obey Drucker Postulate and associated flow rule, and stress slip line field cannot coincide with velocity slip line field.

For metal, velocity slip line is consistent with the stress characteristic line, and their included angle is $\frac{\pi}{4}$ with axis x or y, so the included angle is zero between the slip line and the characteristic line. For geomaterial, the included angle is $\frac{\varphi}{2}$ between the velocity slip line and the Mohr–Coulomb characteristic line. Therefore, at any point in the field, the included angle will be $\frac{\varphi}{2}$ between slip line and characteristic line, rather than superposition.

10.4 Principle of Limit Analysis and Approximate Method

Limit state: if the strengthening of ideal plastic material and the change of geometry size caused by the deformation is ignored, the plastic flow can occur in the case of constant force when the external force reaches a certain value, then it is called that the object is in a limit state, and the load is ultimate load. The limit state of the object is a critical state between a static equilibrium state and plastic flow state. The limit state is characterized as the stress field is static permission field and strain rate field is a kinematic admissible field.

Limit analysis theorem is a general theorem for ideal plastic (rigid-plastic body) under the limit state. This theorem can be used to solve the problem directly, which avoids the integral of the differential equation and the mathematical difficulty in the calculation. Limit analysis theorem is a theory to obtain the solutions of the upper limit and lower limit of ultimate load, and relaxation of some constraints. The scope of the exact solution will be determined by the upper and lower solutions. The most common method is the static method and the kinematic method in the Limit analysis. A static method is required to construct a static permission field, and kinematic method requires constructing a kinematic admissible field.

10.4.1 Limit Analysis Theorem

Static permission solution: The stress field is called static permission stress field which satisfies the equilibrium Eq. (10.2.1), yield condition Eq. (10.2.2) and part of the boundary load condition, and the load is referred to static permission load or solution which corresponds to the stress field.

Kinematic admissible solution: The velocity field is called the kinematic velocity field which meets geometric Eq. (10.2.3), motion Eq. (10.2.4) and the velocity boundary condition. The stress field is called kinematic admissible stress field which is corresponding to the velocity field and meets the energy dissipation rate, the stress boundary condition, and the sliding surface stress is equal to the shear strength. The corresponding load is kinematic admissible load or kinematic admissible solution.

Geomaterial should satisfy equilibrium state under the action of static permission load. Geomaterial should satisfy flow deformation state under the action of kinematic admissible load. Therefore, static permission load should not be greater than the real limit load, and the kinematic admissible load is not less than the real limit load. The real limit load will be the maximum of static permission load and the minimum of the kinematic admissible load. This is just the physical meaning of the principle of upper and lower limits.

10.4.2 Example of Limit Analysis Principle

1. **Example of static admissible solution**

To get the stress field of the limit state which meets the static equilibrium, such as Rankine theory of earth pressure (Fig. 10.6).

Hypothesis: The back is upright and smooth for retaining wall; the surface of the filled soil is horizontal.

Wall back moves away from soil, and the stress state of soil reaches the limit equilibrium. The maximum principal stress is the gravity stress, and the minor principal

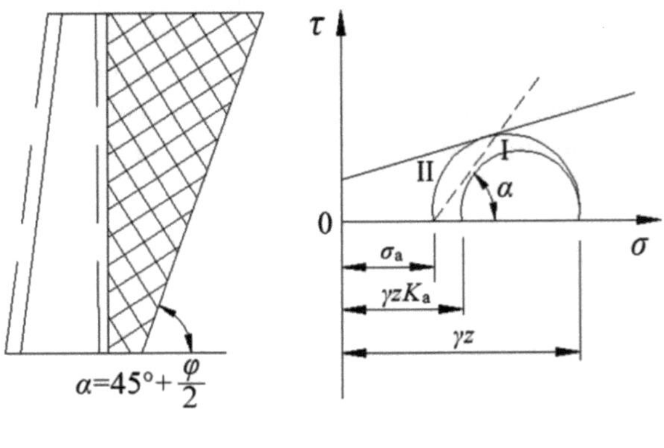

Fig. 10.6 Calculation principle for Rankine earth pressure

stress is the horizontal earth pressure. Earth pressure can be calculated according to Mohr–Coulomb theory.

$$\begin{aligned}\sigma_a &= \sigma_3 = \sigma_1 \tan^2(45^\circ - \frac{\varphi}{2}) - 2c\tan(45^\circ - \frac{\varphi}{2}),\\ &= \sigma_1 k_a - 2c\sqrt{k_a}\end{aligned} \tag{10.4.1}$$

where σ_a is active earth pressure intensity, $\sigma_1 = \gamma z$.

$$k_a = \tan^2(45^\circ - \frac{\varphi}{2})$$

$$E_a = \int_{z_0}^{h} \sigma_a \mathrm{d}z = \frac{1}{2}(h - z_0)(\gamma h k_a - 2c\sqrt{k_a}), \tag{10.4.2}$$

where E_a is just the active earth pressure.

The solutions should be a static admissible solution and the minimum obtained by Rankine theory of earth pressure. So the calculation results based on Rankine theory of earth pressure tend to be minor than the right ones.

2. **Example of kinematic admissible solution**

Provide the limit load by constructing the kinematic admissible stress field, such as Coulomb theory of earth pressure (Fig. 10.7).

Assumption: $c=0$; The sliding failure surface is a plane through the wall heel.

Earth pressure is produced under movement trend of soil along the slip surface.

The wedge ABM meets mechanical limit equilibrium under the active earth pressure. That is three forces of the wedge satisfying the condition of the below closing triangle (G, the gravity of the earth, the reaction on the slip surface R, and the reaction of the wall back E) (Fig. 10.8).

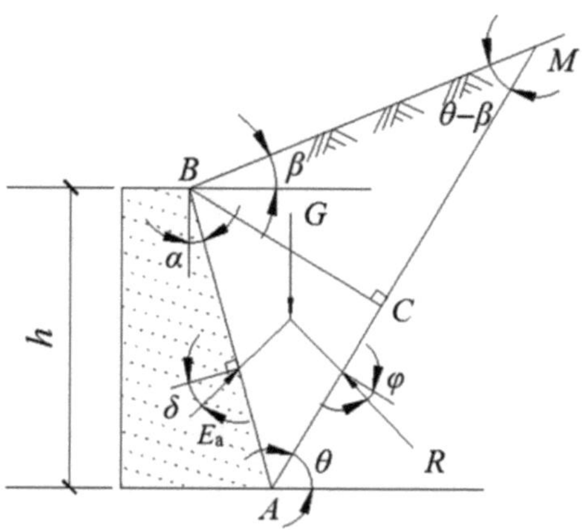

Fig. 10.7 Schematic diagram for Coulomb earth pressure

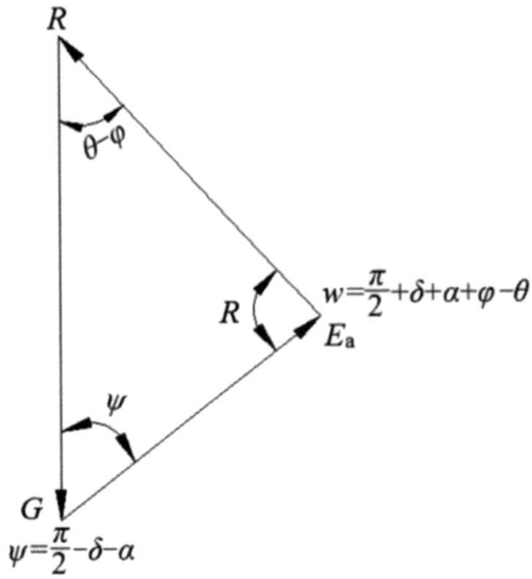

Fig. 10.8 Mechanical equilibrium under active earth pressure

According to sine theorem,

$$\frac{E}{\sin(\theta-\varphi)}=\frac{G}{\sin\omega}$$

$$E=\frac{G\sin(\theta-\varphi)}{\sin\omega} \tag{10.4.3}$$

Through the derivation about θ, the maximum of E which should be the active earth pressure is obtained.

$$\frac{\mathrm{d}E}{\mathrm{d}\theta}=0$$

To obtain the failure angle θ_{cr},

$$\theta_{\mathrm{cr}}=\arctan\left(\frac{\sin\beta\cdot s_{\mathrm{q}}+\cos(\alpha+\varphi+\delta)}{\cos\beta\cdot s_{\mathrm{q}}-\sin(\alpha+\varphi+\delta)}\right), \tag{10.4.4}$$

where $s_{\mathrm{q}}=\sqrt{\frac{\cos(\varphi+\delta)\sin(\varphi+\delta)}{\cos(\alpha-\beta)\sin(\varphi-\beta)}}$.

Substitute Eq. (10.4.4) into Eq. (10.4.3),

$$E_{\mathrm{a}}=\frac{1}{2}\gamma h^2K_{\mathrm{a}}, \tag{10.4.5}$$

where $K_{\mathrm{a}}=\frac{\cos^2(\varphi-\alpha)}{\cos^2\alpha\cos(\alpha+\delta)[1+\sqrt{\frac{\sin(\varphi+\delta)\sin(\varphi-\beta)}{\cos(\alpha+\delta)\cos(\alpha-\beta)}}]^2}$.

It is a kinematic admissible solution.

Coulomb earth pressure theory is based on the stress field of kinematic admissible. The solutions should be a kinematic admissible solution, and the result is the maximum of earth pressure and too big.

Both are a comparatively good skill for constructing static admissible stress field (static admissible solution) and kinematic admissible stress field (kinematic admissible solution). It can also solve the problem with simple boundary conditions and homogenous medium. Great errors will be there if it is applied to complex engineering.

10.5 Numerical Limit Analysis

The theoretic method of limit analysis is always with the aid of theoretical assumptions, dividing the problem space into several separate areas so as to construct the required stress field for each part. The actual physical fields contain an infinite number of points with stress, strain and so on. It should be the simple case through the construction of the stress field in several regions. So, is there a finite method to simulate infinite well? Yes, and there are more mature methods, such as discrete element method. The computational domain is divided into a number of discrete elements, and internal variables within the element are formulated with node variables. The problem with infinite degrees of freedom is transformed into a problem of finite degrees of freedom. The approximate solution of the physical quantity can be obtained directly for the whole domain by a numerical algorithm. The extra assumption is not needed, and if the element number is reasonable, we can get the precision we need. Finite element method is the most commonly used method.

In 1975, British mechanician Zienkiewicz suggested to increase outer load or reduce the strength of geomaterial in the finite element method to calculate the safety factor of geotechnical engineering. It is the essentially finite element limit analysis method. When using the method of reducing strength, it is strength reduction FEM. But because FEM was still in the development stage, the method has not been widely accepted. But now the situation has been fundamentally changed, strength reduction FEM has been accepted by most scholars and engineers. Domestic scholars and the author have done a lot of work in improving calculation theory and calculation precision, and the calculation precision of the method has been improved greatly. This method has been applied to the calculation of rock slope, retaining structure for slope and landslide, and it has been extended to the calculation of the bearing capacity of foundation and the safety factor calculation of tunnel [4, 5]. The application range of strength reduction FEM has been expanded.

10.5.1 The Basic Principle of FEM Limit Analysis

FEM limit analysis can help gain failure mode and the corresponding safety factor of the model by decreasing the strength of geomaterial or increasing load to reach the limit state in the elastoplastic analysis of FEM. Actually, it is the application of a numerical method to solve the limit problems in geotechnical engineering, such as limit load or safety factor and so on. This method can consider the deformation and even failure of geomaterial, interaction of structure and geomaterial, and obtain the internal force of structure. FEM limit analysis is very close to the engineering design, and it will definitely lead the stability analysis of geomaterial into a new era.

10.5.2 Definition of Safety Factor

Take the analysis of slope stability as the example. In the analysis of slope stability, the safety factor is often used to represent the state of slope stability against sliding, and the value is equal to the ratio of the anti-slide force (moment) and slide force (moment).

As shown in Fig. 10.9a, the gravity W, slide force along the sliding surface T, and sliding resistance R. So, the safety factor FS can be represented as

$$FS = \frac{\text{anti - slide force}}{\text{slide force}} = \frac{\int_0^L (c + \sigma \tan\varphi)\mathrm{d}l}{\int_0^L \tau \mathrm{d}l} \tag{10.5.1}$$

where c is the cohesion of slide surface; φ is the internal friction angle of sliding surface; L is the length of slide surface.

Both the sides are divided by the FS at the same time in Eq. (10.5.1).

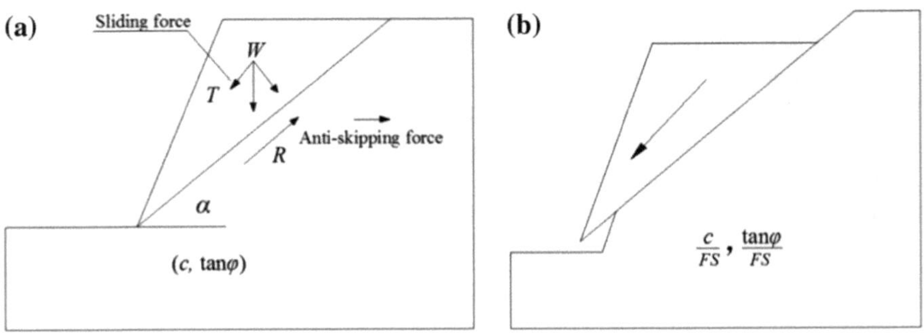

Fig. 10.9 Calculation of slope safety factor

$$1 = \frac{\int_0^1 (\frac{c}{FS} + \sigma \frac{\tan\varphi}{FS})\mathrm{d}l}{\int_0^1 \tau \mathrm{d}l} = \frac{\int_0^l (c' + \sigma \tan\varphi')\mathrm{d}l}{\int_0^1 \tau \mathrm{d}l}, \quad (10.5.2)$$

where

$$c' = \frac{c}{FS}, \tan\varphi' = \frac{\tan\varphi}{FS} \quad (10.5.3)$$

As a result, the safety coefficient can be expressed as

$$FS = \frac{c}{c'} = \frac{\tan\varphi}{\tan\varphi'} \quad (10.5.4)$$

It can be seen from Eq. (10.5.4), the safety factor of the slope is equal to the ratio of the actual shear stress and the shear strength of the limit state of the sliding surface. It can be called the safety factor of strength reduction. This definition of safety factor based on strength reduction is consistent with the definition of the safety factor of slope stability analysis of limit equilibrium. Both of them belong to the safety factor of strength reserve. It means the limit equilibrium and the average safety factor of the total slide surface, rather than the safety factor of stress points. Generally, there is no demand to calculate the sliding surface which can be clearly seen from the nephogram of shear strain or can be determined accurately by mechanical parameters.

10.5.3 Criterion for Limit State

A key problem in FEM limit analysis for slope stability is how to judge whether the slope reaches limit state according to the results of FEM calculation. Generally, the following three criteria are used:

1. Plastic strain is used as a criterion if it is interconnected from the slope foot to the slope top, namely, the plastic zone is used as a criterion of the damage if it transfixes from the inside to the ground or free face. The transfixion plastic zone only means to achieve the yield status, and it not necessarily means the overall destruction of soil. Thus, the transfixion plastic zone is just the necessary condition of failure, rather than a sufficient condition.
2. In the process of FEM calculation, the slope instability occurs at the same time with calculation misconvergence of FEM numerical simulation. Now numerical misconvergence is adopted as a judgment of the instability of slope in international common software.
3. With geotechnical engineering failure, strain and displacement mutates in slip surface, so does the relationship between safety factor (strength reduction factor)

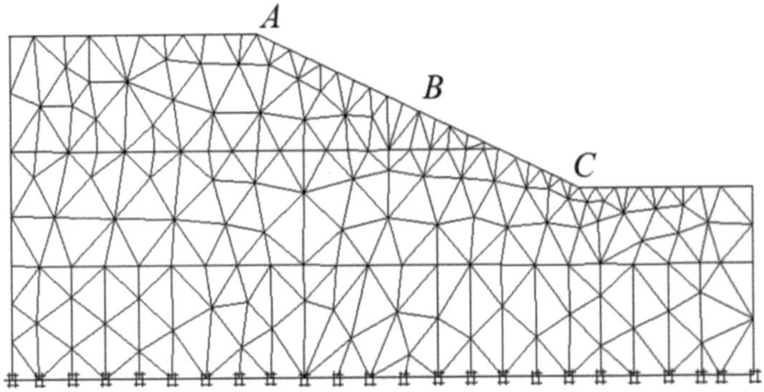

Fig. 10.10 The situation of characteristic point for slope

Fig. 10.11 The mutation of displacement of characteristic point with the increase of reduction coefficient

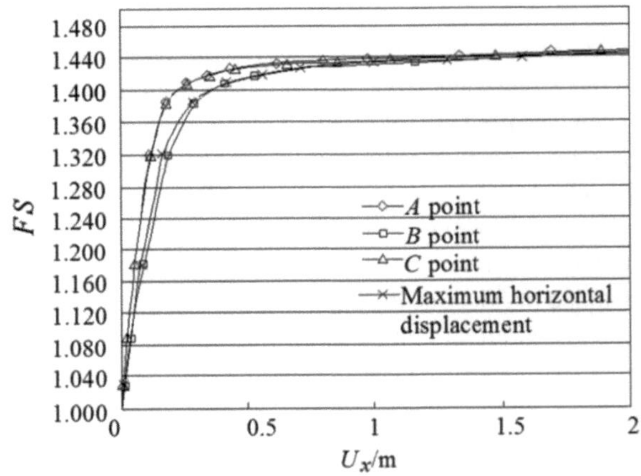

and displacement. Thus, these mutations can also be used as a failure criterion. But we should consider that the displacement will mutate under the earthquake.

Each characteristic point is selected at slope top, mid-slope, and slope foot respectively, as shown in Fig. 10.10. The calculation discloses the horizontal displacement of the characteristic point of slope increases with the increase of strength reduction factor, and mutation of horizontal displacement of characteristic points comes into being when slope reaches the limit state (Fig. 10.11). FEM program cannot find a solution satisfying the static equilibrium, the stress–strain relationship, and strength criterion. The misconvergence of FEM calculation emerges at this time from both the convergence criteria of force, and the convergence criteria of displacement.

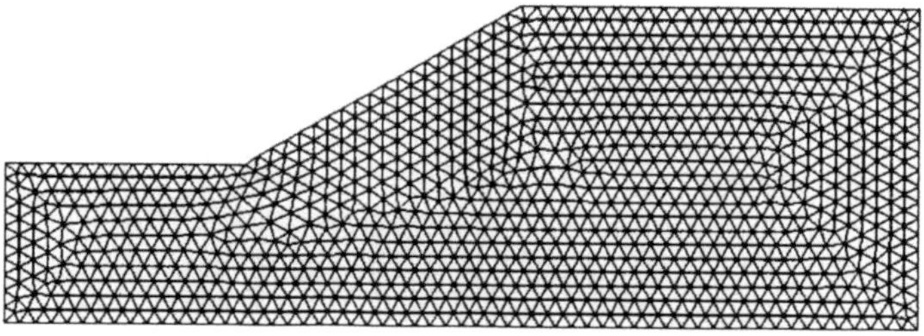

Fig. 10.12 Element grid

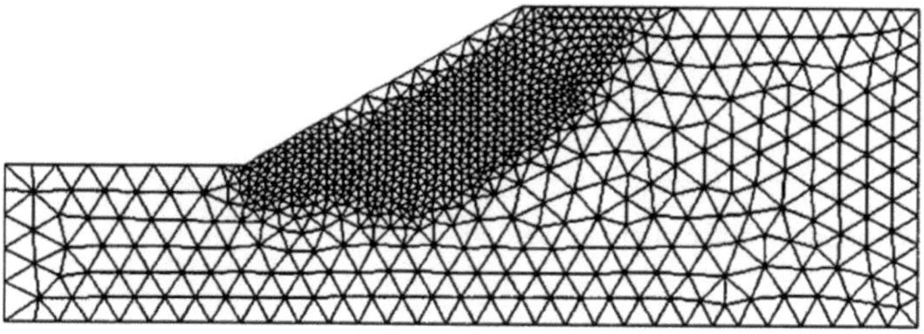

Fig. 10.13 Local dense in element meshing

10.5.4 Example for the Calculation of Slope Safety Factor

This section uses stability analysis of a homogeneous soil slope as an example, introducing the main result of stability analysis of soil slope by the FEM strength reduction analysis.

Free meshing (Fig. 10.12) can be used not only by the mixed element shape, but also only triangle element, while mapping mesh can be used only by quadrilateral element and hexahedron element. In the meshing process, local dense can also be carried out on the important part, and it can be kept loose in some unimportant domain. It's important to note to leave a smooth transition from dense to sparse, and the unit size does not change sharply, as shown in Fig. 10.13.

In this case of the safety factor calculation, the limit state is judged by the criterion of calculation convergence. When the strength reduction factor takes 1.56, FEM calculation converges. When the reduction coefficient is 1.67, FEM calculation misconverges. So safety factor of strength reserve is between 1.56 and 1.57. The safety factor of the slope is determined as 1.56. For this example, the safety factor is 1.55 by using the stability analysis software of Canadian GEO-SLOPE (Spencer method). Obviously, the safety factors are very close for that got from FEM strength reduction and from the traditional limit equilibrium method.

The colored part in Fig. 10.14 is the plastic zone. It can also be demonstrated through the plastic strain distribution of number, but each color should be reasonably given to each quantity of plastic strain, otherwise, it is difficult to distinguish the small plastic strain area from no plastic strain areas (that is, the elastic area), as shown in Figs. 10.15, and 10.16. Although the most part of the slope element is in

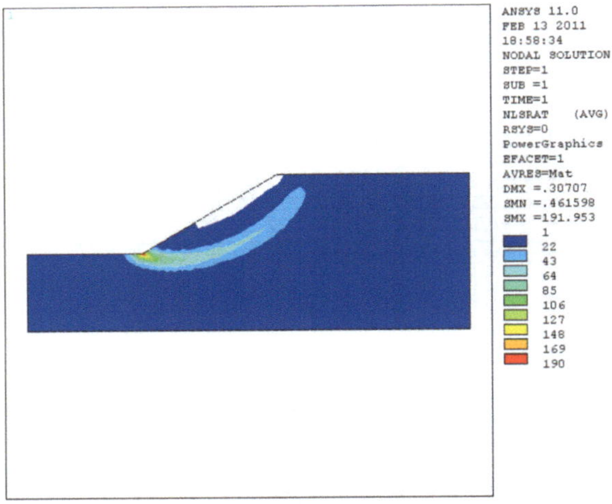

Fig. 10.14 Distribution range of plastic zone

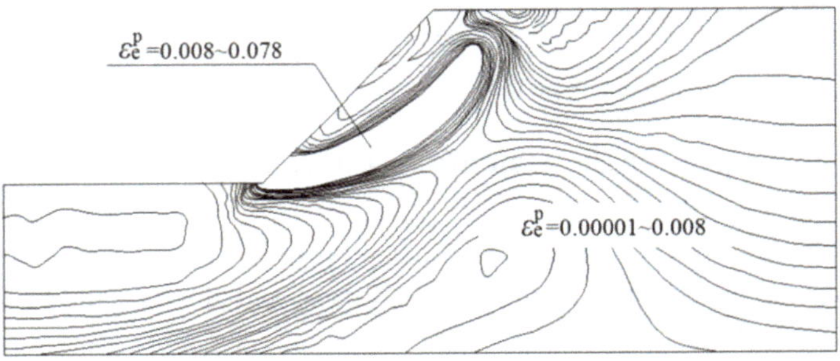

Fig. 10.15 Contour of equivalent plastic strain

plastic state, the elements outside the sliding surface have small plastic strain values (0.00001 ~ 0.008). The plastic strain values of the node on slip surface are relatively bigger, and the values are between 0.008 and 0.078. Figure 10.16 is distribution figure of the color nephogram, and only one color is given to the equivalent plastic strain of the range 0 ~ 0.000264. So it is easy to be mistakenly regarded as the elastic zone, and other obvious color part is wrongly regarded as the plastic zone. Actually, this part is a relatively sufficient development area of plastic strain.

According to the features of slope failure, the slope failure will cause mutation of displacement and the plastic strain of the node in slip surface. The sliding surface is located on the place where there is a mutation of the horizontal displacement and plastic strain, so the sliding surface can be determined by drawing nephogram of horizontal displacement or the equivalent plastic strain in the post-processing of ANSYS software. The sliding surface determined by the above two methods is consistent, and the shape and position of the sliding surface determined is illustrated in Figs. 10.17, 10.18 and 10.19, and the display proportion is 0 for slope deformation.

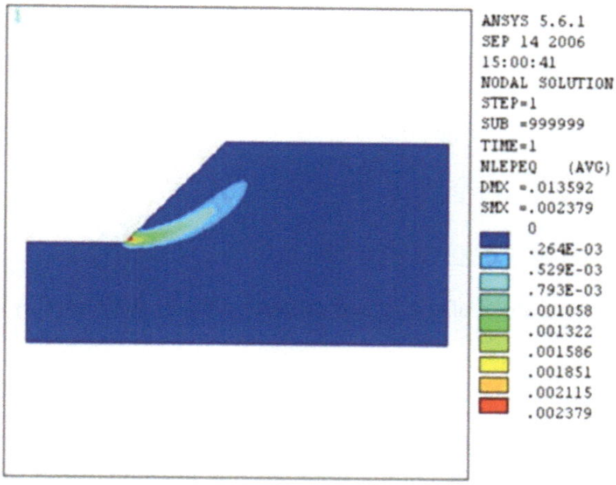

Fig. 10.16 Nephogram of equivalent plastic strain

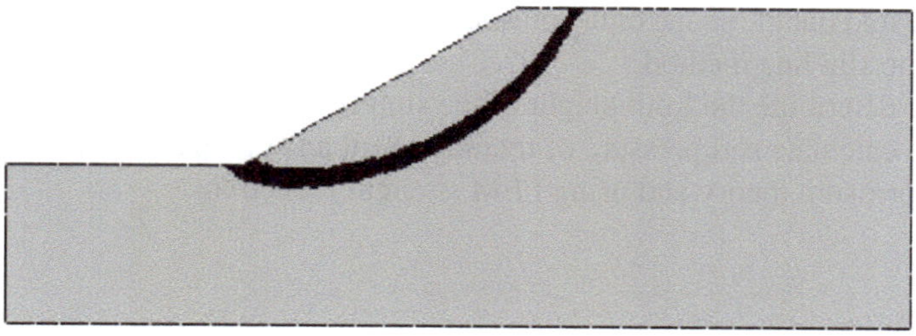

Fig. 10.17 Location and shape of the sliding surface determined by the nephogram of equivalent plastic strain

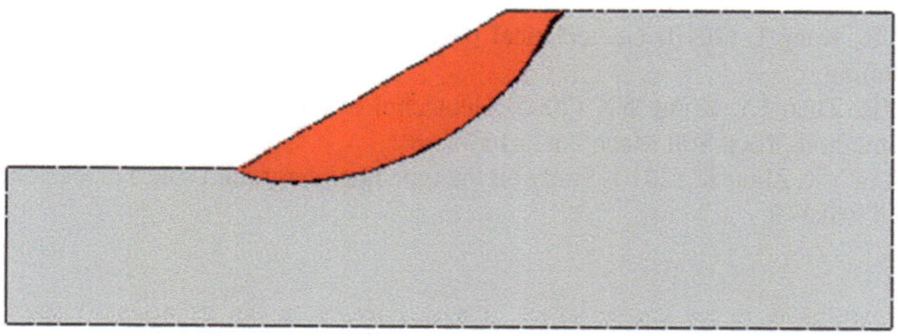

Fig. 10.18 Location and shape of the sliding surface determined by the nephogram of horizontal displacement

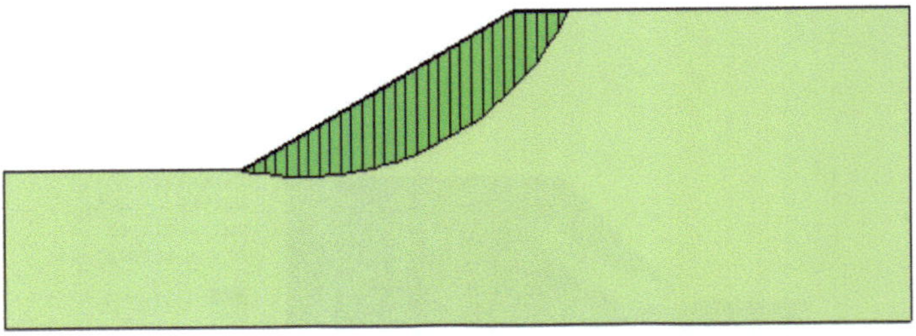

Fig. 10.19 Location and shape of sliding surface determined by slope stability analysis software, Canada slope/w

From the above figures, the positions and shapes of sliding surfaces got from three methods are very close to each other. It shows that the superiority and feasibility of the FEM strength reduction in seeking potential sliding surface.

Questions

1. Please find out the scope of the plastic area for the circular cavity in underground with the slip line method.
2. Please determine the limit height of the simple slope using limit theorem.
3. Please calculate soil pressure of retaining wall and compare it with the classical earth pressure theory and using FEM strength reduction.

References

1. Shen ZJ (2000) Theoretical soil mechanics. China Water Power Press, Beijing
2. Zhen YR, Shen ZJ, Gong XN (2002) Principle of geotechnical plastic mechanics-generalized plastic mechanics. China Architecture & Building Press, Beijing
3. Zheng YR, Kong L (2010) Geotechnical plastic mechanics. China Architecture & Building Press, Beijing
4. Zheng YR, Zhao SY, Kong WX (2005) Geotechnical engineering limit analysis using finite element method. Rock Soil Mech 26(1):163–168
5. Gao G, Liu YX, Zhou JZ (2010) Study on the type in underground space extension. Mod Tunn Technol 47(6):1–9

MIX
Papier aus verantwortungsvollen Quellen
Paper from responsible sources
FSC® C105338

If you have any concerns about our products,
you can contact us on
ProductSafety@springernature.com

In case Publisher is established outside the EU,
the EU authorized representative is:
Springer Nature Customer Service Center GmbH
Europaplatz 3, 69115 Heidelberg, Germany

Printed by Libri Plureos GmbH
in Hamburg, Germany